Kulkarni

Robust Process Development and Scientific Molding

Suhas Kulkarni

Robust Process Development and Scientific Molding

Theory and Practice

3rd Edition

HANSER

Print-ISBN: 978-1-56990-908-9
E-Book-ISBN: 978-1-56990-909-6
ePub-ISBN: 978-1-56990-394-0

Bibliographic information of the German National Library:
The German National Library lists this publication in the German National Bibliography; detailed bibliographic data are available on the Internet at http://dnb.d-nb.de.

Vilshofenerstr. 10 | 81679 Munich | info@hanser.de
www.hanserpublications.com
www.hanser-fachbuch.de
Editor: Dr. Mark Smith
Production Management: le-tex publishing services GmbH, Leipzig
Cover concept: Marc Müller-Bremer, www.rebranding.de, Munich
Cover design: Max Kostopoulos
Cover picture: Suhas Kulkarni
Typesetting: Eberl & Koesel Studio, Kempten

Dedicated to my Parents

Dr. Mohan P Kulkarni

Jayashree M Kulkarni

The Author

Suhas Kulkarni is the founder and president of FimmTech Inc, an injection molding service-oriented firm focusing on Scientific Molding and Scientific Processing for injection molding. He has a bachelor's degree in polymer engineering from Pune, India and a master's degree in plastics engineering from UMass, Lowell, USA. FimmTech is based in San Diego, CA, USA.

Suhas has been associated with injection molding since 1993 with special focus on injection molding processing. He started his career with Republic Tool and Manufacturing in Carlsbad, CA, which is now a part of the Scotts Fertilizer company. He started as a setup and process technician, where he learnt all the nuts and bolts of injection molding. He later joined Molding International and Engineering in Temecula, CA and worked as a Process Engineer for 12 years before he started FimmTech in 2004. The main goal in starting FimmTech was to provide simple and easy-to-use tools for processors on the production floor for process development and troubleshooting. FimmTech has developed several custom tools that help molders develop robust processes and its seminars have trained individuals from hundreds of companies throughout the world. He is a regular contributor to the widely distributed Plastics Technology magazine and is also an adjunct faculty member at UMass, Lowell.

Preface to the Third Edition

It has been six years since the second edition of this book was published. In those six years my approach to process development has become increasingly streamlined and methodical, prompting the need for this third edition. Outdated sections have been deleted or revised and new insights have been incorporated based on my recent findings. My quest to set the value of every molding process parameter based on an experimental data continues. The basic molding principles and theories have remained unchanged but I have experimented and added a few more steps of process development.

A complete chapter on melt preparation has been added. This will help the reader understand the science and procedure of preparing the melt and setting the values for the barrel zone temperatures, the screw rotation speed, and the back pressure. Throughout the book, several existing figures were updated and several new ones were added to enhance the comprehension of the topics.

I extend heartfelt gratitude to my readers who have also encouraged me to take on this task. Special thanks to those who have sent me feedback on the book and mentioning some errors. This not only helps me but helps the readers to understand the subject and not be misguided.

I must express my gratitude to two individuals who were inadvertently omitted from previous editions' acknowledgments. First, Mr. Kirloskar of Kirloskar Kisan, Pune, India. It was in his company that I first saw an injection molding machine. The earliest seeds of my molding future were sown there. I was a summer intern student in my third year of engineering. Mr. Kirloskar was a towering figure but soft spoken with a pleasant smile. The first time I met him he was giving a solution to a molding problem which, at that time, I did not understand a word of. But I loved the technical talk which I am sure ignited the curiosity in me. As the 'Butterfly Effect' explains, an extra flap of a wing of a butterfly somewhere in South America can theoretically create a storm in North America; one of those extra flaps is the internship that has led to the place I am today. I owe part of my success to Mr. Kirlosakar – thank you, sir.

Second is my dear friend Biff Lamkin, with whom I worked for 10 years in my second job. He was a genius who was an expert not only at molding but also in computers, IT, programming, and technology in general. He introduced me to a lot of concepts of molding in my early days. I can still remember him teaching me Pressure Drop Study, which is one of the most important studies in process development. He was also the one who came up with the idea of the melt probe preheater that I use in my courses. Thank you, Biff – I will forever be indebted to you.

A special thanks to my brother Sachin Kulkarni, who has supported and encouraged me all these years and to Ravi Khare of Symphony Technologies, who taught me statistics, DOE concepts, and now theories behind artificial intelligence and machine learning. The continued support of Prof. Basargekar, Tim, and Violeta of Distinctive Plastics, and my friends from UMass Lowell is a constant source of encouragement.

I would have never entered the world of plastics engineering had my dad not exposed me to chemistry as a young kid. I remember doing my high school homework sitting in his lab as he conducted his research. I will be forever indebted to my parents for the encouragement and the exposure at an early age.

A special thanks to my wife Neela for having sacrificed a vacation we were supposed to take as instead I ended up finishing up this third edition. She has always been patiently supportive of all my crazy ideas, work goals, irregular work hours, and last-minute schedule changes. I could not have released this edition on time without her support.

It is my commitment to continue enriching our industry and provide better and easier solutions not only to increase all efficiencies but also to build systems for a sustainable planet and future.

Thank you.

Suhas Kulkarni

June 2024

Preface to the Second Edition

As the saying goes "the only thing that is constant is change." It has been six years since the first edition of this book was published, and it has been very well received. Thank you to all its readers. Since then, I have continued my research to further understand the process of injection molding with the final goal of robust process development. As I kept publishing and teaching this new material, it became time to revise the book.

This second edition has new material in almost all the chapters. Some concepts, which were explained in the first edition, have been expanded upon and rewritten for better understanding. Several figures have been added to complement the explanations. Some of the chapters and text have been split up and rearranged to have a better flow of understanding. A complete chapter on "Basic Quality Concepts" has also been added.

The topic of process development is a complex one, but once the concepts are understood, implementation is easy. The key is to understand the basics first. Over the years, in my consulting business, I often get called on by companies to 'fix' their processes. I always go back to the basics and ask them several simple questions about their molds, machines, and processes to which they sometimes have no answer, or when they do answer my questions, they figure out the solution to the problem on their own. Their process development was probably done by throwing darts on a dartboard and hence the issues. This book is attempting to change that. By using the techniques described in this book, one can establish what I call *cruise control processes*: set the process, start molding, and never touch a setting until the run is done.

The topic of "Design of Experiments" (DOE) has great importance in injection molding. Many companies employ this technique, but not effectively. The reason is not because of their lack of knowledge of DOE, but because of their lack of understanding of the basics of molding, along with their choice of factors and levels for the DOE. This topic has been expanded in the new edition.

I would like to thank Hanser Publications and their staff for this opportunity to write the second edition. Mark Smith and Cheryl Hamilton have been very helpful with the proofing and, moreover, very patient with all the delays from my side. I would also like to thank several other people who have helped me with the second edition. Lorena Castro who took all the bits and pieces of my writing and transformed it into readable flow needs a special mention and acknowledgement.

In the preface to the first edition, I failed to mention a very important place that also helped shape my career and my life. The National Chemical Laboratory (NCL), Pune, India, is where my dad worked all his life as a research scientist. I lived in the shadows of this great institution and its several researchers. My dad would often take me to his lab when he conducted his research, and that is where the seeds of my future were laid. I worked on a couple of projects during my college days in its Polymer Engineering Department, and that was my first personal exposure and involvement with research. It was my experience at NCL, which was one of the contributing factors that pushed me to study further.

My constant sources of inspiration and help include Tim and Violeta of Distinctive Plastics, who have opened their company for my research and seminars, my professor from college, Dr. Basargekar, my colleagues in the industry, Ravi Khare, Atul Khandekar, Vishu Shah, Vikram Bhargava, Randy Phillips, and my family.

To my mom, dad, and siblings, I will be forever indebted to you for all the support and inspiration you have given me over the years.

Suhas Kulkarni

October 2016

Preface to the First Edition

When I interviewed for my second job after I graduated, I was told that if the position was offered to me, I would have to spend my first three days at a seminar on Scientific Molding and Design of Experiments. It was all new to me then. My job was to implement this new technology as a standard across the company. The job was offered to me; I accepted and attended the seminar. Implementing the techniques on the first couple molds was a refreshing change from how I did it before. The scientific method of developing the process left no room for any guess work by applying the theories of polymer science and injection molding. Scientific evidence proved why parts could be or could not be molded consistently within the required specifications. My enthusiasm for the use of these techniques grew as I found more and more evidence of success. Over the next few years, I gave presentations at the local SPE chapter and the attendees wanted to learn more to make their operations efficient. In 2004 I decided to start consulting in the area of Scientific Processing, a term I coined to include all the processes that are involved in the transformation of the pellet to the final product that is shipped out to the customer. My research work on the 'overdrying' of PBT and Nylon was the main driving force to think of the process as being outside of the molding machine and not just what happens in the mold. As my consulting and teaching career expanded, I found many people looking for a resource to learn the basic underlying principles of polymers and plastics and apply them to injection molding. They wanted to understand the why, and then how of Scientific Processing. 'Where can I find this information?' was always a question that was asked. This book is the answer to their question.

Understanding the molding process from the scientific perspective helps in making better decisions to establish the parameters that are involved in controlling the journey of the pellet; from the warehouse to the molding machine and then to its conversion as a molded product. All the parameters are set on the basis of scientific knowledge and experience making the process efficient in terms of productivity.

Higher yield, reduced scrap, robust processes, reduced quality inspection, reduced number of process changes leading to less human intervention are some of the benefits of Scientific Processing. This book details the theory and practice of Scientific Processing. There are a lot of 'rules of thumb' in injection molding. My mission is to eliminate them and present a scientific solution. A good example is the size of vents in the mold.

I hope my commitment to researching and understanding of the molding process will continue to give a better insight to the process. I hope to share those with you in the future editions of this book. There are a number of people who are part of the success of writing this book. Some gave me the knowledge, some inspired me to learn more while others gave me unconditional support in this endeavor. It is impossible to thank all of them individually but without all of them this project would not have been accomplished. First and foremost, special mention must be made of my father who introduced me to the fascinating world of chemical research. It is from here that I get my curiosity, creativity and my analytical abilities of problem solving. Thanks to my teachers and professors who not only imparted the knowledge but also instilled in me the value of education through the dedication to their students. It is from here that I get my inspiration to teach and spread my knowledge. Thanks to my family and friends who have supported me and believed in me. It is from them that I get my will power and courage to get past the current frontiers and take a step into an unknown future.

In the production of this book I would like to thank Christine Strohm and the management of Hanser Publications for publishing the book. The sections on cavity pressure sensing and the chapter on rheology were reviewed by Mike Groleau of RJG and John Beaumont of Beaumont Technologies respectively. Thanks to them for their valuable comments. Thanks also to Dave Hart for proofreading the text and making the matter an interesting technical read. Valuable comments from Ravi Khare of Symphony Technologies were included on the DOE chapter. Without the unconditional help of Tim and Violeta Curnutt of Distinctive Plastics I would have not had the chance to experiment with many of the theories and applications put forward in this book. Special thanks to them for letting me make Distinctive Plastics my home during the book writing process. I am often told I am an effective teacher with clear concepts in polymer science and rheology – I have picked the teaching skills and the knowledge from Prof. Basargekar – my sincere acknowledgements to him. Under the leadership of Vishu Shah I conducted a few successful seminars with the Society of Plastics Engineers. These seminars gave me the fuel and material for this book. Thanks to Vishu not only for the opportunities of the seminars but also for being a professional guide and a personal friend. I would also like to acknowledge the efforts of John Bozzelli and Rod Groleau for their pioneering work in Scientific Molding and raising its awareness in the molding community.

To my alma maters, Maharashtra Institute of Technology, Pune, India and University of Massachusetts, Lowell, USA: Hidden in one of your foundations' bricks are the enriching roots to my success. Thank You.

Suhas Kulkarni

FIMMTECH Inc.

Vista, CA.

January 2010

Contents

The Author ... VII

Preface to the Third Edition ... IX

Preface to the Second Edition ... XI

Preface to the First Edition ... XIII

1 Introduction to Scientific Processing ... 1

1.1 The Evolution and Progress of Injection Molding ... 1

1.2 The Molding Process ... 2

1.3 The Three Types of Consistencies Required in Injection Molding ... 3

1.4 Scientific Processing ... 6

1.5 The Five Critical Factors of Molding ... 9

1.5.1 Part Design ... 9

1.5.2 Material Selection ... 10

1.5.3 Mold Design and Construction ... 10

1.5.4 Machine Selection ... 11

1.5.5 Molding Process ... 11

1.6 Concurrent Engineering ... 12

1.7 Variation ... 12

1.8 The Molding Parameters and a Recommended Process Development Procedure ... 14

2 Properties of Polymers and Plastics That Influence Injection Molding . . . 17
2.1 Polymers . . . 17
2.2 Molecular Weight and Molecular Weight Distribution . . . 19
2.3 Polymer Morphology (Crystalline and Amorphous Polymers) . . . 21
2.4 Role of Morphology in Injection Molding . . . 26
2.4.1 Differences in Shrinkage between Amorphous and Crystalline Materials . . . 26
2.4.2 Melt Processing Range . . . 27
2.4.3 Mold Filling Speed . . . 28
2.4.4 Mold Temperatures . . . 28
2.4.5 Barrel Heat Profile . . . 29
2.4.6 Screw Recovery Speeds . . . 29
2.4.7 Nozzle Temperature Control . . . 30
2.4.8 Cooling Times . . . 30
2.4.9 Mechanical Properties . . . 30
2.4.10 Optical Clarity . . . 30
2.5 Exceptions of Morphology Rules to Polyolefins . . . 31
2.6 Thermal Transitions in Polymers . . . 32
2.6.1 Relationship between the Glass Transition Temperature and Post Mold Shrinkage . . . 38
2.6.2 Importance of Mold Temperatures in Injection Molding . . . 41
2.7 Shrinkage of Polymers in Injection Molding . . . 44
2.8 The Plastic Pressure–Volume–Temperature (PVT) Relationship . . . 46
2.8.1 Importance of Plastic Density in Injection Molding . . . 47
2.8.2 Residence Time and Maximum Residence Time of a Plastic . . . 49
2.8.3 Plastic Datasheets . . . 50
2.9 References . . . 52

3 Polymer Rheology . . . 55
3.1 Viscosity . . . 55
3.2 Newtonian and Non-Newtonian Materials . . . 57
3.3 Viscosity in Polymer Melts . . . 58
3.4 Effect of Temperature on Viscosity . . . 61
3.5 Velocity and Shear Rate Profiles . . . 62

3.6 Application to Injection Molding ... 64
3.6.1 Flow Imbalance in an 8-Cavity Mold ... 64
3.6.2 Racetrack Effect in a Part with Constant Thickness ... 66
3.6.3 Stress Build-Up in Molded Parts ... 67
3.6.4 Warpage Difference between Cavities ... 67
3.7 Solving Flow Imbalances Using Melt Rotation Techniques ... 67
3.8 Fountain Flow ... 69
3.9 Effect of Fountain Flow on Crystallinity, Molecular Orientation, and Fiber Orientation ... 72
3.10 Characterization of Polymer Viscosity ... 73
3.11 References ... 75

4 Plastic Drying ... 77
4.1 Problems in Melt Processing Related to the Presence of Moisture ... 79
4.1.1 Degradation of Plastic ... 79
4.1.2 Presence of Surface Defects ... 80
4.2 Hygroscopic Polymers ... 83
4.3 Drying of Plastics ... 85
4.3.1 Drying Temperatures and Times ... 85
4.3.2 Relative Humidity and Dew Point ... 87
4.3.3 Air Flow Rate ... 88
4.4 Equipment for Drying Plastics ... 88
4.4.1 Oven Dryers ... 88
4.4.2 Hot Air Dryers ... 89
4.4.3 Desiccant Dryers ... 89
4.4.4 Classifications Based on the Location of the Dryer ... 89
4.5 Determination of the Amount of Moisture ... 90
4.5.1 The Glass Slide Technique (TVI Test) ... 90
4.5.2 The Karl Fischer Titration Method ... 91
4.5.3 Electronic Moisture Analyzer ... 91
4.5.4 Measurement of the Dew Point ... 92
4.6 'Overdrying' or Overexposure to Drying Temperatures ... 93
4.7 Cautions ... 100
4.8 Prevention of Overexposure to Longer Drying Times ... 100
4.9 Overdrying Controller ... 100
4.10 References ... 102

5 Common Plastic Materials and Additives. 103

5.1 Classification of Polymers 103

5.2 Commercially Important Plastics 105

5.2.1 Polyolefins 105

5.2.2 Polymers from Acrylonitrile, Butadiene, Styrene, and Acrylate 106

5.2.3 Polyamides (PA) 107

5.2.4 Polystyrenes (PS) 108

5.2.5 Acrylics 109

5.2.6 Polycarbonates (PC) 109

5.2.7 Polyesters 109

5.2.8 Polyvinyl Chloride (PVC) 109

5.2.9 Polyoxymethylene (POM or Acetal) 110

5.2.10 Fluoropolymers 111

5.3 Additives 111

5.3.1 Fillers 111

5.3.2 Plasticizers 112

5.3.3 Flame Retardants 112

5.3.4 Anti-Aging Additives, UV Stabilizers 113

5.3.5 Nucleating Agents 113

5.3.6 Lubricants 113

5.3.7 Processing Aids 114

5.3.8 Colorants 114

5.3.9 Blowing Agents 114

5.3.10 Other Polymers 115

5.4 Closing Remarks 115

5.5 Reference 116

6 Injection Molding and Molding Machines 117

6.1 The History of Injection Molding 117

6.2 Injection Molding Machines and Their Classifications 118

6.3 Machine Specifications 120

6.3.1 Clamp Force (Tonnage) 121

6.3.2 Shot Size 121

6.3.3 Screw Diameter and L/D Ratio 121

6.3.4 Plasticating Capacity ... 122
6.3.5 Maximum Plastic Pressure ... 122
6.4 The Injection Molding Screw ... 122
6.5 Screw Designs ... 125
6.6 The Check Ring Assembly ... 126
6.7 Intensification Ratio (IR) ... 126
6.8 Obtaining Intensification Ratios ... 128
6.9 Selecting the Right Machine for the Mold ... 129
6.9.1 Physical Size of the Mold ... 130
6.9.2 Calculating the Required Machine Tonnage for a Mold ... 132
6.9.3 Percentage Shot Size Used and Number of Shots in the Barrel ... 137
6.9.4 Residence Time of the Material in the Barrel ... 140
6.9.5 Practical Methods to Find Percentage Shot Size, Shots in a Barrel, and Residence Time ... 141
6.9.6 Residence Time Distribution ... 142

7 Scientific Processing, Scientific Molding, and Molding Parameters ... 143

7.1 Introduction ... 143
7.2 What Scientific Molding and Scientific Processing Are Not! ... 145
7.3 Process Robustness ... 145
7.4 Process Consistency ... 146
7.5 The Injection Molding Cycle ... 149
7.6 The Injection, Pack, and Hold Phases of Injection Molding ... 150
7.7 Speed and Pressure ... 153
7.8 The 11 + 2 Plastic Injection Molding Machine Input Parameters ... 154
7.8.1 Barrel Temperatures ... 155
7.8.2 Mold Temperatures ... 156
7.8.3 Shot Size ... 158
7.8.4 Transfer Position ... 158
7.8.5 Injection Speed (Velocity) ... 158
7.8.6 Injection Pressure ... 159
7.8.7 Pack Pressure ... 159
7.8.8 Pack Time ... 159
7.8.9 Hold Pressure ... 160

7.8.10 Hold Time ... 160
7.8.11 Screw Rotation Speed ... 160
7.8.12 Back Pressure ... 161
7.8.13 Cooling Time ... 164
7.9 Additional Settings ... 165
7.9.1 Decompression or Suck Back Position ... 165
7.9.2 Pack Speed ... 166
7.9.3 Tonnage ... 166
7.10 Process Outputs ... 166
7.10.1 Actual Melt Temperature ... 167
7.10.2 Actual Mold Temperature ... 168
7.10.3 Injection (Fill) Time ... 170
7.10.4 Peak Injection Pressure ... 170
7.10.5 Pressure at Transfer ... 171
7.10.6 Cushion Value ... 171
7.10.7 Screw Rotation Time ... 173
7.10.8 Cycle Time ... 173
7.10.9 Injection Only Shot Weight (Including Runners) ... 174
7.10.10 Full Part(s) Weight ... 175
7.10.11 Runner Weights ... 175
7.11 Decoupled Molding℠ ... 176
7.12 References ... 177

8 Introduction to Process Development ... 179
8.1 Introduction ... 179
8.2 Process Development Fundamentals ... 179
8.3 Premolding Setup ... 182
8.3.1 Storage and Drying of Resin ... 182
8.3.2 Plastic Drying ... 182
8.3.3 Machine Selection ... 185
8.4 Importance of Adding Charge Delay Time ... 185
8.5 Filling in Injection: Weight or Volume? ... 187

9 Process Development Part 1: The Melt Preparation 189

9.1 Setting of the Barrel Temperatures 191

9.2 Optimization of Screw Rotation Speeds 195

9.3 Screw Rotation Speed Rule of Thumb That Should Not Be Used 199

9.4 Optimization of Back Pressure 200

10 Process Development Part 2: The 6-Step Study – Exploring the Cosmetic Process 205

10.1 The 6-Step Study 205

10.1.1 Step 1: Optimization of the Injection Phase – Rheology Study 205

10.1.1.1 Procedure to Determine the Viscosity Curve at the Molding Machine 210

10.1.1.2 Interpreting the Data 212

10.1.1.3 Cautions and Exceptions 213

10.1.1.4 Profiling of Injection Speeds 214

10.1.1.5 When a Short Shot Sticks 215

10.1.1.6 Selecting Melt Temperature for Viscosity Graphs 217

10.1.2 Setting the Shot Size and the Transfer Position 217

10.1.3 Step 2: Determining the Cavity Balance – Cavity Balance Study 219

10.1.3.1 Procedure to Determine the Cavity Balance 221

10.1.3.2 Reasons for Cavity Imbalance 222

10.1.3.3 Determining the Cause of Cavity Imbalances 225

10.1.3.4 Calculating Cavity Imbalance 226

10.1.3.5 Acceptable Level of Cavity Imbalance 227

10.1.4 Step 3: Determining the Pressure Drop – Pressure Drop Studies 229

10.1.4.1 Procedure to Determine Pressure Drop 231

10.1.4.2 How to Use the Information from the Study 235

10.1.4.3 Effect of Pressure Drop on the Pack and Hold Phase 236

10.1.5 Step 4: Determining the Cosmetic Process Window – Process Window Study 237

10.1.5.1 Procedure for Determining the Process Window for Amorphous Materials 241

10.1.5.2 Procedure for Determining the Process Window for Crystalline Materials 242

10.1.5.3 How to Use This Information 243

10.1.5.4 Shape of the Process Window 244
10.1.5.5 Cautions and Exceptions 245
10.1.5.6 Relationship between Cavity Balance and Process Windows 245
10.1.5.7 The Pack and Hold Pressure Rule Not to Be Used 246
10.1.6 Step 5: Determining the Gate Seal Time – Gate Seal Study 246
10.1.6.1 How to Use This Information 249
10.1.6.2 Cautions and Exceptions 249
10.1.7 Differentiating between the Pack and the Hold Phase 249
10.1.7.1 Hot Runner and Valve Gated Molds 253
10.1.8 Step 6: Determining the Cooling Time – Cooling Time Study 253
10.1.8.1 Procedure to Determine Cooling Time 255
10.2 Optimization of Tonnage 257
10.3 The Cosmetic Scientific Process 260
10.4 Post-Mold Shrinkage Studies 261
10.4.1 Procedure to Measure Shrinkage 263
10.4.2 How to Use This Information 264
10.5 Completion of Part 2 of the Recommended Process Development Procedure 264
10.6 Recommended Adjustments to Maintain Process Consistency and Robustness 266
10.7 Process Documentation 267
10.8 References 267

11 Process Development Part 3: Exploring the Dimensional Process via Design of Experiments 269

11.1 Why Perform DOEs? 271
11.2 The PVT Diagram and Injection Molding Relationship 271
11.2.1 Design of Experiments: Definition 272
11.3 Terminology 274
11.3.1 Factor 274
11.3.2 Response 275
11.3.3 Level 275
11.3.4 Designed Experiment 276
11.4 Relationships between the Number of Factors, Levels, and Experiments 277

11.5 Balanced Arrays ... 278
11.6 Interactions ... 280
11.7 Confounding or Aliasing ... 282
11.8 Factorial Experiments ... 284
11.9 Data Analysis ... 285
11.9.1 Tornado Charts ... 286
11.9.2 Contour Plots ... 288
11.9.3 Prediction Equation ... 289
11.9.4 Process Sensitivity and Outliers Charts ... 289
11.10 Using the Results from DOE ... 291
11.10.1 Process Selection ... 291
11.10.2 Cavity Steel Adjustment ... 291
11.10.3 Process Adjustment Tool ... 292
11.10.4 Setting Process Change Tolerances ... 292
11.10.5 Reducing Inspection ... 293
11.11 The Dimensional Process Window (DPW) ... 293
11.12 Selection of Factors for DOEs ... 295
11.13 Analysis of Variance (ANOVA) ... 299
11.14 Collecting Shots for a Quality Check ... 301
11.15 Choosing the Highs and Lows for DOEs from the Process Window ... 302
11.16 Randomization ... 302
11.17 DOE Application to Optimize Pack and Hold Times ... 303
11.18 Setting Acceptable Machine Tolerances and Alarms during Production .. 308
11.19 Summary ... 310

12 Mold Qualification Flowchart, Production Release, and Troubleshooting ... 311
12.1 Process Development Flowchart ... 311
12.1.1 Stage 1: Development of the Cosmetic Process ... 312
12.1.2 Stage 2: Mold and Part Quality Qualification Procedure ... 313
12.2 Mold Qualification Checklist ... 313
12.3 Process Documentation ... 314
12.3.1 Process Sheet ... 314
12.3.2 Waterline Diagrams ... 315
12.3.3 Mold Temperature Maps ... 316

12.3.4 Setup Instructions 317
12.3.5 Operator Instructions 317
12.4 Documentation Books 318
12.5 Qualification Production Runs 319
12.6 Mold Specific Troubleshooting Guide 319
12.7 Molding Startup and Shutdown 320
12.7.1 Purging 320
12.7.2 Startup of a Molding Machine 321
12.7.3 Shutdown of a Molding Machine 322
12.8 Troubleshooting 322
12.9 Important Equipment and Tools for Qualifications and Troubleshooting 325
12.10 Common Defects, Their Causes, and Prevention 326
12.10.1 Splay: What Is It? How to Get Rid of It? 327
12.10.2 Defects in Molding 329

13 Role of Mold Cooling, Venting, and Regrind in Process Development 335

13.1 Mold Cooling 335
13.1.1 Number of Cooling Channels 336
13.1.2 Reynolds Number of the Coolant Flow 336
13.1.3 Type of Coolant 337
13.1.4 Series and Parallel Cooling 338
13.2 Venting 339
13.2.1 Dimensions of the Vent 340
13.2.2 Primary Vent Depths 341
13.2.3 Location of Vents 344
13.2.4 Forced Venting or Vacuum Venting 345
13.3 Regrind 346
13.3.1 Effect of the Molding Process on the Part Properties 347
13.3.2 Using Regrind 348
13.3.3 Batch and Continuous Processes of Incorporating Regrind 349
13.3.4 Estimating the Amount of Regrind from Different Generations .. 350
13.3.5 Effect of Regrind on Processing 352
13.3.6 Closing Remarks 353

14 Related Technologies and Topics 355

14.1 Cavity Pressure Sensing Technology 355

14.1.1 Sensors and Output Graphs 356

14.1.2 Types and Classification of Pressure Sensors 357

14.1.3 Use of Information from the Pressure Graphs 359

14.1.4 Controlling the Process with Cavity Pressure Sensors 361

14.1.5 Sensor Locations 363

14.2 Building a Knowledge Base 364

14.3 Concurrent Engineering in Injection Molding 366

14.3.1 The Product Designer 368

14.3.2 The Tooling Engineer 369

14.3.3 The Mold Designer and Mold Maker 369

14.3.4 The Material Supplier 370

14.3.5 The Process Engineer 370

14.3.6 The Quality Engineer 371

14.3.7 The Sales Team at the Molder 371

14.3.8 Mandatory for All Departments 372

14.3.9 Implementing Concurrent Engineering 372

15 Quality Concepts 375

15.1 Basic Concepts 375

15.2 Histogram 376

15.3 Normal Distribution 377

15.4 Standard Deviation 378

15.5 Specification Limits and Standard Deviation 379

15.6 Capability Index 380

15.7 Process Capability 382

15.8 Statistical Quality Control (SQC) and Statistical Process Control (SPC) 383

15.9 References 384

Appendix A Materials Data Sheet 385

Appendix B Conversion Tables for Commonly Used Process Parameters 391

Appendix C Water Flow Tables 393

Appendix D Part Design Checklist 395

Appendix E Mold Design Checklist 397

Appendix F Mold Qualification Checklist 399

Appendix G Regrind Tables – Percentage of Regrind in Total Shot 401

Appendix H Calculation of Melt Density 405

Index 407

1 Introduction to Scientific Processing

1.1 The Evolution and Progress of Injection Molding

Injection molding and extrusion are the most common techniques employed in the manufacture of plastic products. Injection molding of plastics began as an idea by the Hyatt brothers for the manufacture of billiard balls. The idea was borrowed based on a patent by John Smith to inject metal castings. Since then, injection molding of plastics has come a long way. The technique became a popular way to fabricate plastic parts because of the simplicity of the concept, efficiency of production, and the possibility of producing intricate parts with fine details.

The art of injection molding evolved to its present state for a few key reasons. The requirements of the molded parts became more stringent because of the advances in the fields of science and technology. The demand for tighter tolerances and more complex parts increased and is ever increasing. A required tolerance of a few microns on a 10 mm part, or a couple thousandths of an inch on a one inch dimension, is not uncommon these days. Parts requiring innovative designs, especially designed for assembly (DFA) or parts molded from different materials in the same mold (multi-material molding) are now commonplace. As polymer materials were developed for injection molding, the requirements of processing changed. The discovery of the different morphologies of polymers and the need for better melt homogeneity in molding led to the introduction of the injection screw. Various designs for material-specific screws have followed since. The use of high temperature materials that have high melting points and need high mold temperatures have led to the use of high-temperature ceramic heaters and mold temperature controllers providing higher heat capability. Innovations in electrical and electronic technologies paved the road for machines that could be better controlled, accurate, and efficient. Response times for hydraulic valves can be in milliseconds. All electric machines and hybrid machines are gaining popularity because of their consistency and accuracy. The real-time processing parameters of a molding machine can now be viewed from any part of the world via an internet connection and therefore machine production can be monitored or machines can be debugged online. All these features are becoming a common practice among manufacturers. Even some auxiliary equipment can now be debugged and programmed by the suppliers via an internet connection. For the machines tied into the company ERP system, automated messages can be sent to the managers and supervisors about the machine status and quality issues. The need for efficiency and the requirements for advanced product features have dictated the need for innovations in injection molding over the years.

1.2 The Molding Process

The actual molding process has been traditionally defined as the inputs to the molding machine. These are the settings of speeds, pressures, temperatures, times and distances. Examples of these are injection speeds, holding pressure, mold temperature, cooling time and transfer position. These are inputs one would set at the molding machine and record on a sheet, commonly called the Process Sheet. However, the word process now needs to be redefined as the complete operation that encompasses all the activities the plastic is subjected to inside a molding facility – from when the plastic enters the molding facility as a pellet to when it leaves the facility as a molded part. For example, the storage of the plastic, the control of the drying of the plastic, and the post mold shrinkage of the part can have a significant influence on the quality of the part. During this journey of the pellet, every stage can have a significant effect on the final quality of the part or assembly. Naturally, understanding every stage now becomes imperative if we would like to control the quality of the molded part. Molding a part that meets the quality requirements can be a challenge, but the real challenge is molding parts consistently; cavity to cavity, shot after shot, and from one production run to another, meeting all the quality requirements and with the least amount of effort and maximum efficiency. We need to understand not just the inputs but also the outputs of the molding process. In fact, outputs are more important than the inputs.

The goal of this book is to explain to the reader all the molding process inputs, process outputs and provide process optimization techniques. There are several factors that contribute to each of these parameters and these factors will be discussed.

1.3 The Three Types of Consistencies Required in Injection Molding

The aim of developing a molding process should be to develop robust processes that would not need any process modifications once the processes are set. Process consistency leads to quality consistency, see Figure 1.1. We look for three different types of consistencies: cavity-to-cavity consistency (Figure 1.1(a)), shot-to-shot consistency (Figure 1.1(b)), and run-to-run consistency (Figure 1.1(c)). Cavity-to-cavity consistency is required in multicavity molds so that each cavity is of the same quality as the other cavities. Shot-to-shot consistency implies that every consecutive shot would be identical in quality to the previous shot, or the first shot is identical to the last shot of the production run with the process parameters remaining identical during the entire production run. When the process parameters from two different runs are identical and they produce identical quality parts, then this is called run-to-run consistency. Robust and stable processes always yield consistent quality parts with one established process.

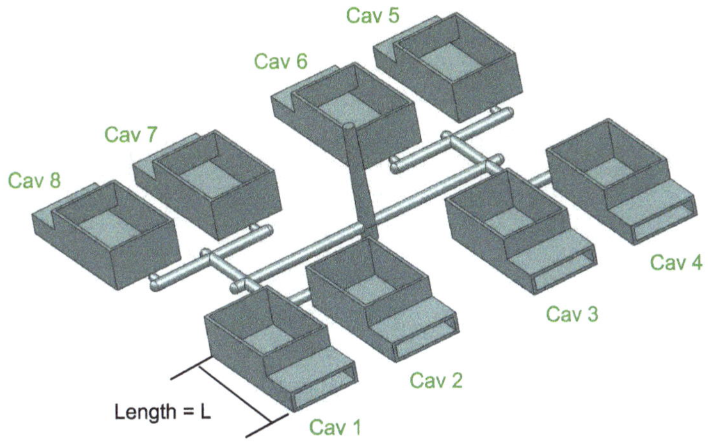

Cavity to cavity consistency:

Part length, $L_{(cav1)} = L_{(cav\ 2)} = L_{(cav\ 3)} = L_{(cav\ 4)}\ L_{(cav\ 5)} = L_{(cav\ 6)} = L_{(cav\ 7)} = L_{(cav\ 8)}$

(a) Cavity to cavity consistency

Figure 1.1 The three types of consistencies required in injection molding

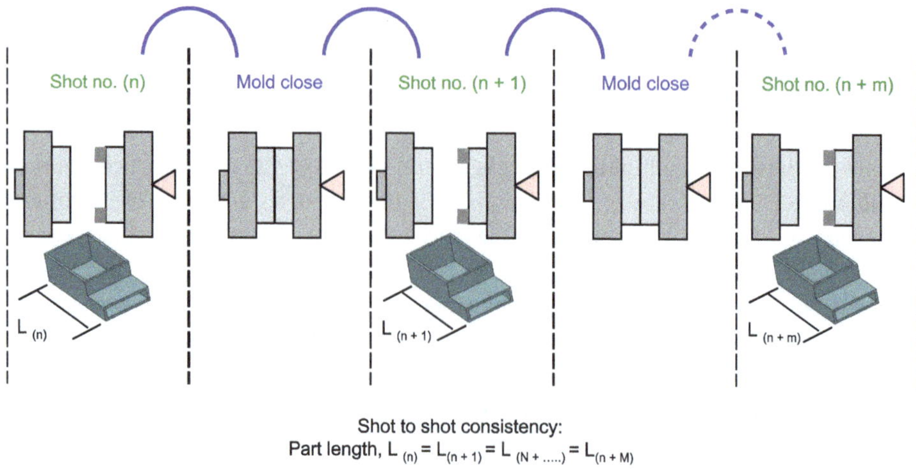

(b) Shot to shot consistency

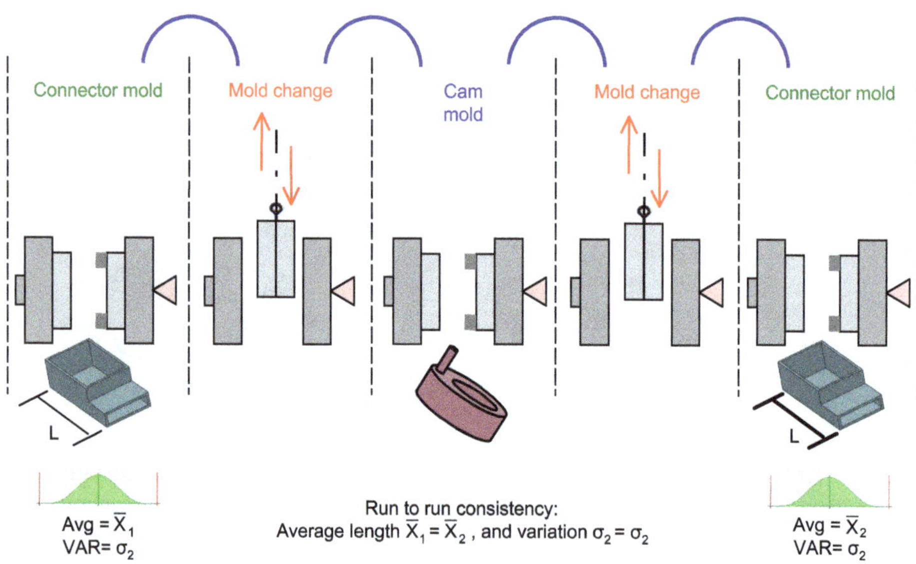

(c) Run to run consistency

Figure 1.1 The three types of consistencies required in injection molding *(continued)*

There can be several reasons for not achieving the three types of consistencies. A cavity-to-cavity inconsistency could be caused because of an error when machining the steel in one of the cavities or by not making the gate sizes of the cavities identical to each other. A shot-to-shot inconsistency could be caused because of a damaged leaking check ring at the end of the molding screw or a problem with the hydraulic injection valve. A run-to-run inconsistency can be caused because of a lack of a robust process or simply because the process was not accurately or completely documented in the previous run. The run-to-run consistency is the one that most companies struggle with. This book deals in depth with process development of robust, repeatable and reproducible processes.

Another reason for inconsistencies and variations in the molded product is the nature of the shrinkage of plastics. When molten plastic is injected inside a mold it cools and freezes to form the product. There is a reduction in the volume of the melt when it cools inside the mold. This is called shrinkage. The magnitude of shrinkage determines the final dimensions of the part. However, this shrinkage is not easily predictable and depends on a number of factors. There is a range of shrinkage values available and that makes it difficult for a mold maker to select a shrinkage value. For example, the shrinkage value for a low-density polyethylene is between 1.3 and 3.1%, which is a wide range. Shrinkage also depends upon the processing conditions. For example, higher the melt temperature, the higher the shrinkage. Every processing parameter can affect the shrinkage to varying degrees. Refer to Figure 1.2, which shows the effect of the molding parameter on the length of the part. To increase or decrease the length of the part, several parameters can be increased or decreased.

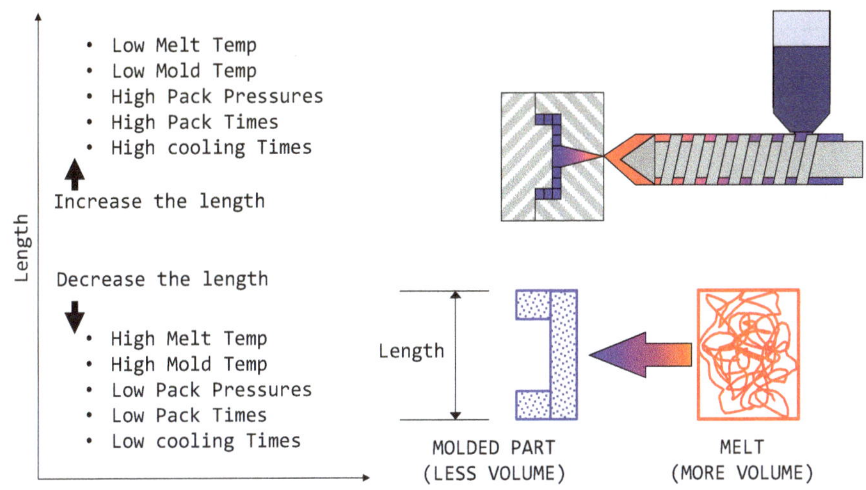

Figure 1.2 Effect of molding parameters on shrinkage and dimension of a part

As seen in the figure, several parameters can have effect on the part dimension and quality. To increase the length of the part, some parameters need to be increased whereas some need to be decreased. Further, the magnitudes of change in length with change in the parameter varies from parameter to parameter. If the molding processes are not developed with these understandings, and in case the dimensions get out of specifications, each processor can work with any one of the parameters. The net result being that processes that were supposedly approved end up having completely different values in a matter of a few runs. When process sheets are compared, for example, from two years ago, there are hardly any numbers that match the current settings.

It should be the goal of every molder to develop an understanding of the molding process for the given mold. A systematic process development approach must be followed. The result of such an approach is a robust, repeatable and reproducible process: the 3 R's. The process also must be such that it can absorb the natural variations that are seen in the production environment such as temperature and humidity changes and even operator changes.

A process shown in Figure 1.3 is not acceptable because there is a lot of inefficiency in the system. Such processes result in defective parts, loss of material, and loss of time, not to mention the time and effort put in by the molding personnel. The parts can be remolded and shipped to the customer; however, the time and efforts lost cannot be recovered. The reputation of the molder is something that can also be permanently affected.

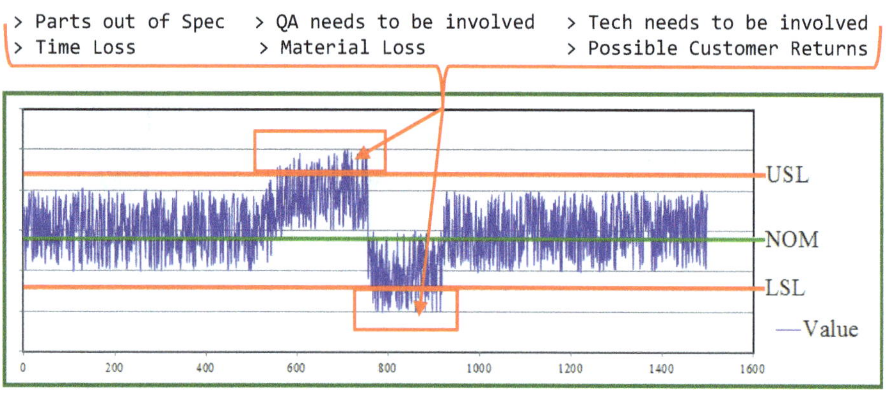

Figure 1.3 Example of an inefficient process

1.4 Scientific Processing

Scientific Processing is the process of achieving consistency in part quality via the application of the underlying scientific principles that control the parameters of the molding process. To achieve this consistency, we must be able to control every activity that is taking place in the process and to control every activity, we must understand the underlying scientific principles. The goal of scientific processing should be to achieve a robust process. Achieving robustness in each of the stages that the pellet travels through automatically translates to an overall robust process. The term consistency must not be confused with the parts being within the required specifications.

A consistent process will produce parts that will reflect the consistency but the parts may be out of specifications. In this case, the mold steel dimensions and/or the part specifications must be adjusted to bring the parts within the required specifications. The process must not be altered if the process is a robust process. Process robustness will be discussed at length in Chapter 8.

Consider that the task on hand is to drive a car in the cruise control mode at 120 km/h (75 mph) for 4 hours with no stops, every day for 365 days a year from location A to location B. There are two choices for the route between the two locations. The first is on a hilly terrain with the road on the edge of a cliff and the second on a wide lane highway as shown in Figure 1.4. Since this is a task to be done on a continuous basis it would be prudent to choose the wide lane highway since any small distractions at any time during the 4 hour daily drive, 365 days a year, could mean the car wavers some but because of the wide lanes the car would still be safe. However, driving the car at the edge of the cliff in the cruise control mode would need constant attention and there would be no chance of any error for the fear of driving off the cliff. For an efficient molding operation it is important to run the machines with no or the least amount of human interaction. In other words, the molding processes must run in a cruise control mode. The available tolerances should also support the natural variations that are seen in a molding environment. Figure 1.5 shows such a process where the part dimensions are well within specifications.

Winding road in the hills

Highway with wide lanes

Figure 1.4 Road options: edge of a cliff or wide lane highway? (Sources: George59/Pixabay; zdrsoft/Pixabay)

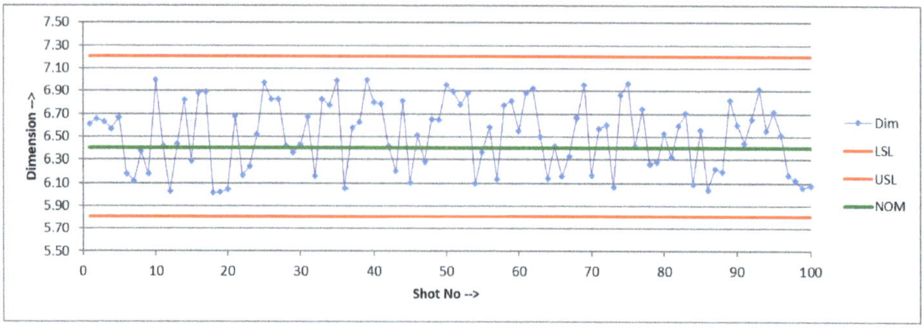

Figure 1.5 Cruise control process

The term *Scientific Molding* was coined and promoted by a two pioneers in the field of injection molding, John Bozzelli and Rod Groleau. Their principles are widely used today and are industry standards. Scientific molding deals with the actual plastic that enters the mold during the molding operation at the molding press. Scientific processing is the complete process from when the pellet enters the facility and leaves the facility as a finished product. Figure 1.6 shows the journey of the pellet and how scientific molding is a subset of scientific processing.

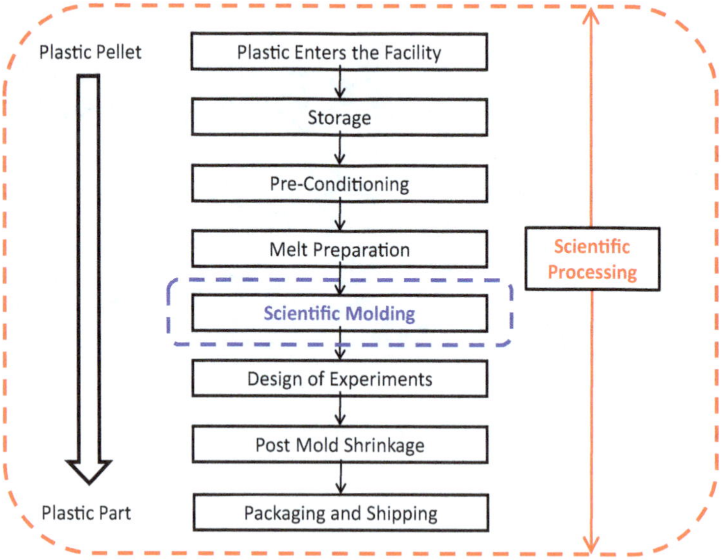

Figure 1.6 The journey of the pellet, Scientific Molding and Scientific Processing

1.5 The Five Critical Factors of Molding

For a part to be molded successfully, there are five factors that have to be considered. These factors, shown in Figure 1.7, are as follows.

1. Part design
2. Material selection
3. Mold design and construction
4. The choice of the molding machine
5. Molding process

Each of these factors plays a very important role in the production of the molded part and therefore every one of them has to be optimized for the production. It is not just the performance of the part but also being able to mold the parts consistently with all the part quality requirements being within the required specifications.

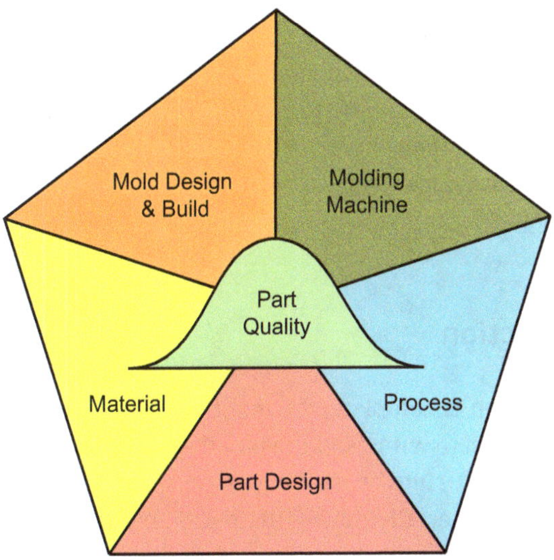

Figure 1.7
The five factors influencing part quality consistency and process robustness

1.5.1 Part Design

The concept of the part starts with the engineer designing it. The part must be designed for molding and all the design rules for plastics must be considered. Rules for plastic part design are considerably different than those used for metal part design because of the inherent nature of the plastic. For example, to avoid sink defects in the plastic part, thick sections cannot be present. Additionally, all corners must have a radius to avoid stress concentration and premature failure. With the growing cost of labor and the need for efficiency in the manufacturing process, the part designers now face the added challenges of designing parts for assembly along with those molded parts that utilize multiple materials, commonly referred to as multicomponent molding or multimaterial molding.

1.5.2 Material Selection

Based on the part design and the part performance requirements, the plastic material must be selected. These are functional requirements. The required tolerances should also be taken into consideration. For example, if a part is around the length of a ruler that is 30 cm (about 12 inches) long and needs to hold tight tolerances then a material of low shrinkage values must be selected. A polyethylene or a polypropylene material will not be able to hold the tolerances since the shrinkage values are high. A glass-filled polycarbonate could possibly work. The part design may require a special plastic material or a special additive to be added to the base plastic for performance. If a thick section must be present, a filled material may need to be selected or if there is a sliding surface, then an additive such as Teflon to reduce the coefficient of friction may need to be added to the plastic. Material selection should typically take place when the basic part design is done. Once the material is selected and the mold is built based on the shrinkage value for the material then changing the material will change the shrinkage and the mold will now not produce the parts to the required specifications. If the shrinkage values are the same, then it will be acceptable.

1.5.3 Mold Design and Construction

Once the part design and material selection is complete, the mold must be designed and constructed such that it is robust enough to withstand the molding process conditions and the plastic material. For example, during the molding process, the mold can be subjected to high mechanical stresses, especially during the plastic injection and the packing phases. The gates are high-wear areas and there are several places where the air needs to vent out for the plastic to enter the mold. Some plastic materials will require special attention and the mold must be specifically designed with the material in mind. Shrinkage may vary considerably from material to material. All these material specific factors must be considered. The required number of parts over the life of the mold is another factor that will dictate the actual materials of construction. Wear on the mold components must be considered, as the materials chosen to build the injection mold and mold cavities will impact the overall life of the mold and associated amount of maintenance required to keep it production worthy.

1.5.4 Machine Selection

Selecting the right machine for the mold should be done once the mold design is complete. It can be done concurrently during the mold construction stage. The machine plays a very important role in the stability of the molding process. For example, machines with large shot sizes must not be used to mold small shots because the part quality consistency will suffer. Vice versa, using a large percentage of the shot size can give rise to problems with melt homogeneity and therefore issues with fill and dimensions. Small molds must also not be mounted in large machines for fear of mold damage due to excessive clamp tonnage being applied.

1.5.5 Molding Process

Process optimization is the last step before the mold is released into production. This book will cover this topic in detail. If the above four factors and activities are not properly selected or performed, process optimization can be a challenge, if not impossible, without incurring significant cost and delay to the project. At this stage, it is usually very late in the project timeline to make any changes to the part design or mold design, especially because of the cost and time involved. An improperly constructed mold can have a very narrow process window leading to a process that will tend to be unstable. If the material selected is not capable of holding the tolerances, no process will be able to produce satisfactory parts. Molding processes should be robust, repeatable, and reproducible. Refer to Figure 1.8, which shows how the cost of making changes to the project changes and increases as the project progresses. For example, if a feature in the part design has to change after the first shots are taken, the cost is substantially higher since the mold design has to change, the mold will need to modified and then trialed again. The figure shows two dotted lines. The bottom line represents the line that shows that the costs will be minimal if all the first four elements are addressed and optimized before the first shots were taken.

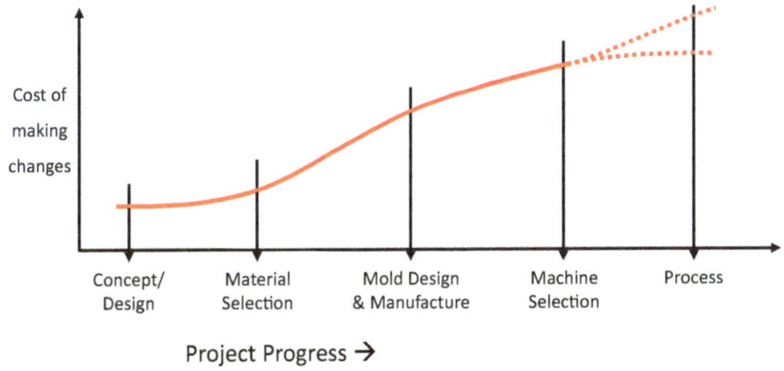

Figure 1.8 Cost of making changes as the each of the phases is completed

1.6 Concurrent Engineering

There are various departments involved in the production of the molded part and therefore regular meetings between the different departments must be held. Each department will have specific knowledge of the selection process and can contribute not just to the process but more importantly predict issues once the mold comes over to their department. For example, getting the process engineer involved in a mold design can help in part orientation in the mold for easy removal, or the mold maker can get help with vent locations based on the process engineer's experience. Involving the quality engineer can help the process engineer understand the required tolerances in the design stage. If the tolerances seem to be unrealistic, they can go back to the product designer for wider tolerances or a material change. There are a lot of benefits associated with implementing concurrent engineering in injection molding. A section is devoted to this topic in this book. Figure 1.9 shows how in the over the wall approach, all the bright people are each working in their own spaces without communicating whereas in the concurrent engineering approach, they are all discussing and contributing the project. The biggest advantage of concurrent engineering is the availability of and access to each member's experience. This contributes to the success of the project.

Over The Wall Engineering Concurrent Engineering

Figure 1.9 Over the Wall and Concurrent Engineering

1.7 Variation

Variation is a natural phenomenon that is present in every process and activity. For example, the time it takes to drive to work has a number, but it can be an average that is collected over a certain period of time. There will be times that are lower than the average and there are times that are above the average. In injection molding, if the lengths of 100 parts are measured, then one could get an average number, but there will be parts below and above this number. Variation can never be eliminated, so the goal should be to minimize it. Variation should be measured in order to predict the quality of the molded parts. As shown in Figure 1.10, a molder could measure the part marked as A and decide that all the molded parts are within specification. However, only when the variation is measured can it be seen that there will be some parts, such as the one marked B, which are out of specification. Once the variation is known, it becomes important to make sure that this variation is always contained within the required specification limits before the mold is released into production.

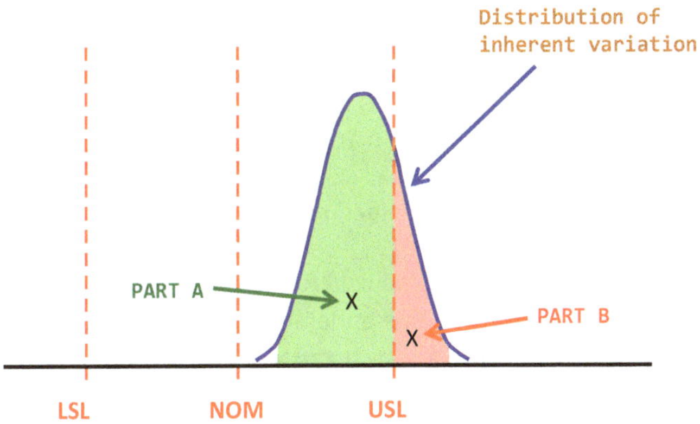

Figure 1.10 Reason to measure variation

A process is said to be capable when all the molded product variation is contained within the specification limits. This is shown in Figure 1.11. The concept of process capability will be discussed in detail in the chapter on quality concepts.

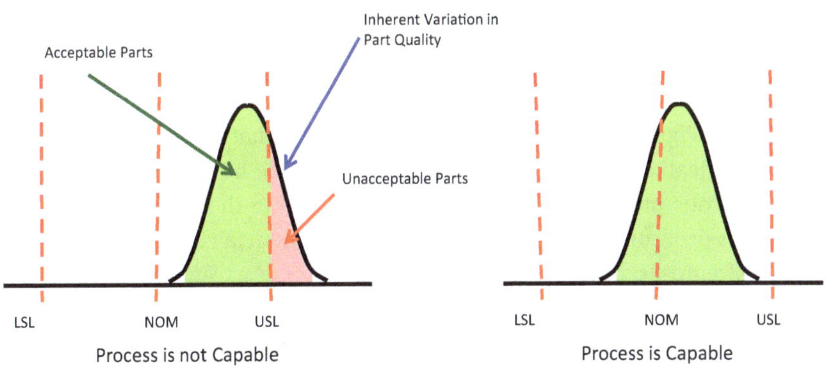

Figure 1.11 Process capability

Variation in injection molding can come from a number of sources, as shown in Figure 1.12. The variation in the molded product is the collective variation from each of these sources, plus many more. Controlling the variation in each source will help the reduction of the overall variation in the final product. The focus of this book will be to discuss the variations that arise from the molding process inputs and how to minimize these variations.

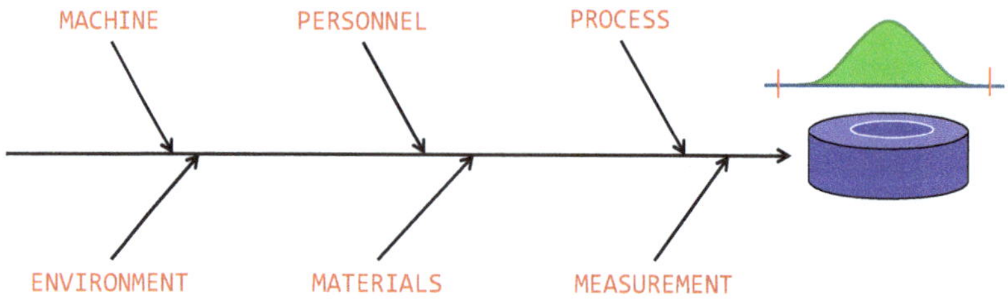

Figure 1.12 Some of the sources of variation in a molded part

In the chapters that follow, the reader will be introduced to the underlying scientific principles to achieve a robust molding process. This understanding will then help in the application of these principles, to develop a robust process and to troubleshoot problems that occur in production. The chapters have been written in a logical sequence to build the readers' knowledge as one would require it or should learn it. However, if the reader is familiar with the topic, he or she can bypass some in favor of other chapters containing the desired information.

1.8 The Molding Parameters and a Recommended Process Development Procedure

There are several process-related molding parameters that affect the part quality. These are shown in Figure 1.13. To achieve a robust process that will produce parts to the required specifications in the most efficient manner, one has to understand and implement a rigorous process development procedure. Note that the phrase is not 'follow' a rigorous procedure, but 'understand and implement' a rigorous procedure. This book will attempt to provide the basics of the molding parameters, what they are, how they can affect the molding process and then go on to explain how they can be optimized. A recommended procedure is shown in Figure 1.14.

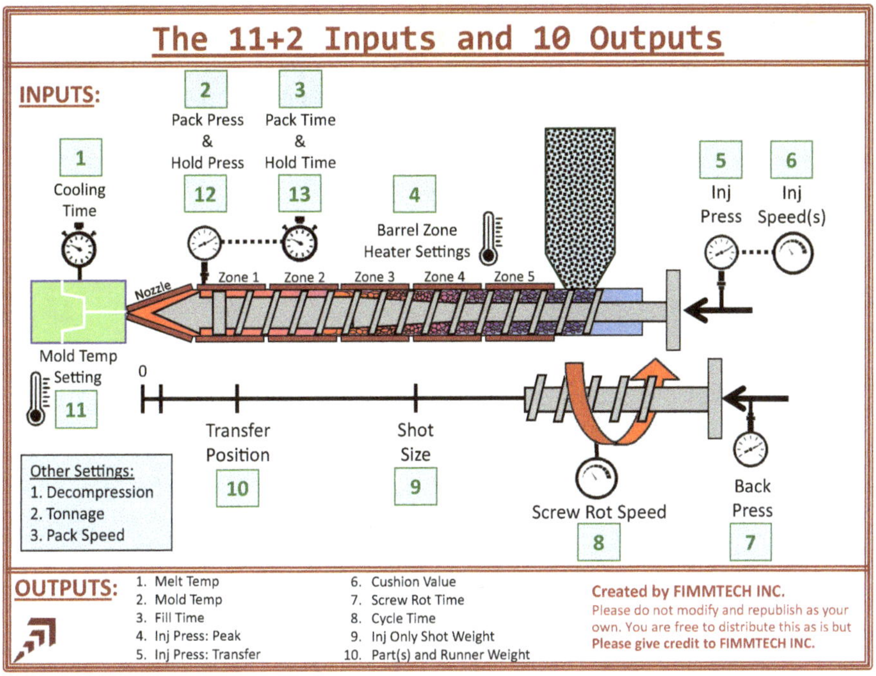

Figure 1.13 The process related molding parameters in injection molding

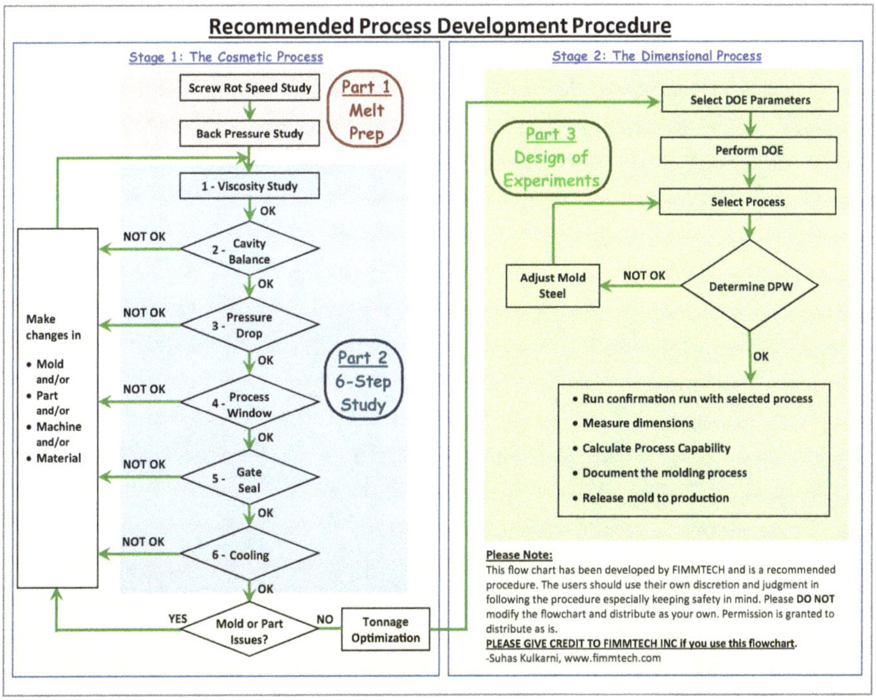

Figure 1.14 Recommended process development procedure

Suggested Reading

Osswald, T. A., Turng, L., Gramann, P. J. (Eds.), *Injection Molding Handbook* (2007) Hanser, Munich

Kulkarni, S. M., *Injection Molding Magazine* (June 2008) Cannon Publications, Los Angeles, USA

2 Properties of Polymers and Plastics That Influence Injection Molding

The term plastic is most commonly used when referring to injection molding materials. Plastics are a class of long-chain molecules called polymers. When polymers have certain properties they are called plastics. Since most of the commercially molded polymers fall under the classifications of plastics, we shall refer to these materials as plastics in this book. The other most commonly molded polymers are thermoplastic elastomers (TPEs) that have the same molding characteristics as plastics but different properties when molded. When referring to these materials, these will be mentioned as TPEs. To understand the concept of injection molding of plastics, a basic understanding about polymers, their properties, and the additives that are incorporated into them is required. This chapter will discuss the topic of polymers and their application to the injection molding process.

2.1 Polymers

Every particle in the universe is composed of atoms. Atoms in turn combine to form molecules. A molecule of water is made up of two atoms of hydrogen and one atom of oxygen. Polymers are very large molecules that have several identical molecules joined together. An ethylene molecule attaches itself to another ethylene molecule and when several thousands or millions of such molecules join with each other, a polyethylene molecule is formed. 'Poly' means many and 'mer' means part. A polymer is many parts chemically joined together. The basic single unit from which it is synthesized is called a monomer, 'mono' meaning one. Polymers are also called macromolecules. The process of converting monomers to polymers is called polymerization. Polymers can also be synthesized from multiple monomers. For example, ABS is synthesized from three different monomers, acrylonitrile, butadiene, and styrene.

Polymers have been around since the beginning of time. DNA, the basic unit of life, is a polymer found in all plants and animals and is a naturally occurring polymer. Today, almost all commercially available polymers are synthesized from natural ingredients. The first commercially synthesized polymers were materials such as ebonite in the late 1800s. Interestingly, the widely used polyolefins gained commercial importance only in the late 1950s, almost a couple decades after the introduction of polyvinyl chlorides, nylons, and polyesters. Recently introduced polymers are based on biomaterials and nanotechnology.

Polymers are synthesized from monomers via a chemical process. There are mainly two types of polymerization processes, the *addition polymerization* process and the *condensation polymerization* process. In the addition polymerization process, a catalyst initiates the polymerization reaction and each monomer adds onto the next monomer until all the monomers are polymerized. A common example of an addition polymer is polyethylene. Polyethylene is polymerized from ethylene monomer, which is a gas at room temperature. The double bond in the ethylene molecule breaks and a bond with an adjacent ethylene molecule is formed. The process continues and the result is a large molecule with high molecular weight. The polymerization process is shown in Figure 2.1.

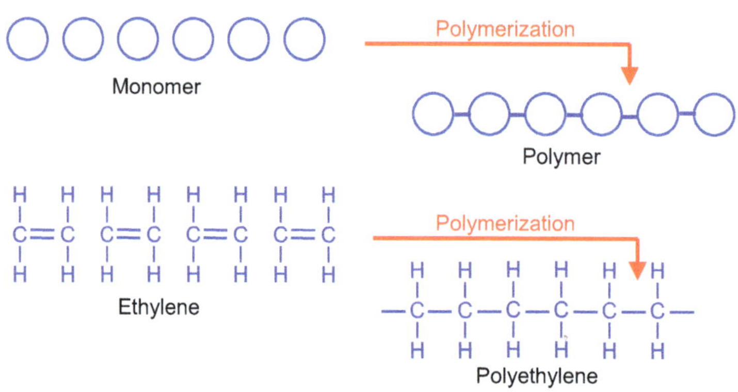

Figure 2.1 The process of polymerization and formation of polyethylene

In condensation polymerization too, each monomer adds on to the next monomer, but this chemical reaction also produces a low molecular weight byproduct that has to be continuously removed out of the system for the polymerization to continue. Condensation polymers are usually polymerized from two or more families of monomers. Nylons and polyesters are examples of condensation polymers. A nylon (chemical name: polyamide) is polymerized from the monomer families of diamines and diacids, as shown in the chemical reaction.

$$nH_2N - R - NH_2 + nHO_2C - R' - CO_2H \rightarrow$$
$$H - (- NH - R - NHCO - R' - CO -)_n - OH + (2n - 1)H_2O$$

R and R′ are the characteristic groups that are present in the monomer. In this case, water is the byproduct. Based on these groups, different types of nylons can be produced. The unit in parentheses repeats itself to form the polymer. If R is $(CH_2)_6$, then the first monomer is hexamethylene diamine and if R′ is $(CH_2)_4$, the second monomer is adipic acid. The polymer that is synthesized from these two monomers is poly(hexamethylene adipamide), commonly called nylon 66.

2.2 Molecular Weight and Molecular Weight Distribution

The repetition of the monomer units causes the molecular weight to increase. The number average molecular weight is the addition of the molecular weights of each of the molecules divided by the number of molecules. Most commercial polymers have a number average molecular weight between 40,000 and 200,000, with some having extremely high numbers. Ultra-high molecular weight polyethylene (UHMWPE) is an example for a molecular weight in the range of 1–6 million. Molecular weights for greases and soft waxes range between 500 and about 3000, whereas some tough and brittle waxes have molecular weights between 3000 and 10,000 [1]. When the attraction between the molecules (intermolecular forces) is high, the materials can gain sufficient mechanical properties at lower molecular weights. Polyamides and polyesters are examples of polymers with strong intermolecular forces. In materials such as polyethylene, where intermolecular forces are low, high molecular weights are required to achieve acceptable mechanical properties. In general terms, molecular weights for polyethylenes are higher those that for nylons or polyesters. UHMWPE was developed for applications in which polyethylenes were the right choice except for their mechanical properties. In most cases, the mechanical properties reach a plateau with increasing molecular weights. Other properties are also affected by molecular weight. Of particular interest to molders is the viscosity of the polymer where an increase in molecular weight results in the increase in viscosity. For melt processing, a certain minimum viscosity is essential for the formation of a processable and homogeneous melt. Processability increases with molecular weight but due to the increase in viscosity, the energy required to process also increases and reaches a point where the increase is not practical for melt processing, see Figure 2.2.

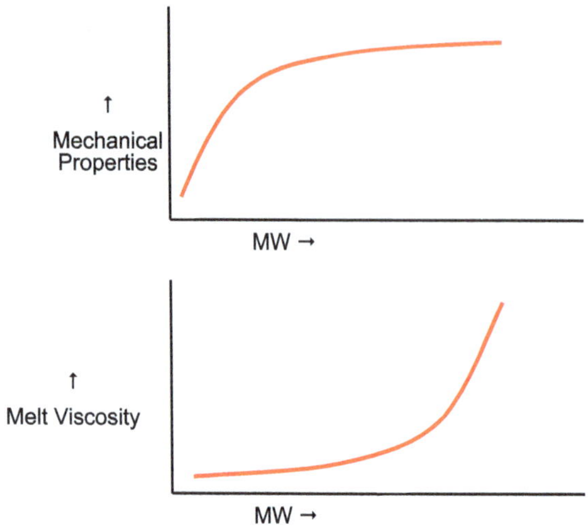

Figure 2.2 Effect of molecular weight on mechanical properties and the viscosity of polymers

The addition of the monomers during the polymerization process (both addition and condensation) is completely random. It is difficult to control the growth of the molecules, which results in molecules of various lengths and therefore varying molecular weights. This in turn leads to a distribution of the molecular weight in the polymer called the molecular weight distribution (MWD) of a polymer. The MWD is an important factor in processing. The lower molecular weight units melt faster than the high molecular weight units. In injection molding, the plastic needs to be injected into the mold as fast as possible to make sure the molecules do not freeze off in the cold mold during injection. If this happens, the part will not fill completely and/or will have built up internal stresses when ejected out of the mold. A narrow MWD ensures that all the molecules are molten during approximately the same period of time. When the residence time in the barrel of the injection molding machine reaches the upper limit, the possibility of molecular breakdown or degradation, resulting in the loss of properties in the final product, increases. This is another reason why a low MWD is desired. However, in the case of extrusion, melt strength is an important element of the process. In this case, the higher molecular weight units have higher viscosity and help to carry the molten lower molecular weight units to form the extrudate. Therefore, a broader MWD is preferred. A narrow MWD would result in the loss of melt strength, and therefore also in the loss of the shape and characteristics of the extrudate profile. The residence time inside the barrel of an extruder is short because extrusion is a continuous process and therefore the risk of degradation is low. This difference in MWD is the decisive factor whether a resin is an injection molding grade or an extrusion grade. Sometimes, extrusion grades are used in injection molding because the viscosity can be low enough to fill the cavities effectively and consistently. The opposite would be less likely, where injection molding grades are used in extrusion. Figure 2.3 shows the difference in the MWD for injection molding and extrusion grades.

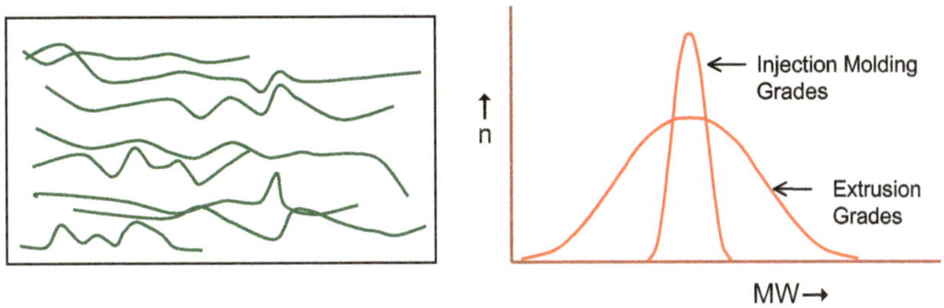

Figure 2.3 Molecular weight distribution for injection molding and extrusion grades

2.3 Polymer Morphology (Crystalline and Amorphous Polymers)

Polymer morphology is the type of arrangement of the molecules in a polymer sample. Based on the different ways the molecules can be arranged, there are two types of polymers: amorphous and crystalline polymers. In amorphous polymers, the molecules are randomly present without any structure or arrangement. Under a high-power microscope, this sample would look like a big bowl of cooked spaghetti. In the case of crystalline polymers, there are certain regions of the sample where the molecules are present in a highly ordered and structured manner. Each of these regions is called a *crystallite*. The difference between the two morphologies is shown in Figure 2.4, and the difference in the properties is mentioned in Table 2.1.

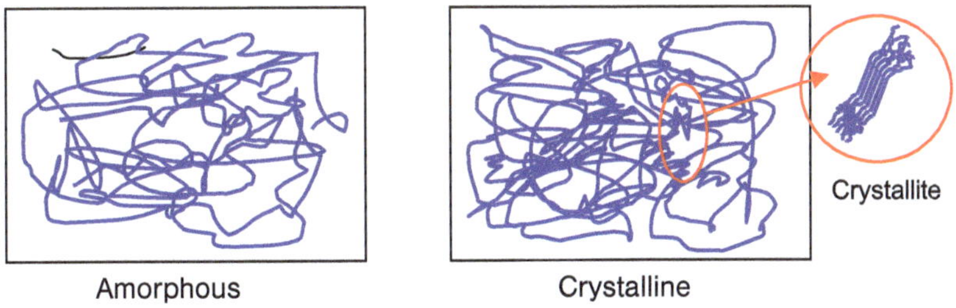

Figure 2.4 Arrangement of molecules in amorphous and (semi)crystalline polymers

Table 2.1 Difference between the Properties of Amorphous and Crystalline Polymers

Amorphous	Crystalline
random structure	ordered structure
broad 'melting' range	narrow melting range
low volumetric shrinkage	higher volumetric shrinkage
lower mechanical properties	higher mechanical properties
transparent/translucent	opaque
e. g., ABS, PS	e. g., nylons, polyesters

No polymer can be completely crystalline and there are always areas in the polymer where the molecules are present in a random manner. A polymer is therefore truly semicrystalline in nature, where crystallites are present in the midst of amorphous regions. Part of a molecule can be present in an amorphous region and part of it can be present in a crystallite, as shown in Figure 2.5. The degree of crystallinity refers to the amount of crystallites present in the sample. Table 2.2 shows the degree of crystallinity for some common polymers.

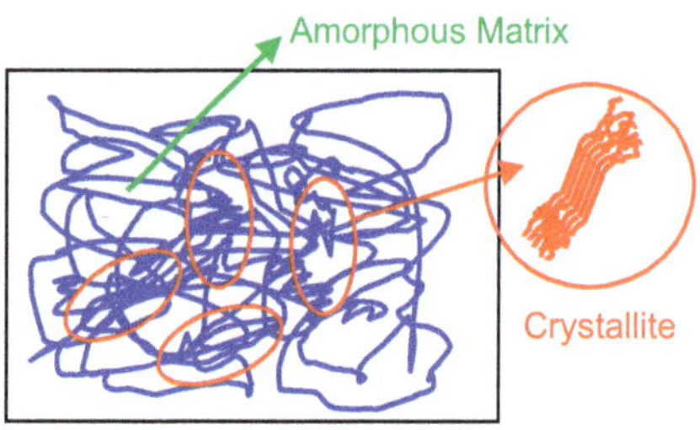

Figure 2.5 Crystallites are present in an amorphous matrix

Table 2.2 Degree of Crystallinity for Common Polymers [2]

Polymer	Degree of Crystallinity
High-density polyethylene	0.80
Isotactic polypropylene	0.63
Poly(ethylene terephthalate)	0.50
Nylon 66	0.70
Nylon 6	0.50

There are two main reasons why a polymer can be amorphous or crystalline: the geometric regularity of the polymers and the strength of the intermolecular forces between them. Geometric regularity is the arrangement of the groups on the main chain. Stereoregular polymers are those where the monomer segments on the main chain are in the same regular configuration. This regularity helps the chains to pack closer and easier together, just as the blocks in the game of Tetris. In the case of atactic polymers, where the monomer segments are randomly oriented, the chance for packing of the molecules is reduced, leading to an amorphous polymer. During the polymerization process of some polymers, it is possible to control the orientation of the monomers and produce stereoregular polymers. Polystyrene is an example of such a controlled polymerization. Regular polystyrene is atactic and therefore amorphous. However, it can be polymerized by using certain metallocene catalysts to form a stereoregular polymer that makes the polystyrene semicrystalline. The properties of crystalline polystyrene are far superior and their mechanical and chemical properties excel. The molecular structures of atactic and syndiotactic polystyrene are shown in Figure 2.6.

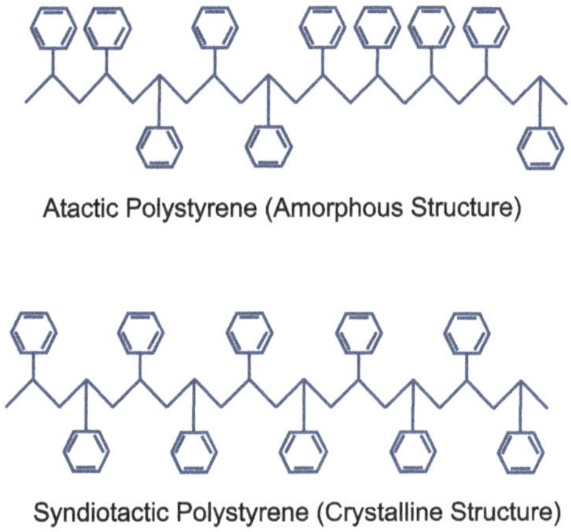

Figure 2.6 Atactic and syndiotactic polystyrene

Another common example is linear polyethylene, which is a highly crystalline and branched polyethylene that is amorphous. Intermolecular forces also play a role in determining crystallinity. The stronger the attraction between the groups of molecules, the higher is the crystallinity. Nylon is an example where the intermolecular forces are high, see Figure 2.7. The hydrogen atom of one molecule has a strong affinity to the oxygen atom of the adjacent molecule, causing the chains to get closer and pack better. This effect can also be observed in the case of polyesters.

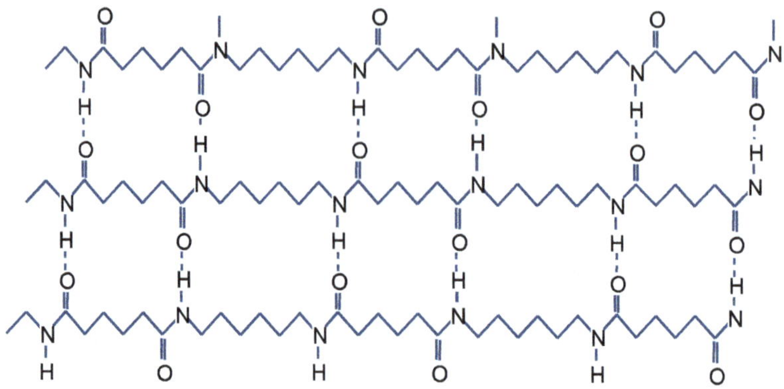

Figure 2.7 Intermolecular attraction resulting in an increase in polymer crystallinity

Table 2.3 lists the morphologies of some common polymers. However, it must be noted that morphologies can be altered during their manufacture and the melt processing of the polymer. For example, mold temperature plays a very important role in the development of crystallinity. Low mold temperatures do not favor the development of crystallites. This is described later in the chapter.

Table 2.3 List of Amorphous and Crystalline Polymers

Polymer	Chemical Name	Amorphous	Semicrystalline
ABS	Acrylonitrile butadiene styrene	Y	
ASA	Acrylonitrile styrene acrylate	Y	
GPPS	General purpose polystyrene	Y	
HDPE	High-density polyethylene		Y
HIPS	High-impact polystyrene	Y	
LCP	Liquid crystalline polymers		Y
LDPE	Low-density polyethylene		Y
PA	Polyamide (nylons)		Y
PAI	Polyamide imides	Y	
PBT	Polybutylene terephthalate		Y
PC	Polycarbonate	Y	
PEEK	Polyether ether ketone		Y
PET	Polybutylene terephthalate		Y

Polymer	Chemical Name	Amorphous	Semicrystalline
POM	Polyoxymethylene (acetal)		Y
PP	Polypropylene		Y
PPS	Polyphenylene sulfide		Y
PSU	Polysulfone	Y	
PVC	Polyvinyl chloride	Y	
SAN	Styrene acrylonitrile	Y	

The relationship between the percentage crystallinity and molecular weight defines the use of the polymer, which defines the use of polyethylene based on this relationship, see Figure 2.8. Low molecular weight and low crystallinity polyethylene is typically used more in soft waxes and greases, whereas high molecular weight and highly crystalline materials tend to be hard plastics.

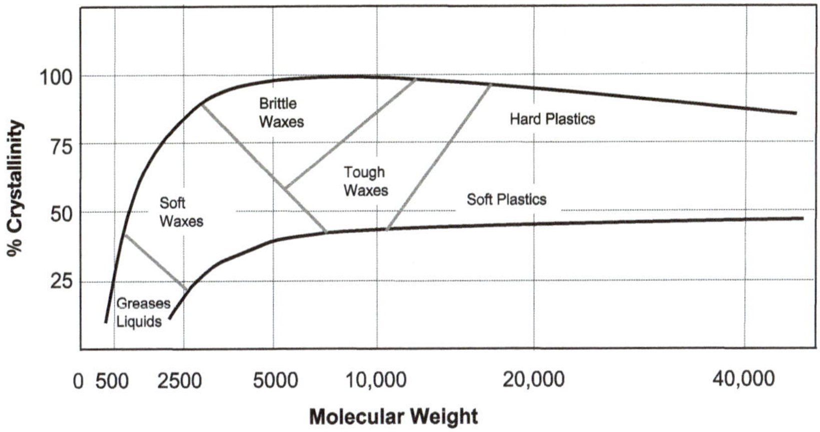

Figure 2.8 Effect of molecular weight and crystallinity on the mechanical properties of polyethylene [1]

Morphology of a Melt

As mentioned in Section 2.3, morphology is the type of arrangement of the molecules. Polypropylenes (PP) are crystalline materials. If a piece of PP is melted, the molecules move away from each other and all the crystallites disappear. There is no systematic arrangement of the molecules placing them in a random arrangement. Since they are random, the PP in its melt form is in an amorphous state. The melt of a crystalline polymer is therefore always in an amorphous state. An exception to this are liquid crystalline polymers (LCP).

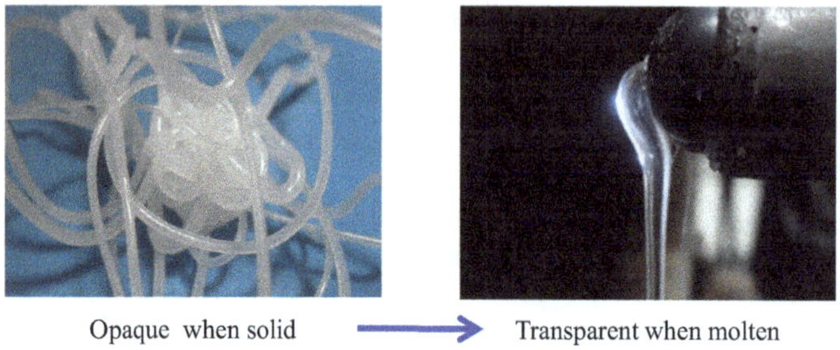

Figure 2.9 The melt of crystalline plastics is always in an amorphous state (except LCPs)

2.4 Role of Morphology in Injection Molding

The molding characteristics of crystalline and amorphous polymers are different. Crystallites are formed because of high molecular attraction and because of the possibility of the chains being unhindered to form the bond. Sometimes just the presence of another molecule or a side chain prevents crystallization. For melt processing, the crystallites must be dissolved and the chains separated from each other in order to reduce the viscosity and inject the melt into the mold. It is this basic nature of forming and dissolving of the crystallites that dictates various differences in processing and melt behavior.

2.4.1 Differences in Shrinkage between Amorphous and Crystalline Materials

Shrinkage is the volumetric change between the melt phase and the glassy or rubbery phase. As the temperature increases, the molecules gain more and more energy, become mobile, and move away from each other. This results in an increase in the volume of the polymer. The intermolecular volume is called free volume. As the polymer cools, the opposite takes place and the free volume reduces, a process known as shrinkage. In crystalline polymers, the movement of the molecules away from each other is much greater compared to amorphous polymers. As the molecules cool, they settle back into a highly structured and closely packed array, which is another contributing reason for their high shrinkage. The absence of such a structure in amorphous polymers negates the need for the molecules to find a definite resting place during the cooling process, resulting in a lower shrinkage value compared to crystalline plastics. Shrinkage values for ABS, an amorphous polymer, are approx. 0.5–0.8% compared to some nylons or acetals that can exhibit shrinkage values of up to 2.5%.

2.4.2 Melt Processing Range

Similar to an ice crystal, the polymer crystallite needs a specific and definite amount of energy to melt. This is called the crystalline melting point and at this particular temperature the crystallite melts. In crystalline plastics, melting occurs over a fairly narrow range of temperatures; this melting range typically covers approx. 20 °C (30 to 35 °F), see Figure 2.10. For example, PBT (Valox 420 from Sabic Innovative Plastics) needs to be processed between 248 and 265 °C (480 to 510 °F), i.e., within a range of 18 °C. The temperature is not as specific and sharp as in the case of simple molecules because of the presence of the amorphous regions and other attraction forces. The lower the percentage of crystallinity, the broader is the processing range.

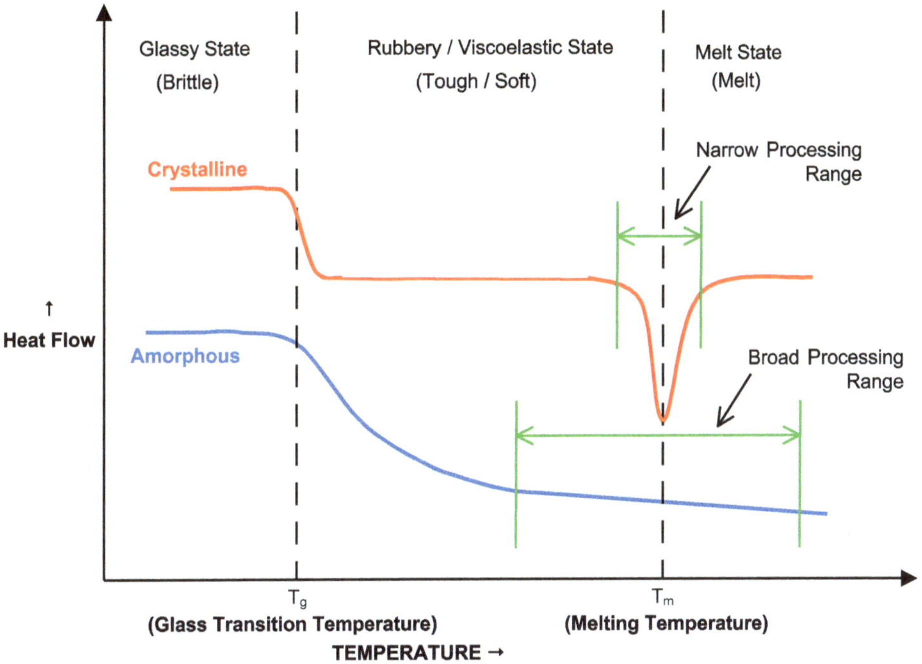

Figure 2.10 Thermal transitions and melt processing ranges for amorphous and crystalline plastics

In the case of amorphous plastics, there is no melting point but the molecules soften over a range of temperatures. There is a recommended processing temperature range for amorphous plastics. In this range, the viscosity of the plastic is low enough to flow and fill the cavities. For example, ABS (Cycolac from Sabic Innovative Plastics) can be processed from 218 to 260 °C (425 to 500 °F), a range of 42 °C.

2.4.3 Mold Filling Speed

Viscosity of the plastic and melt temperature are inversely related. As the plastic temperature increases, the viscosity decreases. As the plastic flows through the cold mold, the temperature of the plastic drops and the viscosity increases. There is also the phenomenon called fountain flow that will be discussed at length in the following chapters. Since for crystalline plastics, the processing range is narrow, the temperature of the flow front must always be higher than the minimum required melt temperature. In the above example of the PBT, the flow front temperature must never drop below 248 °C (480 °F) before the cavity is completely filled. This narrow processing range of crystalline polymers therefore dictates the fact that the plastic must be injected as fast as possible into the mold.

In the case of amorphous plastics, because the plastic stays viscous over a broader range of temperatures, slow injection speeds are permissible, as long as the flow front stays above the minimum processing temperature. This is typical in the molding of lenses and other optical parts.

2.4.4 Mold Temperatures

Similar to the melting of the crystallites at a particular temperature, the crystallites also start to form at a particular temperature, a temperature lower than the melt temperature. This is called the crystallization temperature. This temperature supplies the necessary energy for the formation of the crystallites. If the mold is too cold, the polymer will not be able to receive this energy, preventing the formation of the crystallites. This leads to the loss of the properties in the molded part. Material suppliers conduct extensive research to determine suggested operating mold temperatures. Therefore, it is highly recommended to stay within the recommended specifications for crystalline materials. This concept is further explained in Section 2.4.6 on thermal transitions.

Since crystallites are absent in amorphous plastics, the mold temperature range can be broader and can be especially extended towards the lower temperatures. Because crystallites do not need to be formed, the molecules are not looking for any particular amount of energy; therefore, lower temperatures are acceptable. However, care has to be taken to prevent molded-in stresses when parts are produced using extremely cold molds. Mold temperatures also play a significant role in the cosmetics of the part.

2.4.5 Barrel Heat Profile

In the injection molding barrel, the screw performs the function of conveying and melting the polymer. The base of the hopper is where the polymer pellets first come in contact with the screw. This section is designed to convey and then soften the pellets. In case of crystalline polymers, the crystallites need a lot of energy to melt, so this section, which is usually the second heating zone from the back of the screw, is set at a higher temperature than the next zones in order to initiate the softening of the molecules. But because crystalline plastics can also be heat sensitive or cannot stand high temperatures for long periods of time, the temperature in next heating zone is reduced. This leads to a heating profile that has a hump in the middle, which is typical for crystalline plastics. In case of amorphous plastics, such a profile is not necessary because they need less energy to soften and can tolerate longer residence times in a heated barrel, as is shown in Figure 2.11.

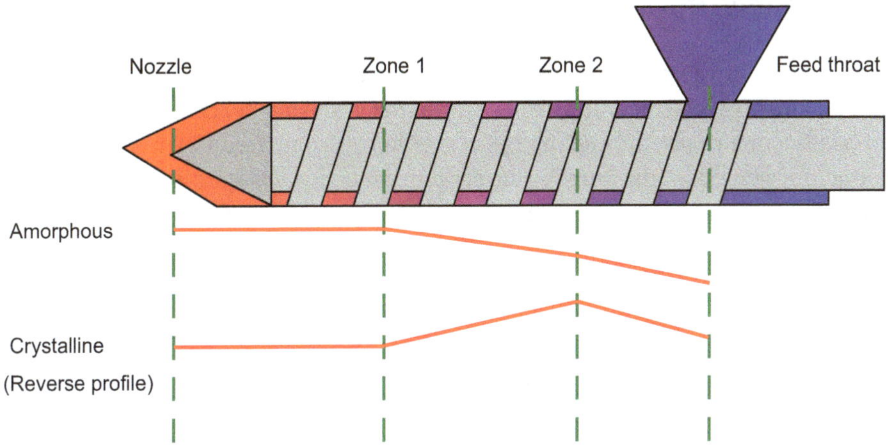

Figure 2.11 Barrel heat profiles for amorphous and crystalline plastics

2.4.6 Screw Recovery Speeds

The barrel heaters provide heat on the outside of the cylinder. Because plastics are bad conductors of heat, the plastic closer to the screw and furthest from the wall of the barrel requires additional heat to become plasticized. This additional effective melting energy comes from the shear friction caused by the rotating screw. High screw speeds generate high amounts of shear, helping the crystallites melt and ensuring melt homogeneity. For amorphous plastics, due to the low energy required to melt the plastic, high screw speeds are not critical. In fact, high screw speeds can degrade the material and cause defects such as splay.

2.4.7 Nozzle Temperature Control

Nozzle temperature control is critical while processing crystalline polymers. Not only must the temperature of the nozzle be maintained within the processing range, it should be maintained within a temperature range that is much narrower than the processing range. This is especially important during the static phases of the plastic flow, which follows the holding phase of one cycle and precedes the injection phase of the next cycle. Lower temperatures tend to freeze off the plastic in the nozzle tip making it impossible to inject the next shot. At higher temperatures, all the crystallites will have disappeared making the viscosity very low and causing what is commonly called nozzle drooling. There are some innovative designs for nozzle tips available in the market to help prevent this problem. With amorphous plastics, the broad processing range helps maintain the viscosity in the nozzle and prevents freeze off or drool.

2.4.8 Cooling Times

Once the crystallites are formed, the molded part gains sufficient strength and only needs a small drop in temperature to be ejected out of the mold. Therefore, in the case of crystalline materials, the cooling times are shorter compared to amorphous plastics (for the same part thickness). Nucleating agents are added to some crystalline plastics to accelerate crystallization, thus reducing the cooling time even further. Nucleating agents have no effect on amorphous materials because there are no crystallites to be formed.

2.4.9 Mechanical Properties

The crystallites provide mechanical strength to the polymer. They are like a rope, rather than a bundle of grass, providing strength. Generally speaking, crystalline plastics have higher mechanical properties than amorphous plastics. Therefore, most are referred to as engineering resins. However, with the advent of new technologies and the discovery of new additives, the properties of amorphous materials can be easily modified to match those of crystalline materials.

2.4.10 Optical Clarity

Most amorphous polymers in their natural and unmodified state are optically clear. The distance between the molecules is large, allowing the wavelengths of light to pass through and making them transparent: polystyrene is an example of this. For crystalline polymers, the packing of the molecules does not allow the passage of light and therefore they are usually opaque. As the degree of crystallinity decreases, the materials tend be translucent. As will be discussed in the next section, melts of any polymer, crystalline or amorphous, are always amorphous and therefore a purge of an unfilled melt will always look clear. Polyethylene, a crystalline plastic, is opaque but a small amount of melt drool out of the nozzle tip will always be transparent.

A summary of the processing differences between amorphous and crystalline materials is summarized in Table 2.4. These are only broad comparisons because part and mold design can affect every one of these parameters.

Table 2.4 Summary of the Processing Differences between Amorphous and Crystalline Materials*

Process	Amorphous	(Semi) Crystalline
Mold fill rates	Slower rates are acceptable	Fast rates are required
Importance of mold temperatures	Mainly for improved cosmetics and stress relief	For mechanical properties, cosmetics and stress relief
Molding barrel settings	Regular profile	May require a reverse profile
Melt heat stability	Fair	Low
Molding screw rotation speeds	Lower are acceptable	Require high
Nozzle tip temperature control	Easy	Difficult because of freeze off/drool
Cooling times	Higher	Lower

* These are broad comparisons only. Part and mold design can affect every one of these parameters.

2.5 Exceptions of Morphology Rules to Polyolefins

Polyolefins are mostly crystalline materials. It would therefore be logical to follow the processing guidelines that are given for crystalline materials. However, there are some properties that these crystalline materials possess that belong to the amorphous category:

- They have sharp melting points but by nature they are soft and can therefore be processed for a small range below their melting temperatures. These materials are also thermally stable for a wide range above the melt temperature, and in some cases, even up to 200 °F above the melt temperatures. Therefore, the overall processing range for these materials is very wide although they are crystalline materials.

- Because they have a wide processing range, slower injection speeds can be used to inject the material into the mold.
- Crystallization temperatures dictate mold temperatures; therefore, mold temperatures would have to be set between 70–110 °C for polyethylenes and much higher in case of polypropylene. However, in most cases, it is perfectly acceptable to deviate from this range and mold parts are lower mold temperatures or even using chilled water as the cooling medium. The reason for this is twofold. First, since the T_g of olefins is lower than room temperature, complete crystallization can happen as a post molded phenomenon. Therefore, ultimately the molded part will achieve the required properties. The associated danger is the presence of molded in stresses, dimensional variation, and warpage in the parts. The majority of the parts molded with olefins are commodity products, and therefore the drawbacks are usually overlooked. As an example, a trash can molded from polypropylene does not have tight tolerances that have to be maintained. In most cases it is mostly cosmetic with some load testing. The end user could use it for a few years during which if it get defective it is thrown away. In some cases these drawbacks are remedied by using fillers and additives. Depending on the quality requirements, the mold temperatures must be set appropriately.
- Screw speeds need not be on the higher end since the crystallites are soft and will ultimately melt because of the smearing action. Screw speeds must be optimized to create a good homogeneous melt. If there are colorants or other additives that may burn and degrade at the higher screw speeds, then it is acceptable to reduce the screw speeds.

Because of the soft nature of the plastic, ejection temperatures are closer to room temperatures. This increases the cooling time and therefore cycle times are usually long. As a related side note, part design can play an important role and reduce cycle times significantly.

2.6 Thermal Transitions in Polymers

Refer to Figure 2.12, which shows the lid of a container that is not closing correctly after a few months of use. The lid as has shrunk and therefore does not fit on the container. This is called post mold shrinkage. Post mold shrinkage is a common phenomenon that occurs to a varying degree in all molded components and must be accounted for. This phenomenon is related to the thermal transitions of polymers. Thermal transitions are also responsible for several other molding-related topics such as mold temperatures, melt temperatures, and surface finish of the product. There is one main thermal transition for amorphous and crystalline materials, which occurs at the *glass transition temperature*. In the case of crystalline materials there are three main transitions. They occur at the glass transition temperature, the crystallization temperature and the melt temperature.

Figure 2.12 Lid of the container does not close after a few months of use

Although there are no polymers that are completely crystalline, for the sake of the following discussion, let us assume that they do exist. Therefore, in the following discussion, a crystalline polymer would mean a polymer that is 100% crystalline, an amorphous polymer would mean a polymer that is 100% amorphous, and a semicrystalline polymer would be a polymer that is partially crystalline with the crystallites present in the amorphous regions.

First consider an amorphous polymer submerged in a very cold liquid such as liquid nitrogen. The temperature of liquid nitrogen is anywhere between −210 to −195 °C (−345 to −320 °F). At this temperature, the different molecular energies are almost nonexistent and therefore the molecules are not free to move, resulting in a brittle polymer. A sheet of flexible plastic, when quenched into liquid nitrogen, becomes brittle. If dropped on a concrete floor, it would sound like glass and shatter into bits and pieces. However, if the temperature of the sheet is gradually increased, the thermal energy from the increasing temperature provides energy to the molecules. Depending on the polymer the sheet is made of, at a given temperature the sheet becomes flexible. There is a transition from a brittle state to the flexible state. The temperature at which this transition occurs is called the *glass transition temperature* or the T_g of the polymer. At T_g, the molecules have sufficient energy to move and the polymer is flexible. A further increase in the temperature provides more energy, the flexibility increases, and eventually the molecules become soft enough to form a viscous mass suitable for melt processing. An increase in the energy of the molecules results in the increase in the specific volume of the polymer. Specific volume is the volume per gram of the polymer. A plot of specific volume versus temperature, like the one shown in Figure 2.10, shows a linear increase in the volume until an inflection point is reached, where the slope of the line changes. The inflection point reflects the glass transition temperature of the amorphous polymer. Further increase after the glass transition shows a steady increase in the specific volume.

Next, let us look at a crystalline polymer and perform the same experiment. Recall that we are considering a 100% crystalline material. In crystalline polymers, the intermolecular forces are very high and therefore a considerable amount of energy is required to move the molecules away from each other. As the temperature increases, the molecules gain more and more energy but the intermolecular forces prevent any movement. At the temperature where the molecules move away from each other and become mobile, the molecules are very flexible and the polymer is now a molten mass. The crystallites need a definite amount of energy to melt. When they receive this in the form of thermal energy they melt all at once. The phenomenon is similar to the melting of low molecular molecules, such as water where ice turns into water at 0 °C. For this reason, crystalline materials show a sharp melting point. Because the transition from a solid to a liquid is sudden, the transition is referred to as melting and the transition temperature is called the *melt temperature* or T_m. A plot of specific volume versus temperature shows an inflection point at the T_m, as can be seen in Figure 2.13. Since there are no crystallites in amorphous materials these materials do not have a melting temperature.

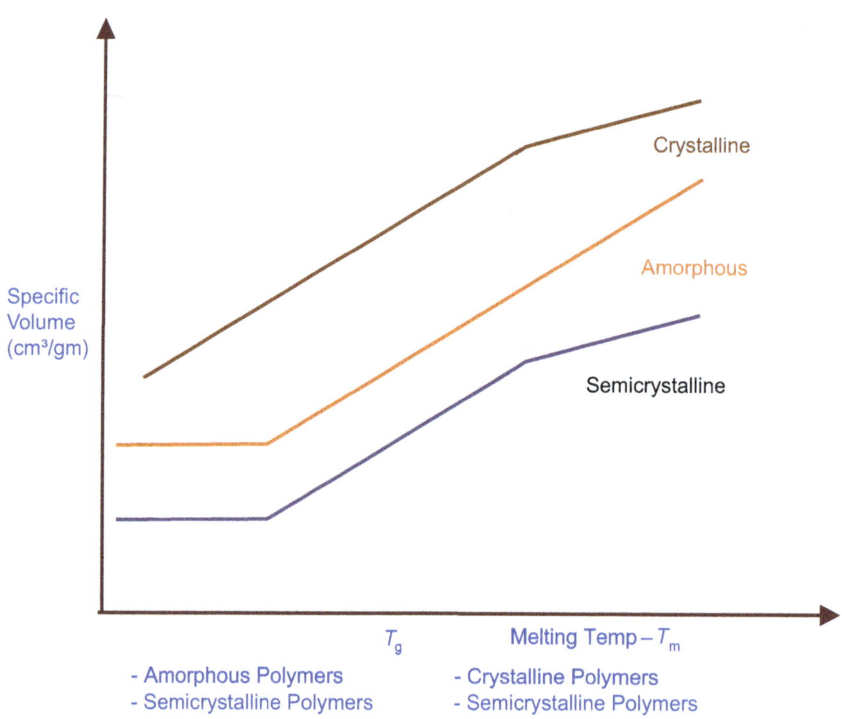

Figure 2.13 Specific volume versus temperature for polymers

Based on the thermal responses of amorphous and crystalline polymers, it is now easy to predict the response of a semicrystalline polymer. As state previously, a semicrystalline polymer can be thought of as crystallites present in an amorphous matrix of molecules. This sample will exhibit the properties of the crystalline and the amorphous polymer and will therefore have a glass transition temperature and a melting point temperature. On the specific volume plot, there will be two inflection points representing each of the transitions, as shown in Figure 2.11. However, in case of semicrystalline polymers, the inflection point is not as sharp as one would see in the amorphous or crystalline samples. Rather, there is a range where the inflection starts and ends. This happens because of the different sizes of the crystallites and differences in molecular lengths. Once the molecules of a semicrystalline material have melted, they now have a lot of thermal energy and any additional energy can start breaking down the molecules causing degradation of the polymer.

This difference between the crystalline and amorphous materials leads to the fact that semicrystalline materials have a much narrower melt processing window compared to amorphous materials. For example, nylon (a semicrystalline material) will have a melt processing window between 248 and 265 °C (480 to 510 °F), while an ABS (amorphous material) has a window between 218 and 260 °C (425 to 500 °F). The processing window is 17 °C for nylon, compared to 42 °C for ABS.

Now, let us consider the reverse process of phase transitions. When the temperature of a melt of a crystalline polymer is gradually reduced, the molecular energy starts to reduce. The viscosity of the melt starts to increase and the crystallites begin to reappear. The temperature at which the crystallites occur is called the *crystallization temperature* or the T_c. For crystallization to occur, the melt must be subjected to the crystallization temperature for a finite amount of time. If the melt of a crystalline polymer is rapidly quenched below the glass transition temperature, the solidified polymer will not exhibit any crystallites and will be completely amorphous. It is for this reason that mold temperature is of greater importance when molding semicrystalline polymers as compared to amorphous polymers.

The above discussed temperatures affect the properties of the polymers and therefore decide the final applications of the polymers. For example, for the product designer, the T_g is one of the important factors of consideration. For a product to be flexible at room temperature, its T_g must be below room temperature, such as for elastomers. For a product to be rigid at room temperature, the T_g must be above the room temperature. For the molder, the knowledge about the crystallization and melting temperatures is important because these dictate the processing conditions. The T_c is used to determine the mold temperature range to start and promote the crystallization of the molecules. The T_m is used to determine the melt temperature ranges. A processing data sheet does not provide these as typical values but they are reflected under the processing conditions. The material manufacturer performs the analysis and uses the results as one of the many tests to recommend the processing conditions.

The differential scanning calorimeter (DSC) is an instrument used to determine the thermal transitions in polymers. A typical graph generated by a DSC is shown in Figure 2.14. For crystalline polymers, two scans are performed. During the first scan, the polymer is taken from a low set temperature all the way past the melting temperature, and the transitions are recorded. The sample, which is now in the molten form, is immediately quenched in liquid nitrogen. Most polymers are below the T_g at this temperature. Since the melt is amorphous and is immediately quenched, all the energy is taken away and the frozen polymer is also completely amorphous. The DSC scan is repeated. As the temperature increases, the molecules gain more and more energy. The glass transition is noticed and when the crystallization temperature is reached the crystallites begin to form. Crystallization temperatures are recorded on the second scan. For amorphous polymers, a second scan is not necessary. Note that the DSC curves look flipped based on the type of DSC instrument used.

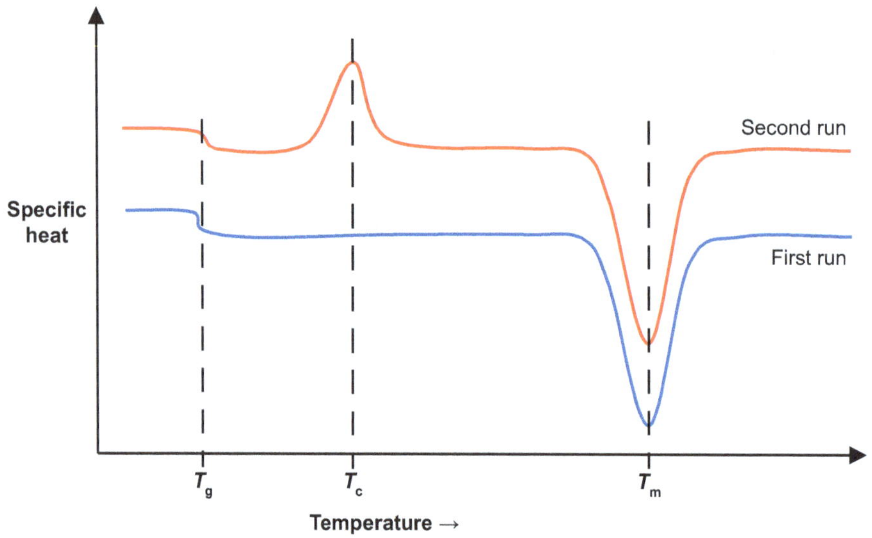

Figure 2.14 Typical representation of a DSC plot for polymers

The relationship of mold temperature and crystallinity is explained in Section 2.4.4. Refer to Figure 2.15; the figure shows the difference in the resulting morphologies when a polymer sample is quenched below the T_g and when one is subjected to the crystallization temperatures for a given time.

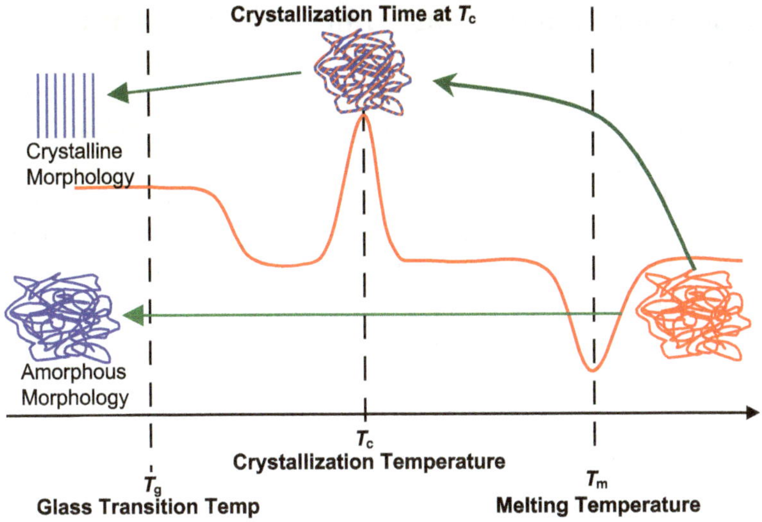

Figure 2.15 Resulting morphologies depending on polymer conditioning

The % crystallinity also depends on the mold temperature. Refer to Figure 2.16, which is a DSC scan for a nylon that was molded at mold temperatures of 82 °C (180 °F) and 32 °C (90 °F). The nylon molded at 32 °C had a % crystallinity of 20.95% and that molded at 82 °C had a % crystallinity of 31.5%.

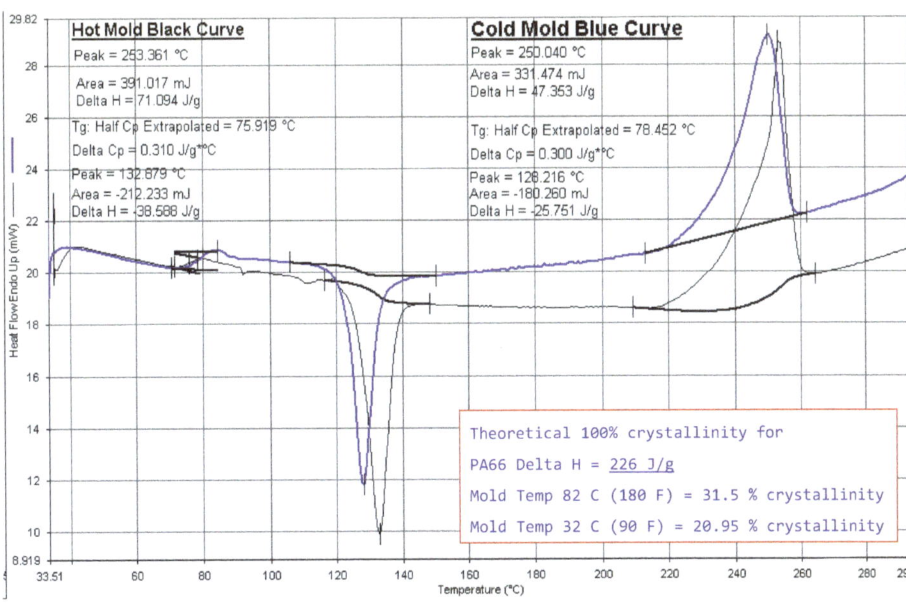

Figure 2.16 DSC graph for nylon molded at mold temperature of 82 °C (180 °F) and 32 °C (90 °F) (previous schematics shown are for a flux DSC)

2.6.1 Relationship between the Glass Transition Temperature and Post Mold Shrinkage

Below the T_g, the molecules do not have sufficient energy to be mobile. The plastic is brittle at temperatures below the T_g. Above the T_g, the plastic molecules have the energy to be mobile and the plastic is therefore flexible. Referring to Figure 2.17, consider the room temperature at 25 °C. Both parts shown in the figure are compact disc covers. One is molded out of polystyrene (PS) and the other out of polypropylene (PP). At room temperature, the PS cover breaks when one tries to bend it because the T_g of PS is about 100 °C. However, the PP cover flexes but does not break because, depending on the grade of the PP used, the T_g of PP can be anywhere between 0 and −30 °C.

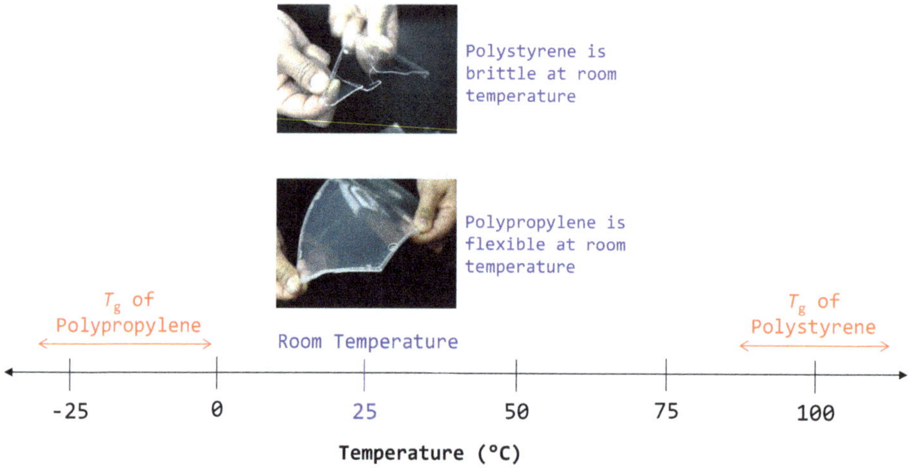

Figure 2.17 PS breaks but PP flexes at room temperature because of the position of their glass transition temperatures, with respect to room temperature

Consider a molding process with the following conditions:

- Ambient temperature is 25 °C.
- Temperature of the molded part when ejected out of the mold is 65 °C.
- The time that the part takes to cool from 65 °C to the ambient temperature of 25 °C is 7 minutes (see Figure 2.18).
- A critical dimension is the length of the part.

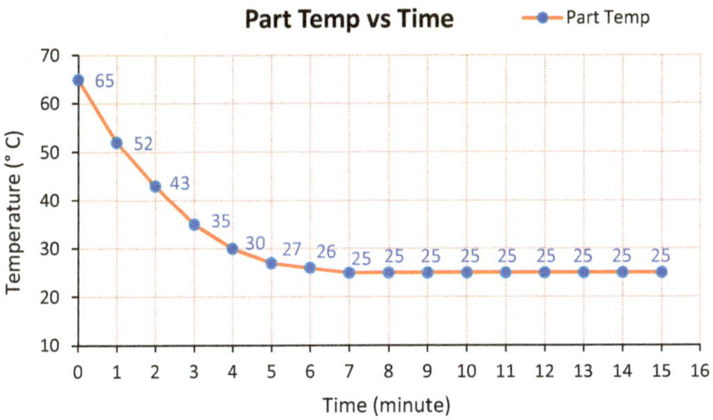

Figure 2.18 Change in part temperature with time

Two materials, Plastic A and Plastic B, were molded. Plastic A has a T_g of 35 °C. This means that above 35 °C, the molecules are mobile and have the energy to move, and below 35 °C, the molecules do not have sufficient energy to move. Plastic B has a T_g of 10 °C. This means that above 10 °C, the molecules are mobile and have the energy to move, and below 10 °C, the molecules do not have sufficient energy to move.

Plastic A: refer to Figure 2.19. The part that is molded with Plastic A has a T_g of 60 °C. From the graph, we can observe that the time the part takes to reach 60 °C is 3 minutes. Because molecules are mobile above the T_g, the material will continue to shrink until it reaches the T_g. Therefore, in this case, the length of the part will keep changing until the temperature of 60 °C is reached, or in other words, the length will keep changing for the first 3 minutes and it will then will stay constant. Therefore, post molding shrinkage after the part has reached its T_g is nonexistent, and the dimensions do not change once the parts have been molded and are always subjected to temperatures below the T_g. In the this example, the length of the part stays constant over time.

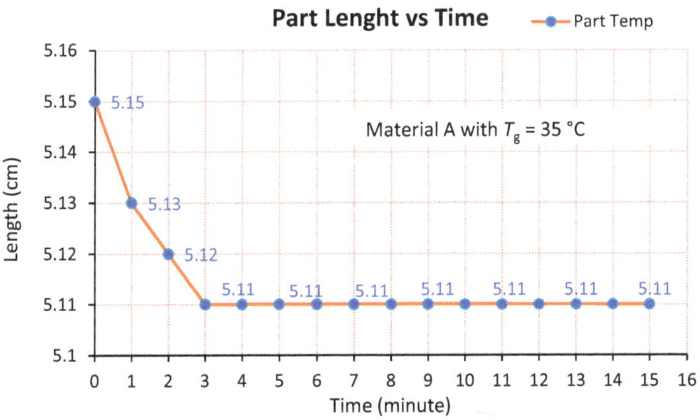

Figure 2.19 Change in part length with time for Plastic A with a T_g of 35 °C

Plastic B: Refer to Figure 2.20. The part that is molded with Plastic B has a T_g of 10 °C. Because the ambient temperature is above the T_g of the plastic, the molecules in the molded part will never reach the T_g. Therefore, in this case, the molecules will always have enough energy to keep moving and reach their equilibrium positions in ambient temperature. This movement causes shrinkage, leading to a continuous change in part length until an equilibrium is reached. This shrinkage can occur over a longer period of time, which could be in hours or even days. In the this example, the length of the part molded from this Part B with a T_g of 10 °C will continue to change over time until it reaches the final equilibrium.

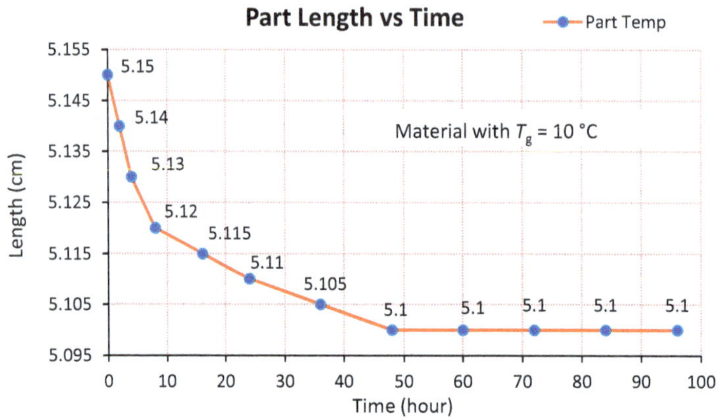

Figure 2.20 Change in part length with time for Plastic B with a T_g of 10 °C

Commodity plastics, such as polyethylenes and polypropylenes, have T_g values that are lower than room temperatures and, therefore, they are susceptible to higher degrees of post mold shrinkage. Often one notices that over time, lids of plastic containers do not fit well and become warped. This is the result of post mold shrinkage. CD discs and DVD discs are molded from polycarbonate, which has a T_g value of about 150 °C. Therefore, one will probably never come across a CD or a DVD that is warped. The T_g is reached very quickly as the disc gets ejected from the mold. It also helps that CDs and DVDs are even-walled and thin.

The T_g is what is called as an *alpha transition*, and it has the most effect on the post mold shrinkage phenomenon. Every polymer also has other similar transitions below the alpha transitions called the *beta* and *gamma transitions*. In reality, there is still a very small movement of the molecules below T_g, but is not as significant as compared to the movement above the T_g. Therefore, a small amount of post mold shrinkage will always continue to happen even below the T_g. Sometimes, this can be confused with and/or added to the stress relief that could be also occurring at the same time. Stress relief can happen below or above the T_g.

2.6.2 Importance of Mold Temperatures in Injection Molding

In addition to the formation of the crystallites for crystalline materials, there are two other reasons for both amorphous and crystalline materials why mold temperatures are important. They are for releasing stress in the part and to achieve the required surface finish of the part.

Refer to Figure 2.21, which shows a music cassette that was processed under conditions that led to in-molded stress. The cassette was lost in the interior of the car and was found several months later after a hot summer. The heat from inside a parked car relieved the in-molded stress and caused the warpage. In this case, the molder, probably to reduce cycle time set the mold temperature too low, quickly freezing the incoming molecules in place, leaving them in a stressed position. This stress was released due to the heat in the car. Setting the mold temperatures higher and in the recommended range would have avoided this problem.

Figure 2.21 Warpage due to release of in-molded stress

The common stress relief technique of annealing is based on the concept of T_g. During the injection molding process, because of the high melt pressures, stresses can develop in the part. If the part is taken slightly above the T_g, these stresses can be relaxed. There can be a dimensional change associated with the annealing process for the reasons previously described in this section.

The surface finish, also called the texture of the part, is a specification that is present on every part print. The plastic picks up the texture from the mold surface as shown in Figure 2.22. The mold is colder than the plastic, which begins to cool as soon as it touches the mold. The rate of cooling will depend on the temperature of the mold. The lower the temperature, the faster the cooling rate. At lower temperature the plastic could cool off before it gets a chance to get packed into the texture, resulting in a part with a texture that is not acceptable. Therefore a higher mold temperature will always provide a better surface finish. This is illustrated in Figure 2.23.

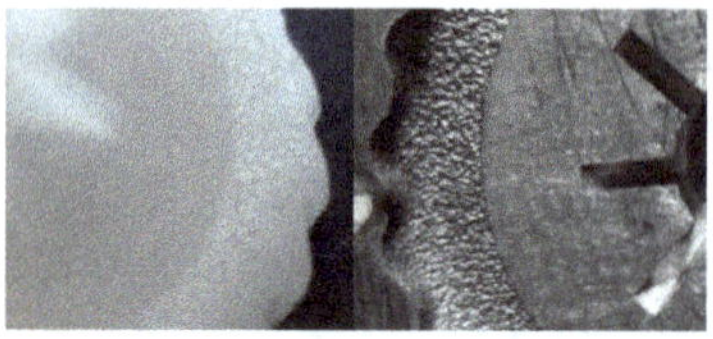

Figure 2.22 Texture of the mold replicated on the part

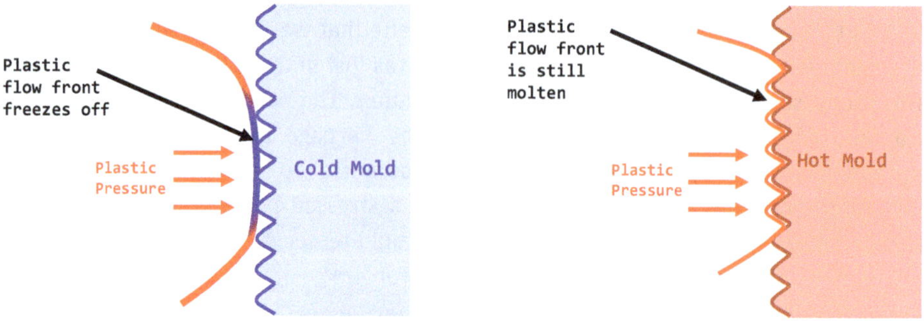

Figure 2.23 Effect of mold temperature on the temperature of the plastic flow front and part texture

Sometimes, it is not practical to set the mold to the high temperature settings that are required to achieve the surface finish. Setting the temperatures high will not let the plastic cool down to the ejection temperatures and/or will increase the cooling time and therefore the cycle time. Technologies have been developed whereby the surface of the mold is heated just before injection and then the mold is cooled down after the injection, pack and hold phases are done. An example of a toy car molded with similar technology is shown in Figure 2.24. The material was a glass-filled nylon. The side of the mold that molded the front part of the windshield of the car was heated and cooled, producing the mirror surface and picking up the fine texture of the mold. The other side was kept at a lower temperature. This cooled down the melt almost instantaneously and froze the plastic–glass mix on the surface, producing the rough surface as seen in the figure. In the case of the windshield side, the molten plastic had the time and viscosity to filter through the glass and come out on to the surface, giving the mirror finish.

Figure 2.24 Effect of mold temperature on surface finish (Courtesy: Trexel, Inc.)

Table 2.5 Glass Transition and Melting Temperatures*

Polymer	T_g (°C)	T_m (°C)
Natural rubber	−73	36
Nylon 6	50	250
Polybutadiene (trans-)	−54	47
High-density polyethylene (HDPE)	−125	146
Polypropylene (syndiotactic)	−8	204
Polystyrene (PS)	100	250
Polyvinyl chloride (PVC)	−18	191
Polycarbonate (PC)	150	243
Acrylonitrile butadiene styrene (ABS)	104	–

* Because there are several grades of the same base polymers, and moreover, because the plastic purchased from suppliers always has additives blended in, the values here should be used for reference only. As in the case of ABS, which is a ter-polymer, the T_g will drop with increasing amounts of the butadiene content.

2.7 Shrinkage of Polymers in Injection Molding

Shrinkage occurs when the melt begins to cool and the molecules start to return to their desired equilibrium states. The distance between the molecules is higher in the melt than when they are cooled. As the melt cools, the molecular distance decreases, reducing the free volume and causing shrinkage. The higher the increase in volume during the melting phase, the higher is the shrinkage. The shrinkage of the plastic during the injection molding process can be easily affected by the various molding parameters. This poses the biggest challenge to mold makers when sizing the core and cavity steel and to the processors in order to maintain the part quality during production. It is therefore important to develop a robust and stable process that will produce parts with consistent shrinkage and that is least affected by natural variations. The shrinkage of injection molded parts also depends on the direction of plastic flow, see Figure 2.25. The plastic is injected through the gate and the molecules get oriented in the direction of flow. Given a chance to relax and with sufficient available energy the molecules will get back to their original non-oriented equilibrium state. But this is not always the case, because molding cycles need to be fast and wall thicknesses are small, inhibiting relaxation. In addition, typically the molecules are under mechanical stress from the injection, pack and hold forces. This causes a variation in shrinkage between the direction of flow and the direction perpendicular to flow. The effect is more pronounced for crystalline materials because they have a high shrink value caused by the formation of the crystallites. Amorphous materials too exhibit a difference in shrinkage depending on flow direction, but the difference is not as pronounced. Materials exhibiting varying shrinkage values in the cross and parallel flow directions are known as anisotropic materials; materials with identical shrinkage values are called isotropic materials. For example, polyesters are anisotropic materials while ABS is an isotropic material. For Valox 357, a polyester manufactured by Sabic Innovative Plastics, the parallel flow shrinkage values range from 1.0 to 1.4% and the cross flow values range from 1.2 to 1.6%. Although there is an overlap, the average values for the cross flow are higher. For Starex AB-0760, an ABS manufactured by Samsung, the shrinkage values in the cross and parallel flows are both 0.30 to 0.60%.

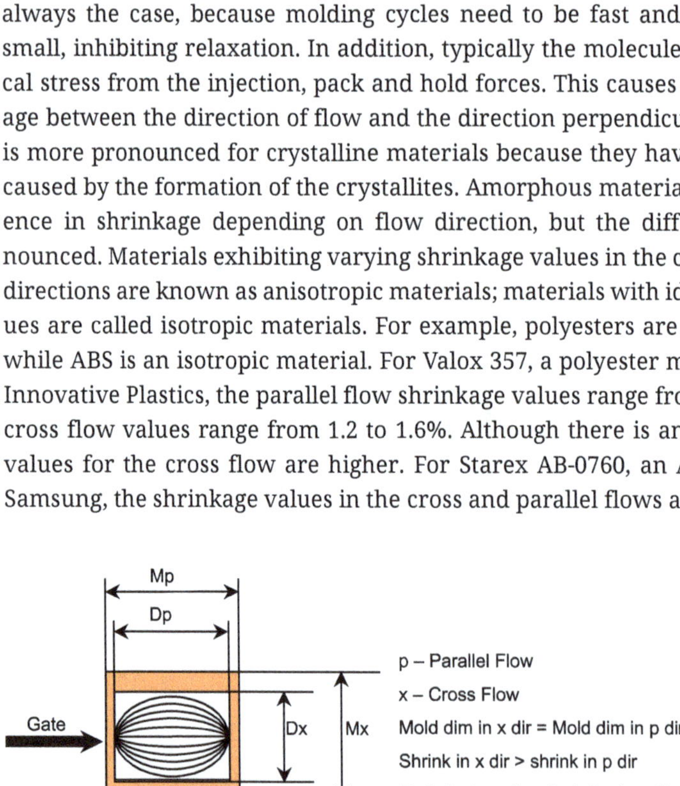

Figure 2.25 Parallel flow and cross flow shrinkage

Exact shrinkage values can never be predicted because shrinkage is a function of various parameters. Some manufacturers publish the shrinkage values for different sample thicknesses because, as stated earlier, in thicker parts heat can be retained for a longer time allowing the molecules to relax and causing more shrinkage. For this reason it is difficult to determine the exact mold cavity dimensions that will yield a molded part with the exact finished dimensions. It is very rare to build an injection mold cavity and cores that are the exact size of the desired finished part or CAD model. Mold makers will typically build the mold 'steel safe' and adjust the steel after the initial sampling of the mold. Another factor that complicates this situation is the fact that final shrinkage is a function of both the cross and parallel flow shrinkages. Rarely will a required dimension be perfectly parallel or perfectly perpendicular to the melt flow. The final shrinkage value is a combination of the two numbers. In this case, leaving the mold cavity and cores steel safe is always a good idea. It is not possible to do so in all cases and the tooling engineer must rely on past experience and make the best decision.

The term shrinkage in injection molding always compares the part dimensions to the dimensions of the mold. Shrinkage of the melt in a free state is always positive. This means that the volume will always decrease and the part dimension will therefore always become smaller. The term free state is used to describe a noninhibited state of movement where the molecules are not held by any other forces. However, in some parts, such as long tubular parts, we find that as the part shrinks the length will decrease, forcing the diameter to increase, see Figure 2.26. This can be visualized similar to Poisson's effect where contraction due to external force in one direction leads to the expansion in the perpendicular direction. This negative shrinkage in the diameter of the part is due to the force that is exerted because of the overall reduction in the length of the part which can be significant. Stress can easily build up in such parts and cause premature product failure.

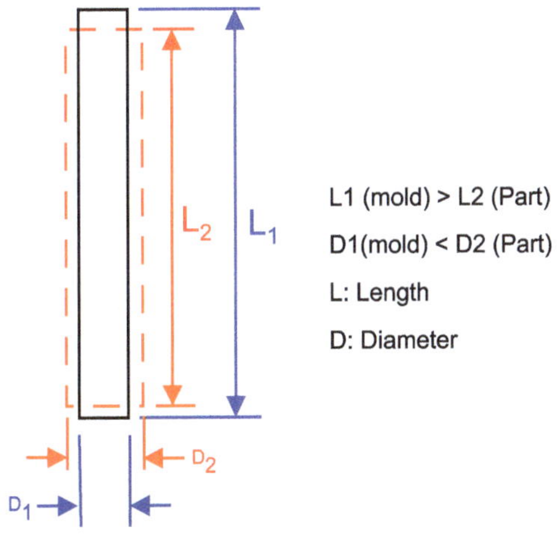

Figure 2.26 Positive and negative shrinkage in molded parts

The phenomenon does not necessarily lead to a negative shrinkage value, and can also result in a reduced shrinkage value, depending on the amount of mechanical stress in the part. This poses another challenge for the mold maker in sizing of the cavities. Mechanical stresses can cause a change in dimension in parts with other geometries, such as square boxes with deep pockets. The deep pockets create unsupported walls. Sometimes reduction in the distance between the walls causes an increase in the distance in the other direction. This is represented in Figure 2.27.

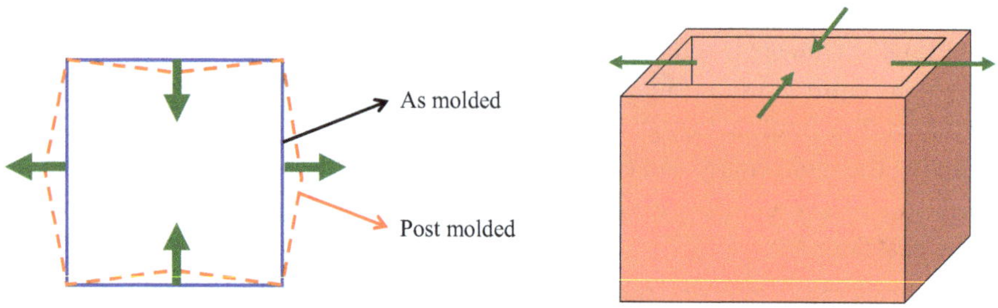

Figure 2.27 Mechanical stresses affecting part dimensions

2.8 The Plastic Pressure-Volume-Temperature (PVT) Relationship

Figure 2.28 shows that at 0 MPa of pressure, as the temperature of the plastic is increased, the specific volume also increases. This increase is the result of the increase in the intermolecular distance and the free volume. At any given temperature, if an external pressure is applied, the melt gets compressed and the specific volume decreases. Specific volume is directly proportional to temperature but is indirectly proportional to the pressure. In injection molding, there is a combination of factors involved as the melt is transformed into the product. The melt is collected as a shot (see Section 2.8.1) in front of the molding screw. There is some compression of the melt because of the pressure (back pressure) during the shot build-up process. The molten plastic is then injected with high pressure into a cold mold. During this process, and after the plastic is inside the mold, the melt is cooling down. As the melt is being transformed to the final molded product, the pressure, volume, and temperature of the melt are all continuously changing, and the final contour in the graph is a combination of the curves in the Figure 2.28. Cavity pressure sensing equipment can track these curves and predict shot-to-shot part quality and consistency.

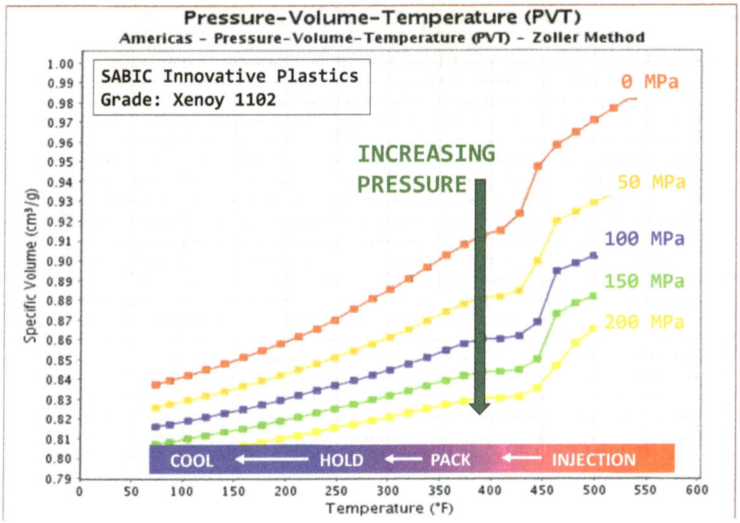

Figure 2.28 Pressure-volume-temperature relationship in injection molding

As a processor, one of the most important concepts to keep in mind is that plastic melt is compressible and elastic, which can be visualized in the example of a rubber band ball (Figure 2.29). It can sometimes be seen that in the molding process, at the end of the second stage pressures and just before the screw recovery process, the screw bounces back. This happens as the melt decompresses when the pressure is released. As will be discussed in Chapter 7, cushion values must also be kept at a minimum in order to avoid shot-to-shot variations.

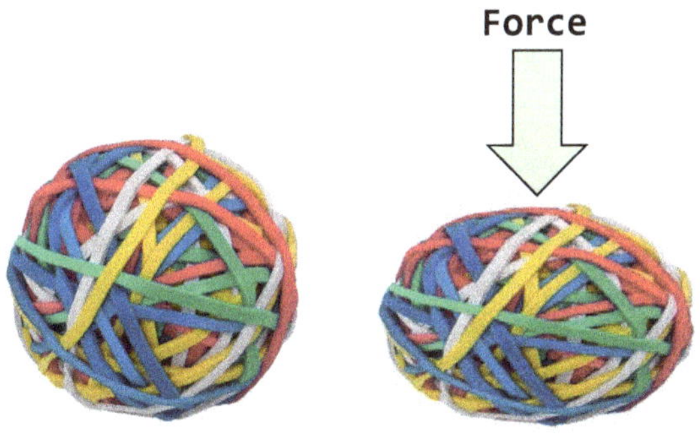

Figure 2.29 Plastic melt is compressible, similar to a ball of rubber bands

2.8.1 Importance of Plastic Density in Injection Molding

Density is defined as the weight of unit volume of a cube of a given material and is expressed as grams per cubic centimeter (g/cc or g/cm^3) in the metric system. For example, the density of polystyrene (PS) is 1.06 g/cc. This means that a 1 centimeter PS cube will weigh 1.06 grams or a volume of 1 cubic centimeter (cc); if filled with PS, will weigh 1.06 grams. If the same volume is filled with lead, it will weigh 11.36 grams because the density of lead is higher than PS. It will weigh 0.789 grams if 1 cc is filled with ethanol because the density of ethanol is lower than that of PS.

By definition, the shot capacity of a molding machine is defined as the maximum amount of PS that can be held in front of the screw. For example, if the shot capacity is specified to be 100 grams, then the barrel will hold 100 grams when the screw is completely in its retracted position. The density of PS is 1.06 g/cc. If the same volume in front of the screw that holds 100 grams is now filled with a plastic with a higher density, such as a 30% glass filled PBT with a density of 1.53 g/cc, the shot capacity will now be 144 g. On the other hand, if the PS is replaced by polypropylene (PP), which has a density of 0.90 g/cc, the shot capacity will now be only be 85 g. This implies that even though the barrel is rated for 100 g, it will be impossible to mold a 100 g PP part on the machine. Refer to Figure 2.30.

Machine shot sizes are frequently used in injection molding calculations. It is important that the shot sizes are therefore always recalculated for the material being molded. The formula is as follows: normalized shot size = (published shot size in GPPS/1.06) × density of the material under consideration. GPPS is a general purpose polystyrene.

To avoid confusion, several manufacturers have started specifying the machine shot sizes in volumetric units. In such cases, all part weights and runner weights will also have to get converted into volumes to perform the appropriate calculations.

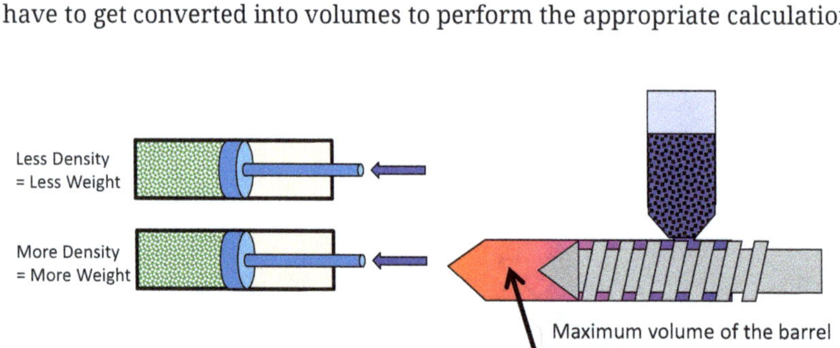

No	Material	Sp Gr	Max Shot Capacity (gms)
1	Polystyrene	1.06	100
2	Nylon 6	1.15	108
3	Acetal	1.42	134
4	30% GF PBT	1.53	144
5	LDPE	0.92	87
6	PP	0.90	85

Figure 2.30 Max shot weight dependence on material densities

2.8.2 Residence Time and Maximum Residence Time of a Plastic

At a given melt temperature, a polymer can be held without degradation only for a finite amount of time. If the polymer is subjected to longer lengths of time, the polymer chains break and start to degrade. This time is defined as the *maximum residence time*. Figure 2.31 is a residence time graph for ULTEM 1000, a polyether imide from Sabic. If the melt temperature is set to approximately 700 °F (371 °C), the plastic can be held in the barrel for a maximum of about 12 minutes. As the melt temperature increases, the maximum residence time number decreases. From the same graph, if the melt temperature is 770 °F (410 °C), then the maximum residence time is 6 minutes.

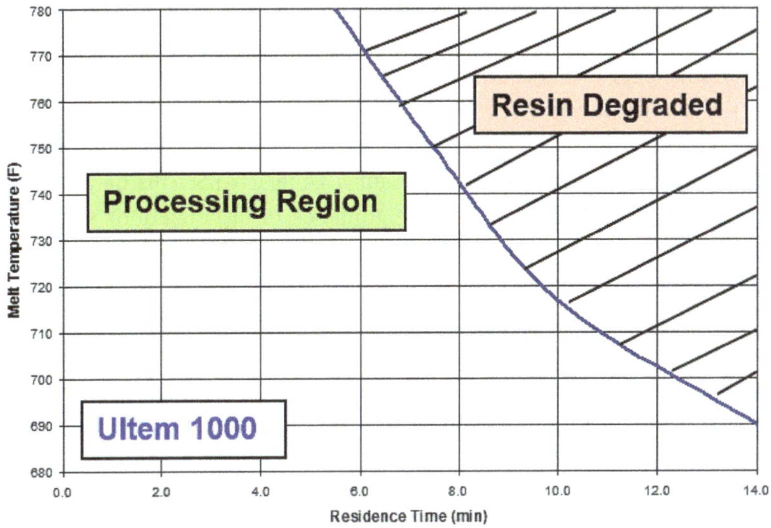

Figure 2.31 Maximum residence time for Ultem 1000 (PEI from SABIC Innovative Plastics)

In the molding process, the plastic enters the feed throat, gets conveyed over the screw, collects in front of the screw, and is then injected through the nozzle into the mold. The section of the barrel that houses the screw is heated with the assistance of the heater bands, which help to melt the plastic. The shear from the rotating screw also contributes to the heat and aids the melting. The plastic enters into the barrel as pellets and leaves as a melt. The amount of time the plastic takes to travel from the feed throat to the nozzle tip is called the *residence time of the plastic in the barrel*. Because injection molding is a batch process, depending on the shot weight of the mold, the cycle time of the mold, and the maximum shot size of the machine, the residence time of the plastic will change.

During the injection molding operation, the melt should always have a residence time that is less than the maximum residence time for the given material. If the mold is a hot runner mold, then the residence time of the melt in the hot runner system must be added to the residence time in the barrel to give the total residence time. See Section 6.8.4 for calculations and more information.

2.8.3 Plastic Datasheets

Datasheets for plastics provide detailed information about them. There are different types of datasheets. The most common datasheets, which the material manufacturers provide, are the property data sheets, the processing datasheets, and the material safety datasheets (MSDS).

The property datasheets provide information about the properties of the plastics. These properties can be physical, mechanical, electrical, and so on. The categories of the properties will depend on the intended end use of the material. An example of the properties datasheet is shown in Figure 2.32 for a generic ABS. This datasheet provides the properties of the plastic under the guidance of organizations such as the ASTM, ISO, DIN, and JIS. These organizations provide specifications of the test sample, the conditions, and the procedure of testing. These values should be used as guidelines. When comparing two materials, care should be taken that the test method used for both is the same because there are no conversion factors available between the test methods, for example, between an ASTM test and a ISO test. The final manufactured product is almost certainly of another geometry and therefore the numbers on the datasheet should only be a guide. Generic datasheets contain only a single point data. Processing parameters can have a substantial impact on the properties. As was shown in this article, the impact of the resistance for a given ABS ranges from less than 2 N-m (1.4 ft-lb) to almost 50 N-m (36.5 ft-lb), as a result of changing melt temperatures between 218 and 271 °C, and mold temperatures between 29 and 85 °C.

TYPICAL PROPERTIES			
MECHANICAL	**Value**	**Unit**	**Standard**
Tensile Stress, yld, Type I, 0.2 in/min	7500	psi	ASTM D 638
Tensile Modulus, 0.2 in/min	450000	psi	ASTM D 638
Flexural Stress, yld, 0.05 in/min, 2 in span	17600	psi	ASTM D 790
Flexural Modulus, 0.05 in/min, 2 in span	325000	psi	ASTM D 790
Hardness, Rockwell R	90	-	ASTM D 785
IMPACT		**Unit**	**Standard**
Izod Impact, notched, 73°F	4.1	ft-lb/in	ASTM D 256
Izod Impact, notched, -40°F	1.2	ft-lb/in	ASTM D 256
Gardner Impact, 73°F	8	ft-lb	ASTM D 3029
THERMAL	**Value**	**Unit**	**Standard**
HDT, 264 psi, 0.250", unannealed	220	°F	ASTM D 648
CTE, flow, -40°F to 100°F	4.21E-05	1/°F	ASTM E 831
Relative Temp Index, Elec	60	°C	UL 746B
Relative Temp Index, Mech w/impact	60	°C	UL 746B
Relative Temp Index, Mech w/o impact	60	°C	UL 746B
PHYSICAL	**Value**	**Unit**	**Standard**
Specific Gravity	1.05	-	ASTM D 792
Mold Shrinkage, flow, 0.125"	0.5 - 0.8	%	GE Method
Melt Flow Rate, 230°C/3.8 kgf	5	g/10 min	ASTM D 1238
OPTICAL	**Value**	**Unit**	**Standard**
Gloss, untextured, 60 degrees	88	-	ASTM D 523
ELECTRICAL	**Value**	**Unit**	**Standard**
Dielectric Strength, in air, 62 mils	952	V/mil	ASTM D 149
Arc Resistance, Tungsten {PLC}	6	PLC Code	ASTM D 495
Hot Wire Ignition {PLC)	3	PLC Code	UL 746A
High Voltage Arc Track Rate {PLC}	1	PLC Code	UL 746A
High Ampere Arc Ign, surface {PLC}	0	PLC Code	UL 746A
Comparative Tracking Index (UL) {PLC}	0	PLC Code	UL 746A
FLAME CHARACTERISTICS	**Value**	**Unit**	**Standard**
UL Recognized, 94HB Flame Class Rating (3)	0.060	in	UL 94

Figure 2.32 Example of a properties datasheet

An example of a processing datasheet is shown in Figure 2.33. Resin manufacturers conduct tests on the materials to arrive at the values on the datasheet. There are no standards available for these tests. For the most part, the mold temperatures, the melt temperatures, the drying times, and other temperatures can be relied upon. However, the injection pressure, the injection speed, the back pressure, the screw rotation speeds in revolutions per minute (rpm), and the shot to cylinder size cannot be relied upon because these are always a function of the part design and the mold design. A part that is about 10 mm thick and 50 mm long will require a lot less pressure and speed to fill as compared to a part that is 3 mm thick and 200 mm long. Some material manufacturers will also give guidelines to set the pack and hold pressures to a certain percentage of the injection pressure, as shown in Figure 2.34. This too is incorrect since thicker parts will require more pressure to pack the shrinkage. They also provide other recommendations that are incorrect. This data is from a material manufacturer with the name hidden to protect their identity. A 30 rpm on a 20 mm molding screw is not the same as a 30 rpm on a 50 mm molding screw because the circumferential speed on a 50 mm will be higher; therefore, it will generate more shear heat. There have been no studies done to prove that a 50 to 70% value of shot to cylinder size provides the best quality parts. A general rule in industry is a 20 to 80% size, but that can also be challenged based on the part and material that is under consideration for molding. The cycle time should also figure into the calculation.

Injection Molding Parameter	Value	Unit
Drying Temperature	200–210	°F
Drying Time	2–4	hr
Drying Time (Cumulative)	8	hr
Maximum Moisture Content	0.01	%
Melt Temperature	450–500	°F
Nozzle Temperature	450–520	°F
Front - Zone 3 Temperature	450–470	°F
Middle - Zone 2 Temperature	420–440	°F
Rear - Zone 1 Temperature	370–390	°F
Mold Temperature	120–180	°F
Back Pressure	50–100	psi
Screw Speed	30–60	rpm
Shot to Cylinder Size	50–70	%
Vent Depth	0.0015–0.002	in

Figure 2.33 Example of a processing datasheet

Manufacturer: XXXXXX
Material: Polypropylene Impact Copolymer
Injection Notes:

- Injection Pressure: 40 to 80% of max machine pressure
- Pack & Hold Pressure: 30 to 60% of max pressure
- Back Pressure: 10% of max machine pressure
- Cooling Time: 25 to 40 secs

Figure 2.34 Misleading processing guidelines from the material manufacturer

2.9 References

[1] Turner, A., Gurnee, E. F., *Organic Polymers*, Prentice-Hall, 1967, p. 51

[2] Zweifel, H., Maier, R., Schiller, M., *Plastics Additives Handbook* (2009) Hanser, Munich

[3] Sepe, M., The importance of melt & mold temperature, *Plastics Technology*, (Nov. 2011); *http://www.ptonline.com/columns/the-importance-of-melt-mold-temperature*

Suggested Reading

Deanin, R. D., *Polymer Structure, Properties and Applications* (1972) Cahners, Boston, MA

Tager, A. A., *Physical Chemistry of Polymers* (1978) Mir Publishers, Moscow

Odian, G., *Principles of Polymerization* (1991) Wiley Interscience, USA

Gowariker, V. R., Viswanathan, N. V., Sreedhar, J., *Polymer Science* (1996) New Age International (P) Limited, Delhi

Billmeyer, F. W., *Textbook of Polymer Science* (1984) Wiley Interscience, NY

Brydson, J. A., *Plastics Materials* (1995) Butterworth-Heinemann Ltd, Oxford, UK

3 Polymer Rheology

Polymer rheology is the science of flow of polymers. The study of polymer flow is essential to understand the melt processing of plastics. In any melt processing technique, the plastic has to be melted and then deformed to conform to the final product specifications. In injection molding, the melt must be injected into a mold and then cooled to get the final part. In extrusion, the melt is shaped through a die and then cooled down to get the required profile. During the injection of the melt into the mold and until it reaches its final form, the melt is subjected to various forces including mechanical and thermodynamic forces. In this chapter, some of these concepts are explained in brief to give the reader an understanding sufficient enough to help apply them to injection molding. The concepts also have been simplified and the general mathematical details have been left out. For those readers who would like to get an in depth understanding on this topic, a number of books are mentioned in the references at the end of the chapter.

3.1 Viscosity

Viscosity is the resistance to flow. The higher the resistance to flow, the higher is the viscosity. Honey or corn syrup do not flow easily and therefore have high viscosities. On the other hand, water flows very easily and therefore has low viscosity. Gasses have even lower viscosities as compared to water and therefore flow easier. Viscosity depends on a number of factors and is an inherent property of the fluid but can be influenced by external forces. Based on the types of forces the polymer is subjected to and/or on the type of medium the polymer is present, various types of viscosities are defined. Of particular interest in injection molding is the viscosity of the melt as defined by the apparent viscosity. As will be discussed later in the chapter, the viscosity of the melt is dependent on the applied force and therefore is not constant. Hence the adjective ‘apparent’ is used to describe the viscosity at a particular given shear rate.

Consider a liquid sandwiched between two metal plates as shown in Figure 3.1. The distance between the plates is H. The area of the plates is A. The bottom plate is held stationary and the top plate is displaced in the X direction with a force F and a velocity $V = V_H$, see Figure 3.2.

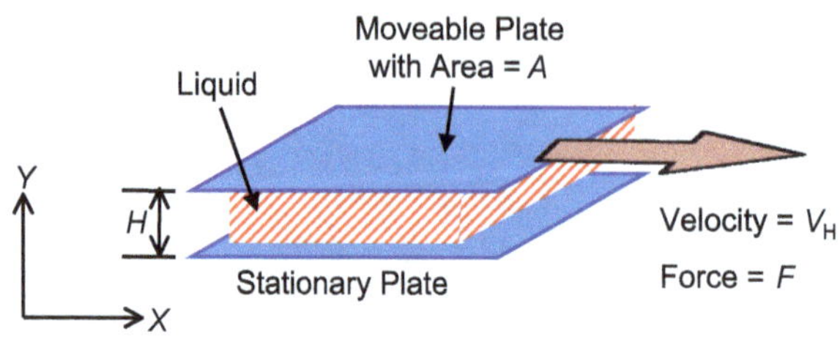

Figure 3.1 Experimental setup to determine the velocity profile

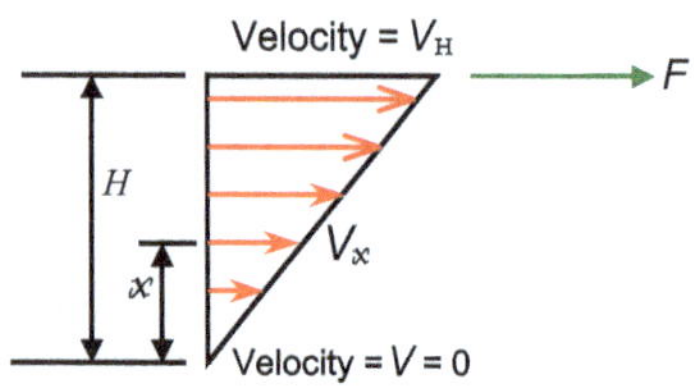

Figure 3.2
Velocity profile for a liquid under shear force

The layer of liquid just below the top plate will also move with the plate at a velocity V_H and the layer just above the bottom plate will be stationary with a velocity equal to $V = V_0 = 0$. At any layer between the two plates, at a distance of x, the velocity is V_x and is proportional to the distance from the bottom plate. The shear rate is the differential velocity between the layers. Since each of the layers is moving at different velocities, the shear rate is also a function of the distance x. The shear rate of the top layer is given by the equation

$$\dot{\gamma}_H = V_H/H \tag{3.1}$$

and at any given layer at a distance x is given by

$$\dot{\gamma}_x = V_x/x \tag{3.2}$$

The units of shear rate are reciprocal seconds or s^{-1}.

Each layer is subjected to shear forces as it is being pulled in the direction of the applied force. Since stress is force divided by area, the shear force τ at the top layer can be defined as

$$\tau = F/A \tag{3.3}$$

Any layer at a distance *x* from the bottom layer will be subjected to a force F_x, which is between zero and *F*. The area on which the force is applied stays the same. Therefore the shear stress for this layer will be

$$\tau_X = F_X / A \tag{3.4}$$

F_x and V_x are both proportional to the distance of the layer from the bottom plate. That means the greater the distance, the higher the velocity and the higher is the shear force. The increase in shear stress is directly related to the increase in shear rate. The two are governed by the equation

$$\tau = \eta \dot{\gamma} \tag{3.5}$$

where η is the constant of proportionality called the viscosity of the liquid. The viscosity is the slope of the line when the shear stress is plotted versus shear rate.

3.2 Newtonian and Non-Newtonian Materials

In the above discussion, the shear stress was linearly proportional to the shear rate with viscosity being the constant of proportionality, see Figure 3.3. In some fluids, the relationship between the shear stress and the shear rate is not linear. The viscosity is not a constant and is dependent on the shear rate or the time the fluid is subjected to the shear. Such fluids are not non-Newtonian fluids. Based on the response to shear, non-Newtonian fluids can be classified into two types, dilatant and shear thinning fluids. In dilatant fluids, the viscosity increases with increasing shear rate and in shear thinning fluids, the viscosity decreases with increasing shear rates, see Figure 3.4(a). Based on the response to the time of shear at constant shear rate, non-Newtonian fluids are classified into rheopectic and thixotropic fluids (see Figure 3.4(b)). All plastics show shear thinning behavior.

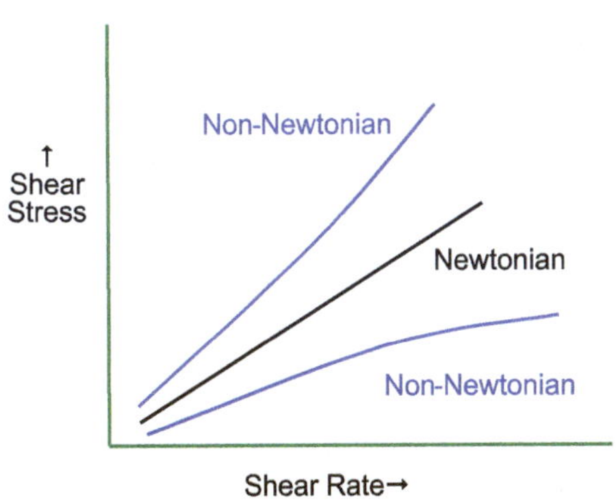

Figure 3.3
Shear stress versus shear rate for Newtonian and non-Newtonian liquids

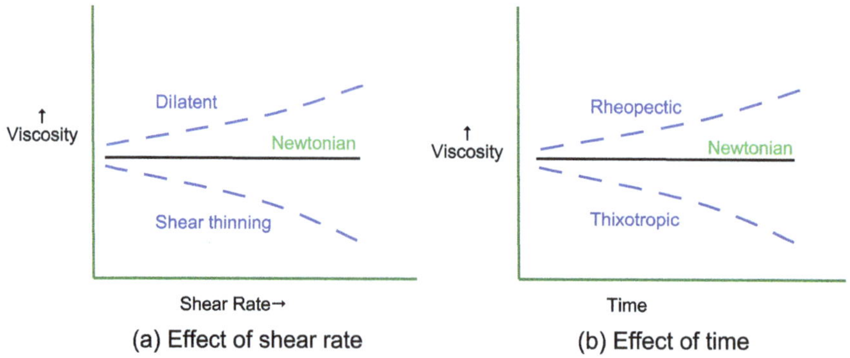

Figure 3.4 Change in viscosity as a function of (a) shear rate and (b) time for Newtonian and non-Newtonian liquids

Rheology is the science of the flow of non-Newtonian materials. All polymers are non-Newtonian, specifically shear thinning, as the shear rate increases, the viscosity drops.

3.3 Viscosity in Polymer Melts

Polymer melts are non-Newtonian. Research has shown that the velocity profile, which is the line that joins the velocities of each layer, is never a straight line as shown in Figure 3.2. A typical velocity profile for a polymer melt, called fountain flow, is shown in Figure 3.5. The parabolic velocity profile is a result of a pressure driven flow through a closed channel. The drag from the channel walls slow the material, resulting in a higher velocity in the center. At the flow front, this condition causes the faster flowing material in the center of the channel to "fountain" to the slower flowing regions. Fountain flow conditions only exist at the flow front and cannot occur behind the flow front. However, the velocity profile conditions exist throughout the flow, at the flow front and behind the flow front., i.e., the center is always flowing faster than the perimeter. Hot runners, where the channel is already full, will not experience fountain flow, except when they are first filled. Fountain flow will exist in the cavities and in cold runners. An incompletely filled runner or a short shot will always exhibit their flow front velocity profiles as shown in Figure 3.5.

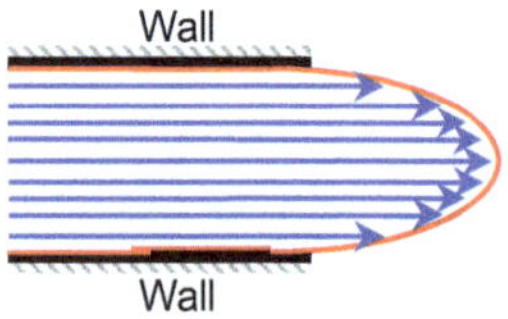

Figure 3.5 Velocity profile of plastic flow

Because the profile is not linear, the shear rate equation in Equation 3.1 now needs to be rewritten as

$$\dot{\gamma} = (x/h)\,(dx/dt) \tag{3.6}$$

Where t is time, h is the thickness or diameter of the flow channel, and x is the distance along the flow channel. The units do not change and are reciprocal seconds or s^{-1}.

The relationship between viscosity and shear rate is described by various mathematical models. The most popular model, the power law viscosity model proposed by Ostwald and de Waele, is widely applicable to injection molding. This equation accurately represents the shear thinning region found at the high shear rates in injection molding. The power law viscosity model is given by the equation

$$\eta = m\dot{\gamma}^{n-1} \tag{3.7}$$

where m is a constant called the consistency index and n is the power law index. A representation of a typical graph of viscosity versus shear rate is shown in Figure 3.6 for a polyester. Note that the shear rates represented here are of the order of those that are experienced in injection molding.

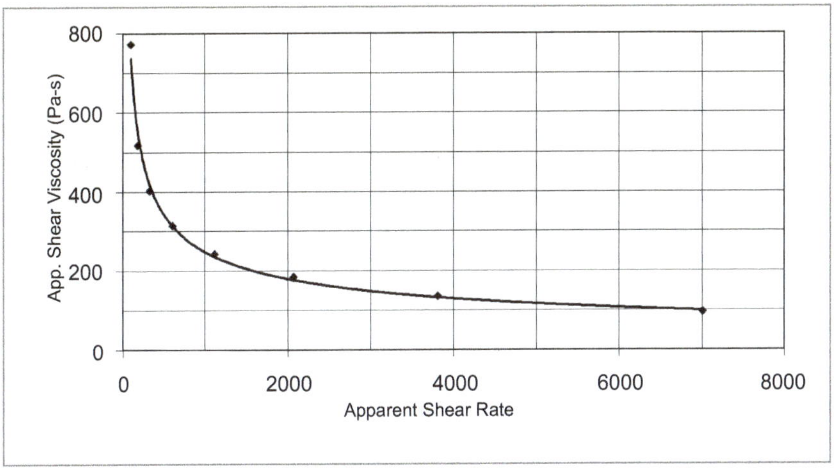

Figure 3.6 Effect of shear rate on viscosity represented on the linear scale

For polymer melts subjected to very low shear rates, the viscosity is essentially unchanged. There is very little or no effect of the shear rate. However, these shear rates are not encountered in injection molding. This is the reason why there are different viscosity models to represent viscosity.

Equation 3.7, $\eta = m\dot{\gamma}^{n-1}$, can be rewritten as

$$\log\eta = \log m + (n-1)\log\dot{\gamma} \tag{3.8}$$

This equation now becomes a linear equation and if a graph of log η versus log $\dot{\gamma}$ is plotted, we get a straight line as shown in Figure 3.7. This is the typical graph that is obtained from a melt rheometer test. The graph in Figure 3.7 is generated from the same data used for the PBT graph in Figure 3.6. The slope of the line is $(n-1)$. Polymer melts are shear thinning and therefore the viscosity drops, resulting in a negative slope of the line. Therefore, the value of n is always less than 1 but greater than zero.

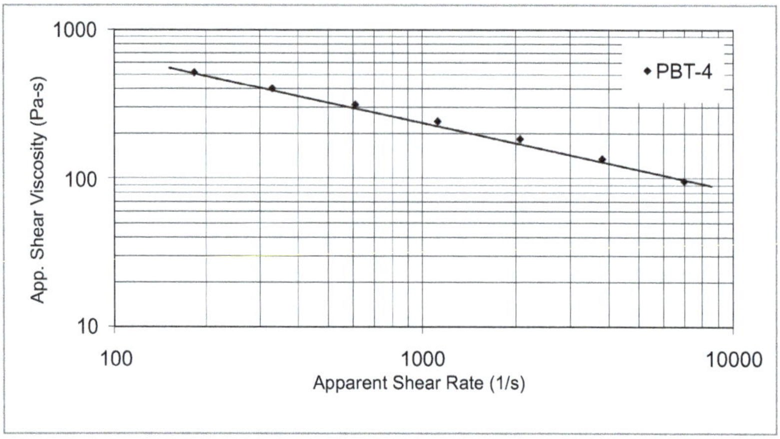

Figure 3.7 Effect of shear rate on viscosity represented on the logarithmic scale

In polymer melts, as the shear rate is applied, the molecules start to align themselves in the direction of flow, moving away from their equilibrium intertwined states. Increasing shear rates stretch and align more and more molecules in the direction of flow. This alignment facilitates the easy movement of the flow layers past each other, thus reducing the resistance to flow or viscosity. At a certain point, all the molecules become aligned in the flow direction and increasing the shear rate further has little or no further effect on the viscosity.

For the sake of practical injection molding, we could consider the region of higher shear rates as a Newtonian region where the viscosities become consistent. Since the viscosities are a result of the fill speeds, the corresponding regions of fill speeds are now treated as the consistent region. Injection speed is analogous to shear rate and an in-mold viscosity curve appears similar to the one shown in Figure 3.8. Shear rate can be calculated as the reciprocal of the fill time, where fill time is the time the screw takes to travel from the set shot size to the holding phase transfer position of the screw on the molding machine. Simply put, it is the time for which the screw moves in the injection phase.

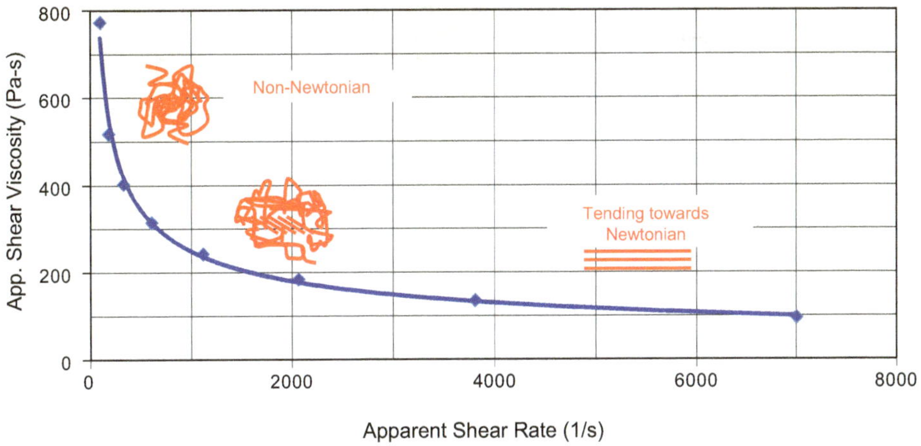

Figure 3.8 Effect of apparent shear rate (injection speed) on the viscosity of the polymer

3.4 Effect of Temperature on Viscosity

In the solid state the molecules have very little thermal energy and therefore are almost immobile. Depending on the ambient temperature and the glass transition temperature (T_g) of the polymer, the polymer can be either brittle or soft and tough. The thermal regions are explained in Chapter 2. Below the T_g, the plastic is very brittle and is said to be in the glassy state. Above the T_g the plastic is soft and is said to be in the rubbery or viscoelastic state. Above the melting temperature (T_m), for crystalline plastics, the plastic is in the melt form. In general, as the temperature is increased, the thermal energy reduces the weak intermolecular attraction that holds the molecules together, making them more mobile. Amorphous polymers continue to soften and crystalline polymers show a sharp melting point. Increase in temperature increases the mobility of the molecules, thereby reducing the viscosity of the polymer. Temperature and viscosity are inversely related. Figure 3.9 shows the effect of temperature on the viscosity of the melt. It is also clear that the effect of shear rate is higher than the effect of temperature.

In injection molding, it is common practice to increase the melt temperature to make the plastic flow easier. However, Figure 3.9 demonstrates that increasing the injection speed will have a greater effect on part fill. This advantage is discussed in Chapter 7 and a procedure to generate the in-mold rheology curve is also discussed.

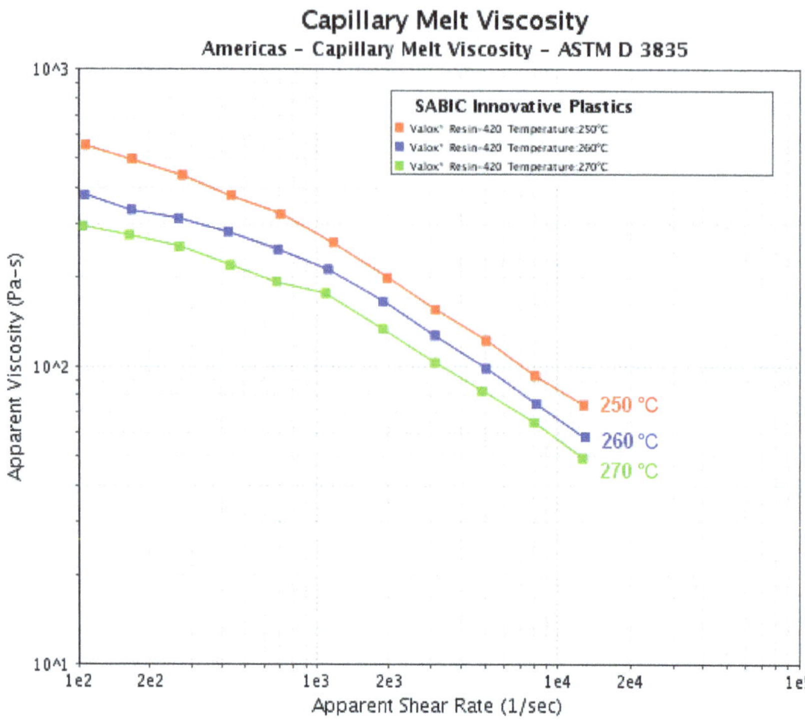

Figure 3.9 Effect of temperature on the viscosity of the melt (Source: Sabic Innovative Plastics)

3.5 Velocity and Shear Rate Profiles

The velocity profile shows the relative velocities of the different layers in the polymer melt as it flows through a channel. The length of the arrows in the velocity profile represents the velocity of each of the layers. The velocity at the wall is zero and therefore the shear rates are very low near the wall. The velocity increases towards the center of the channel, following a parabolic profile and it is highest at the center. As discussed earlier, shear rate is the difference in the velocities of the adjacent layers. In Figure 3.10, the difference in the velocities in the first two layers near the wall is very high compared to two layers near the center of the channel. Therefore, the shear rate is higher near the wall, compared to the center of the channel. The shear rate profile, which is a derivative of the velocity profile, is plotted below the velocity profile in Figure 3.10. These studies are relatively recent [1].

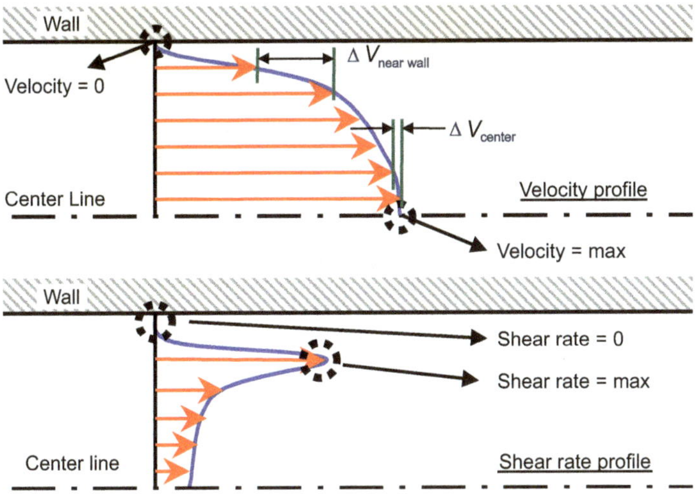

Figure 3.10 Velocity and shear rate profiles in polymer melts

Figure 3.11 [1] shows the outputs from a finite difference flow analysis program. It shows the velocity, shear rate, melt temperature, and viscosity of the cross section of the melt. The *y*-axis represents the direction perpendicular to flow.

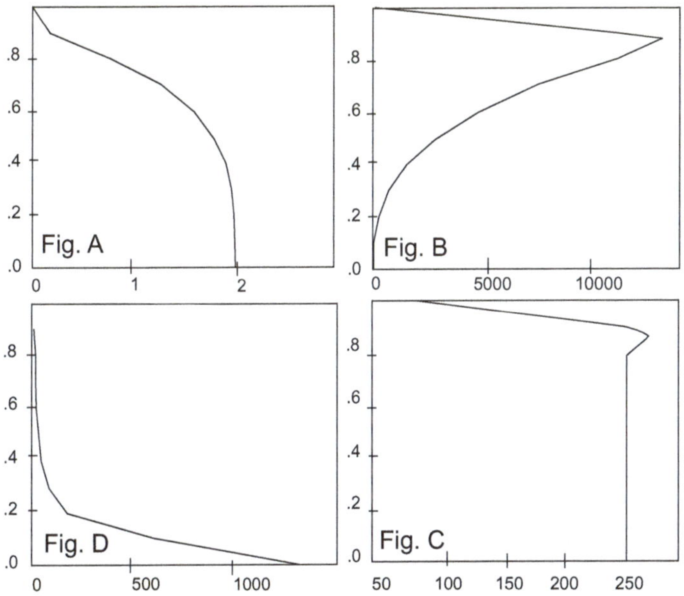

Figure 3.11 Outputs from a finite element, finite difference flow analysis program providing information on the melt conditions through the cross-section of a cold runner; *y*-axis is from the center line of the channel to the channel wall: (A) velocity; (B) shear rate; (C) melt temperature; (D) viscosity [1]

3.6 Application to Injection Molding

The direct effect of the shear rate distribution discussed above is visually evident in injection molding, particularly in the filling pattern of multicavity molds. As the plastic begins to flow in the runner, the shear layers are formed as shown in Figure 3.10. Since the melt flow is always laminar, these layers split and/or flow into the various flow channels in laminas. Each of these laminas has their own characteristic properties, such as shear rates and temperatures. The high shear laminas just below the wall of the flow channel create a low viscosity region changing the velocity of flow in some cavities and causing cavity-to-cavity flow imbalances. Some of the typical and common characteristics of such flows and their effect on the parts are described in the following.

3.6.1 Flow Imbalance in an 8-Cavity Mold

Consider an 8-cavity mold as shown in Figure 3.12. As the plastic flows through the primary runner, the shear layers are developed.

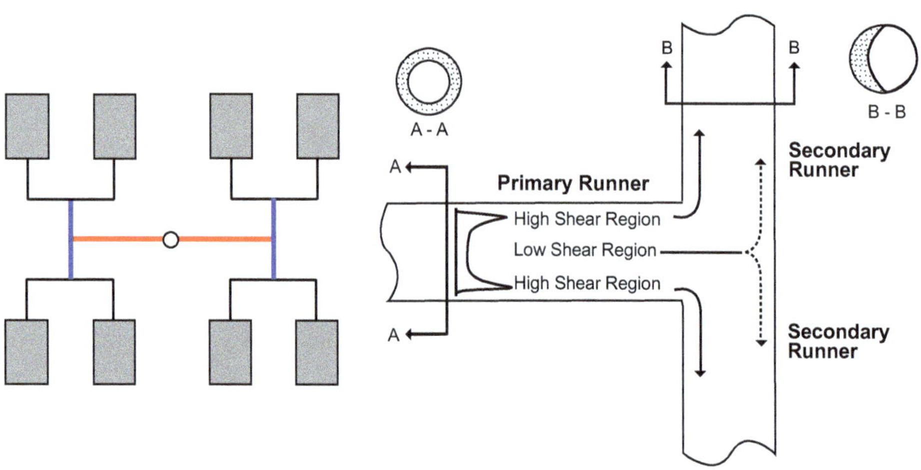

Figure 3.12 Split of different shear rate regions in an 8-cavity mold [2]

If we disregard the frozen layer (in a cold runner mold), we can distinguish between two distinct layers. The outside layer is the high-shear layer and the inside layer is the low-shear layer. In the diagram, the high shear layer is the shaded area. Section A-A is the cross section of the primary runner and shows the two layers concentric to each other. Since the flow through the mold is laminar, the variations in shear, temperature, and viscosity across the runner proceed into the secondary runner. The inner low-shear laminas hit the far wall of the secondary runner and the high-shear laminas on the outer perimeter continue to flow along the near wall of the branching secondary runner. The cross section shown in Section B-B illustrates this distribution of high- and low-sheared material in the branching secondary runner.

The laminar flow continues through the secondary runner, the tertiary runner, and then into the cavities. The result is that inside cavities (those close to the sprue) are filled first, as they are fed by the hotter, high-sheared, lower viscosity material developed earlier in the runner. This is shown in Figure 3.13. A visual proof of the high-shear lamina is shown in Figure 3.14.

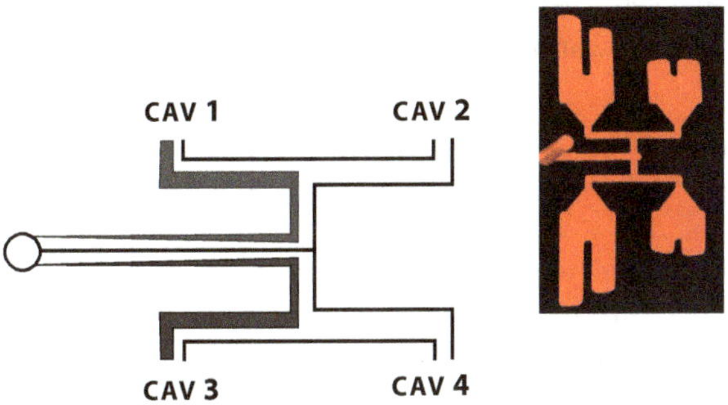

Figure 3.13 Flow imbalance between the inside and outside cavities. The photo on the right is an actual short shot of a polycarbonate material molded at optimal process conditions [3]

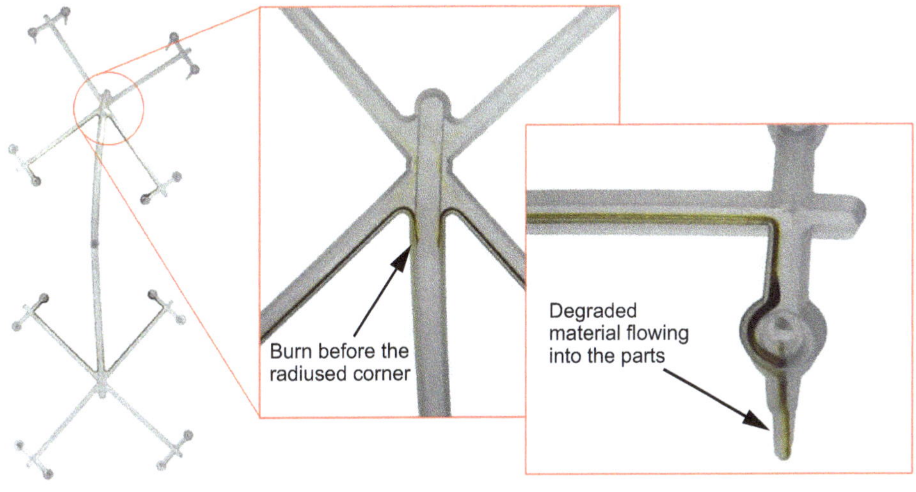

Figure 3.14 Visual evidence of the high-shear lamina [3]

Here, a runner used in the molding of a PVC part clearly showed degradation caused by the high shear. The black streak on the inside is the high-shear lamina in the figure. Note that the burning/streaking of the material develops prior to the actual corner. We point this out in particular, because shear induced imbalances are sometimes misrepresented as being caused by a sharp corner in a runner. Here, the evidence dispels this theory as the runner is not only radiused (no sharp corners) but the burning begins before the corner.

The cavity-to-cavity imbalance described above is called a rheological imbalance. The runner is said to be rheologically imbalanced despite the fact it is geometrically balanced. When the distance from the sprue to the gate is the same for each cavity, the mold is said to be geometrically balanced. This geometrical balance in a runner is still commonly incorrectly referenced as a "naturally balanced" runner.

3.6.2 Racetrack Effect in a Part with Constant Thickness

The part shown in Figure 3.15 is made with a single-cavity mold, producing a 100 mm × 2 mm thick flat part with a constant wall thickness. The shear effect on viscosity can be seen in the flow pattern developed in this cavity. The high-sheared low-viscosity material developed in the perimeter of the runner splits and concentrates along the perimeter of the flat disc, causing the racetrack effect. Note that the photo in the right shows that the effect is significant enough to create a gas trap, opposite the gate, in this flat part.

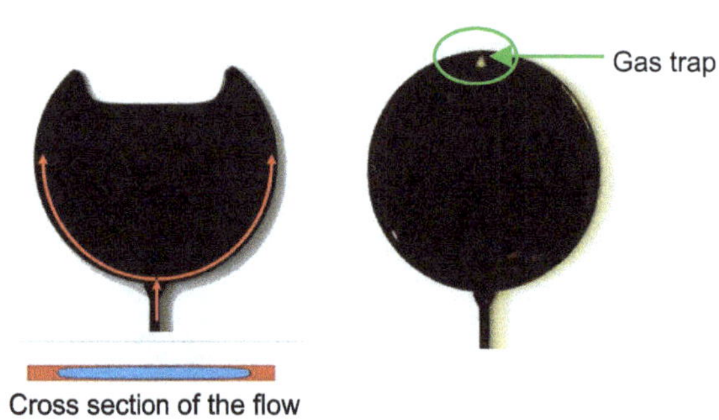

Figure 3.15 Racetrack effect causing a gas trap in a part with uniform thickness [3]

3.6.3 Stress Build-Up in Molded Parts

The part shown in Figure 3.16 is made from a transparent material in a two-cavity mold. The parts are packed out and examined with a polarizing lens after molding. The build-up of stress can be seen on the inside of the parts. This is the area where the hotter laminas flow, causing differential cooling and therefore stress.

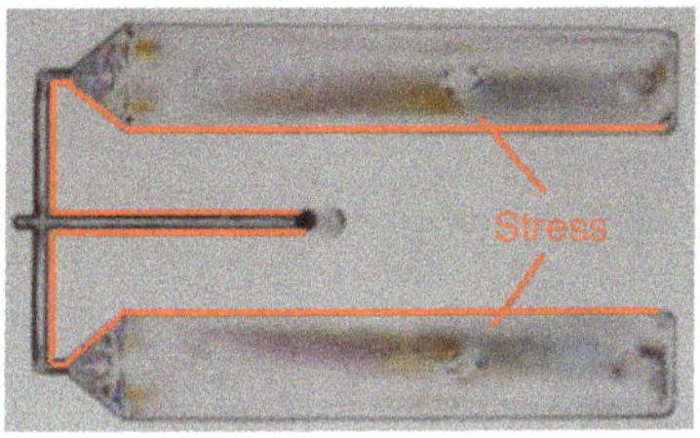

Figure 3.16 Stress build-up observed under a polarizing lens [3]

3.6.4 Warpage Difference between Cavities

Figure 3.17 shows parts molded in a 4-cavity mold. Because the location of the hot laminas in the cavities is different, two of the cavities are warped while the other cavities are perfectly flat.

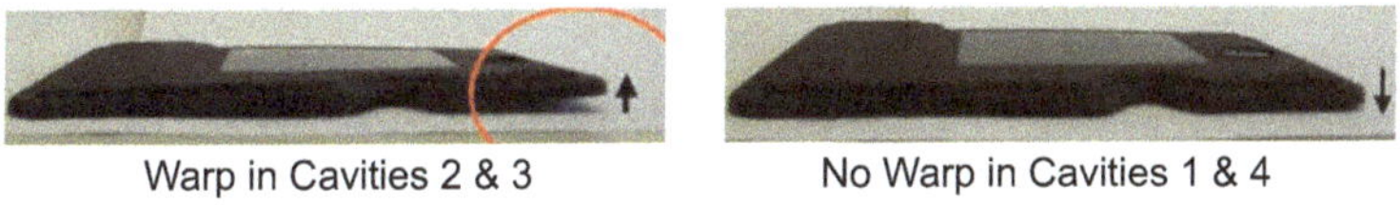

Figure 3.17 Warpage differences between cavities from the same mold caused by a melt imbalance [3]

3.7 Solving Flow Imbalances Using Melt Rotation Techniques

The solution for balancing the flow and creating rheologically balanced molds was developed and patented by John Beaumont of Beaumont Technologies in Erie, Pennsylvania (please note that the use of this technology requires a licensing agreement from Beaumont Technologies Inc.) Beaumont's varied methods of melt management, commonly known as MeltFlipper® technology, can be used in applications that include the rheological balance of mold and part filling, control of intra-cavity filling, warpage, part property, and cosmetic manipulation. One of the more common applications is shown for an 8-cavity mold in Figure 3.18(a). This is a conventional H-shaped runner with eight cavities. The cross sections of the flows are also shown. In the primary runner, the high shear and the low shear areas are concentric to each other. As the flow splits at the secondary runner, the high-sheared material stays on the inside. At the split at the tertiary runner, the high-shear material ends up in the inside cavities, causing these cavities to fill before the outside cavities. Inside and outside cavities exhibit different melt conditions. The result is that the parts formed in these two cavity groups will be different in size, weight, and properties.

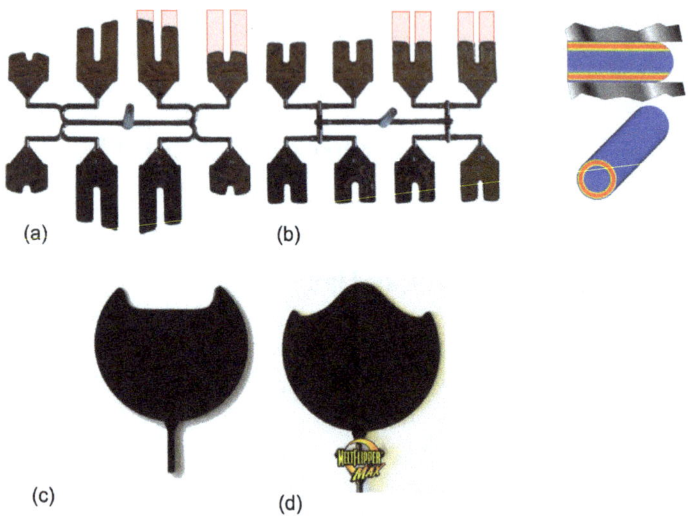

Figure 3.18 Before and after examples using the melt rotation technology [3]

Beaumont's patented melt rotations technologies use a variety of methods to manage the position of the high- and low-sheared laminates to achieve the desired balancing effects. In this 8-cavity example, the melt would be "flipped" or rotated, 90 degrees prior to entering the tertiary runner to the position shown in Figure 3.18(b). This rotation is commonly created at the intersection of the primary and the secondary runner. When this melt exits the secondary runner, the high-shear area is now on the top rather than on the inside, as is shown in Figure 3.18(a) without the flip. When this reoriented material enters the split of the tertiary runner, the high- and low-sheared melt splits up evenly into the two branches and each of the cavities receives melt with equal amounts of high-shear and low-shear material. This creates a fill and rheological balance between all cavities in the mold.

In the above case, the melt was rotated at one location, the intersection of the primary and the secondary runner. In case of a 16-cavity mold, the concept can be extended and the melt will need to be rotated at two locations for balancing all 16 cavities. In addition, by applying similar melt rotation techniques, some of the other problems in the parts can be solved. An excellent treatise on this subject with detailed explanation is provided in [1]. Some of the 'before and after' examples are shown in Figure 3.18. Figure 3.18(a) shows the filling pattern resulting in a conventional geometrically balanced runner. Filling is not only unbalanced from cavity to cavity, but each side of the Flow #1 cavities (inside four cavities) is different. Figure 3.18(b) shows the filling pattern after melt rotation was applied. Note that a balanced filling results not only between cavities but also within cavities. Not only will all eight parts be almost similar, but the use of pressure transducers (or thermocouples) for controlling and monitoring the process can be significantly improved. Figure 3.18(c) shows the filling pattern resulting in a simple flat disk, where the high-sheared material from the runner is causing a racetrack effect around the perimeter. Figure 3.18(d) is the same disk, except the melt has been conditioned using Beaumont's multi-axis rotation technology (MAX™ technology). Here, the majority of the high-sheared laminates have been repositioned to flow across the center region of the cavity.

Figure 3.19 shows the temperature distribution before and after the use of melt rotation technology. The photos of these parts were taken with an infrared camera immediately after molding. The left hand side part is the inside cavity fed by the high-sheared lamina. In the conventional runner there is clear evidence that the inside cavity is fed with hotter material (white). Note that this evidence of thermal variation exists even after the parts were partially cooled in the mold.

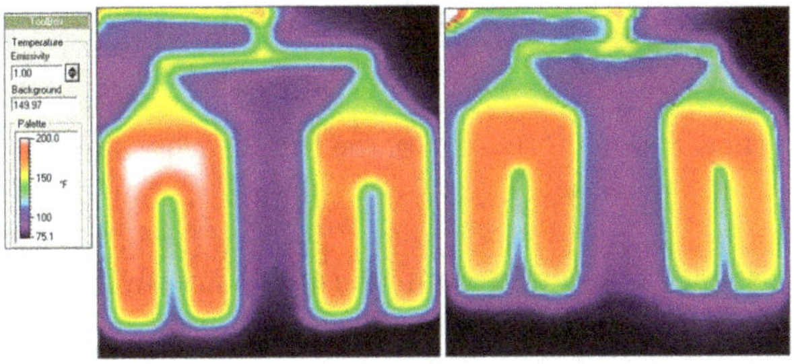

Figure 3.19 Temperature distribution before and after using melt rotation technologies [3]

3.8 Fountain Flow

In thermoplastic injection molding the mold temperature is set lower than the melt temperature because the melt needs to cool down to at least the ejection temperature. To the touch of the human hand, the mold may be hot but the mold is always considered to be cold. For example, the mold temperature for a polyetherimide (PEI) material is in the range of around 165 to 180 °C. Personnel are required to wear safety gloves when working with these molds. In the following sections, a cold mold therefore does not mean the mold is cold to human touch, but that it is colder than the plastic.

Plastic flow is laminar. Refer to Figure 3.20. When the plastic injection starts and the plastic first touches the cold mold, it freezes at the walls of the mold. The plastic that is injected behind this melt pushes its way through the center, moves ahead of the frozen layer, and ends up against the mold wall just ahead of it. The result is an inside-out flow, where the flow front looks similar to a water fountain, and therefore is called fountain flow.

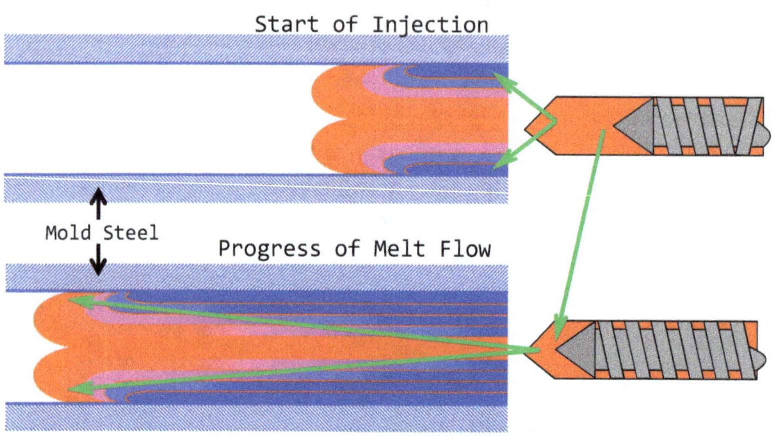

Figure 3.20 Depiction of fountain flow and the formation of the frozen layer with melt progression

If we divide the plastic melt that is ready to be injected into the mold into various sections, it is interesting to see where each of these sections end up. The plastic that is in the nozzle tip forms the skin of the sprue whereas the center of the sprue contains the plastic from the very back of the melt that was in front of the check ring. Refer to Figure 3.21, which shows a schematic cross section a mold with the progression of the melt marked by the numbers. For example, in the barrel, Section 1 is in the nozzle tip. When this gets injected into the mold, it will end up on the skin of the part that is closest to the machine nozzle. Section 2 will get pushed further and will form the skin on the part that is ahead of Section 1.

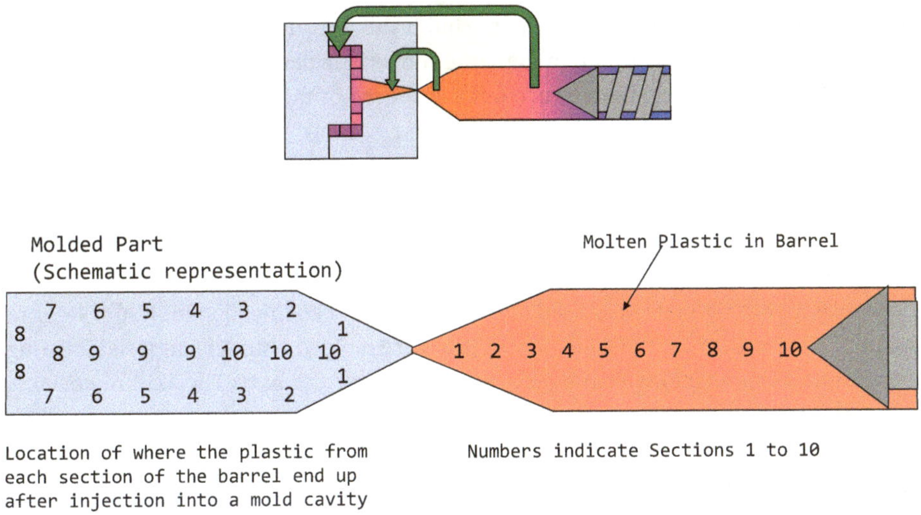

Figure 3.21 Relationship between position of the molecules in the barrel and their position in the final molded part

Fountain flow can be seen during color changes. If a molder is molding parts in blue and then switches to yellow, and then the yellow shows up at the end of fill, and the blue shows up in the sprue or the gate area of the part. Refer to Figure 3.22, which shows the molded part and the runner at a color change from a clear plastic to a yellow plastic. The skin of the sprue and the part are both formed with clear plastic and the inside of the part is formed with the yellow plastic. The cold slug well at the end of the sprue is also filled with clear plastic since that is the first plastic that enters the mold. The gate area of the part is filled with the yellow plastic whereas the end of fill and the skin of the part are filled with the clear plastic.

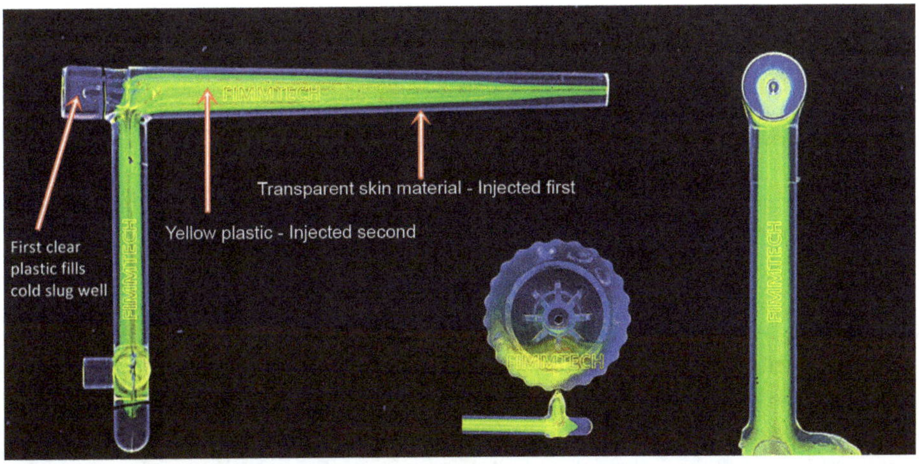

Figure 3.22 Evidence of fountain flow during color change

Understanding fountain flow can help in troubleshooting defects in molded parts. If a cosmetic defect is always located in the same spot on the part, one could backtrack the possible location of the melt to figure out the source of the problem. For example, a molder had a part that failed in one particular section. The problem was tracked down to a long nozzle body with the thermocouple at one end and the heater band at the other. The plastic was getting degraded and during the injection ended up in the section of the failure. Changing the nozzle body to a shorter one helped solve the problem.

The concept of fountain flow is used in the technique of co-injection molding. Refer to Figure 3.23. In the process of co-injection, parts are molded with one material forming the skin and the second material forming the core of the part. The skin material is selected for cosmetic reasons and/or other reasons such as environmental protection. The core material is used as a filler to drop the cost or to use as regrind material or just as a foaming matrix to reduce weight and prevent sink. In this technique, the skin plastic is injected first immediately followed by the core plastic. When the required amount of skin plastic is injected, the mold is only partially filled, but since the core material immediately follows, it pushes the melt of the skin plastic further to the walls, and the core plastic fills the part from the inside.

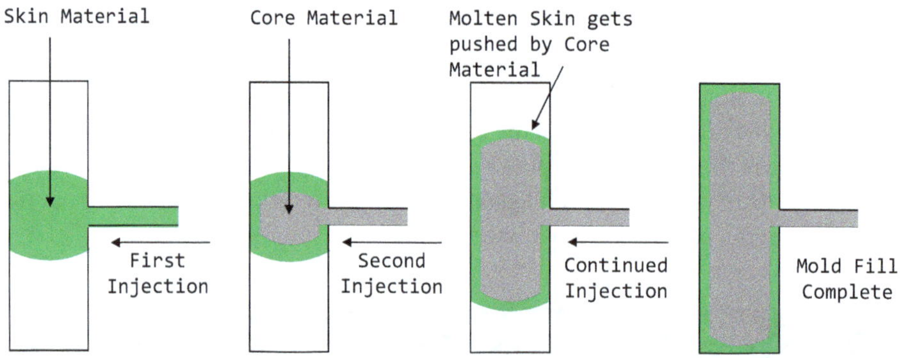

Figure 3.23 Co-injection molding; an application of fountain flow

3.9 Effect of Fountain Flow on Crystallinity, Molecular Orientation, and Fiber Orientation

Consider the flow of a crystalline material into a mold. During the formation of the skin, the plastic touches the cold mold and forms the skin almost instantaneously. Therefore, the molecules in the skin layer have no time nor the energy to crystallize completely. The layers underneath this skin are insulated from the mold surface and have a little more time before they cool down to a no flow condition. Therefore, the layers progressively have more time and energy to complete the crystallization process. If a cross section of a molded product is examined for crystallinity, it can be found that the center of the part has the most crystallinity and that the skin has the least.

For these same reasons mentioned in the previous paragraph, and because the skin layers freeze off almost immediately, the molecules and the fibers that are oriented in the direction of the plastic flow freeze off in the same orientation. With the progressive layers, the heat lets the molecules and fibers relax, and reach an equilibrium position. Therefore, the molecules on the surface have the most orientation but the ones on the inside have the least orientation. Refer to Figure 3.24.

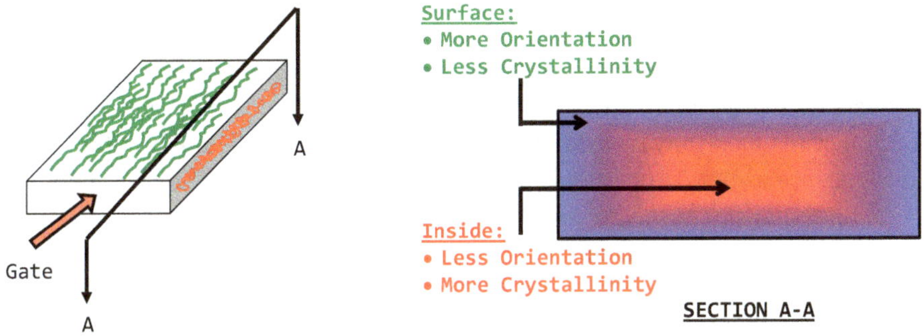

Figure 3.24 Molecular orientation, fiber orientation, and crystallinity in injection molded parts

3.10 Characterization of Polymer Viscosity

Polymers are found in a variety of applications ranging from their use in a solution (as in the case of paints) to being processed as melts. In this range of applications, the shear rates range from very low to very high. There is a need to employ various techniques to characterize the viscosity for a couple of reasons. First, there is no one viscosity model that fits the wide range of shear rates. A low shear model will not fit the viscosity-shear rate relationship at a high shear rate and vice versa. Second, the presence of another medium, such as a solvent or a plasticizer, alters the models and the properties of these mediums have characteristic effects on the viscosity. Based on these different characterization methods, various viscosities are also defined. *Solution rheology* is the study of polymers in solvents and *melt rheology* is the study of rheology of melts. Of particular interest related to injection molding is melt rheology at high shear rates. This is done with the help of the *capillary rheometer*. The shear rates that are observed in the machine nozzle, sprue, runner, gates, and the part can all be duplicated in this rheometer. Typical shear rates can range from 100 s^{-1} to almost 100,000 s^{-1}, with the shear rates in the gates potentially going into the millions per second. Figure 3.25 shows a schematic of a basic rheometer. It consists of a barrel and plunger arrangement. On the bottom end of the barrel is a die. The barrel is heated to a desired temperature based on the material being tested and is then filled with the plastic to be tested. Once molten, the plunger travels at various shear rates and the plastic is extruded through the die. The geometry of the die is very important and is used in the calculations of the shear rates. The plastic is extruded at various shear rates and the viscosity is measured. Typically, the viscosity is measured at three different temperatures. These data are sometimes made available by the material suppliers for mold design and flow simulation purposes.

Another common test is called the *melt flow test*. This test is carried out using a melt flow indexer that has a basic arrangement similar to the capillary rheometer (Figure 3.25). The difference is that this is a low shear test, with shear rates not commonly seen in injection molding. The advantage of simplicity and a general correlation with the high shear viscosity makes this popular. The output number is called the melt flow index (MFI) or the melt flow rate (MFR). The melt flow indexer is similar to the capillary rheometer, except that instead of the load cell, a known weight is placed on the plunger. The arrangement consists of a plunger, barrel, and die. The barrel is set to a desired temperature, then loaded with the plastic. Once a predetermined time has lapsed, the weight is placed on top of the plunger. The weight forces the plastic to be extruded. The weight of the plastic extruded in 10 minutes (in grams) is called the MFI or simply the melt index of the plastic. For example, if 25 g of plastic was extruded from the die, the MFI is 25. In the industry, it is common to say that the material is a 25 melt material. The units for MFI are always grams regardless of the measurement system. The MFI number is commonly found on material certifications as an incoming quality control parameter. A number of companies perform their own test to confirm these numbers and to keep a log of the incoming lots of material. This is an easy and inexpensive way of quality control, though care must be taken with filled compounds. In this case, the fillers do not melt and can cause an obstruction to flow and bias the results making them inconsistent.

The procedures for the capillary rheometer and the melt flow indexer tests are both defined by ASTM or other test organizations. They also define the various temperatures, weights, times, and other test conditions that may vary based on the type of plastic being tested.

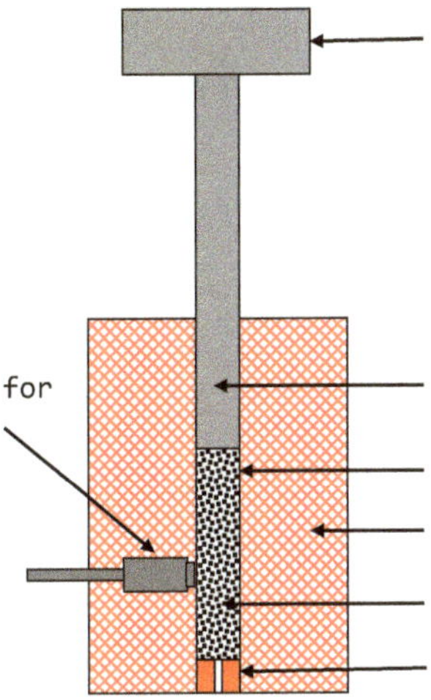

Figure 3.25 Basic schematic for rheometers and melt flow indexers

3.11 References

[1] Beaumont, J., *Runner and Gating Design Handbook* (2007) Hanser, Munich

[2] Beaumont, J., et al., Solving Mold Filling Imbalances in Multicavity Injection Molds, *Journal of Injection Molding Technology* (June 1998) Vol 2, No. 2, p. 47

[3] Beaumont Inc., *Technical Presentation* (2009)

Suggested Reading

Beaumont, J., Nagel, R., Sherman, R., *Successful Injection Molding* (2002), Hanser, Munich

Aklonis, J. J., *Introduction to Polymer Viscoelasticity* (1983) Wiley Interscience, NY

Billmeyer, F. W., *Textbook of Polymer Science* (1984) Wiley Interscience, NY

Cogswell, F., *Polymer Melt Rheology* (1981) John Wiley, NY

Dealy, J., Wissbun, K., *Melt Rheology and its Role in Plastic Processing Theory and Applications* (1990) Van Nostrand Reinhold, NY

Wilczyński, K., *Rheology in Polymer Processing* (2020) Hanser, Munich

4 Plastic Drying

Most plastics tend to absorb moisture when exposed to humidity. This is true with plastic in any form, whether as pellets before processing or in a finished product, such as an injection molded product. Such plastics are called hygroscopic or hydrophilic plastics. Plastics that do not absorb moisture are called hydrophobic plastics. Nylons are common examples of hygroscopic plastics. Nylon parts will absorb moisture and alter the dimensions of the molded part, depending on the humidity. As a nylon part absorbs moisture, it can swell in physical size, causing dimensions to change beyond the required specification limits. Although moisture absorption is inevitable in the molded part, excess moisture should be removed from the plastic resin to an acceptable level before molding in order to produce an acceptable part. Every plastic has an acceptable maximum moisture level above which melt processing problems can occur. Moisture levels must be below this recommended value before processing. A list of maximum moisture levels for various materials is provided in Table 4.1. The numbers mentioned are for nonfilled plastics. Most fillers are nonhygroscopic and therefore do not absorb any moisture. Consider as an example nonfilled nylon. Nylons typically need to have a moisture level of less than 0.20% before they are processed. If a particular nylon resin is 50% glass filled, then the amount of nylon is 50%. Therefore, the amount of allowable moisture will also be 50% of 0.20 or 0.10%. The amount of filler must always be taken into consideration when conducting a moisture test. Most material manufacturers overlook this when providing the datasheets.

Table 4.1 Unfilled Materials and the Recommended Maximum Moisture (Courtesy: IDES.com)

Long name	Short name	Suggested maximum moisture (%)
Acetal (POM) copolymer	Acetal copolymer	0.15–0.20
Acetal (POM) homopolymer	Acetal homopolymer	0.2
Acrylic, polymethyl methacrylate	PMMA	0.097–0.10
Acrylonitrile butadiene styrene	ABS	0.010–0.15
Polyamide 6	Nylon 6	0.095–0.20
Polyamide 66	Nylon 66	0.15–0.20
Polyamide 66/6 copolymer	Nylon 66/6	0.099–0.20
Polyphthalamide	PPA	0.045–0.15
Polycarbonate	PC	0.019–0.020
Polybutylene terephthalate	PBT	0.020–0.043
Polyethylene terephthalate	PET	0.0030–0.20
Polyetherimide	PEI	0.020–0.021
Polyethylene, high density	HDPE	NA
Polyethylene, low density	LDPE	NA
Polyethylene, linear low density	LLDPE	NA
Polyphenylene sulfide	PPS	0.015–0.20
Polypropylene homopolymer	PP homopolymer	0.050–0.20
Polystyrene, general purpose	PS (GPPS)	0.02
Polystyrene, high impact	HIPS	0.1
Polyvinyl chloride	PVC	NA
Styrene acrylonitrile	SAN	0.020–0.20

In materials such as nylon, moisture plays an extremely important role during processing by acting as a viscosity regulator for the melt. Therefore, there is also a minimum required level of moisture for such materials. This topic will be discussed further later on in this chapter. Drying plastic resins before processing is a critical step. Plastic resins that are hygroscopic must be subjected to elevated drying temperatures for a specified time, in order to effectively remove excess moisture. However, excessive drying of the plastic resin, beyond the manufacturer's recommended times and temperatures, can also create problems. Until recently, this has been an overlooked condition although it can have a very negative potential impact on the mechanical properties and appearance of the finished molded part. Exposure of plastic resins to drying temperatures above the manufacturer's recommended levels for excessive periods of time is typically referred to as "overdrying."

If the process of drying is not done correctly, it can result in losses in production in the form of scrap parts and also lost production time that cannot be recovered. Drying is accomplished with the help of dryers in the molding facility. Some plastics come pre-packaged in vacuum-sealed bags and do not need to be dried as long as they are processed immediately after being removed from their packaging. Any opened and unused portion must be dried before it is processed, if later stored in an open environment.

4.1 Problems in Melt Processing Related to the Presence of Moisture

There are several problems that can result due to the presence of moisture during the melt processing of the plastic.

4.1.1 Degradation of Plastic

In the presence of moisture, the plastic can degrade in the injection molding barrel. A chemical reaction called hydrolytic degradation can take place at molding temperatures and attack the long-chain molecules. The water molecule acts as a catalyst and initiates the degradation. The degradation can itself produce more water and increase the rate of the degradation reaction. This breakdown of the molecules causes the loss of properties in the final product. There are two types of hydrolytic degradations that can occur: If the chain ends are broken, the loss in molecular weight is not significant and the effect on the end product is negligible. This type of degradation is called end-chain degradation. In random degradation, the polymer molecules are broken at random bonds along the molecule's length and this can cause a significant drop in molecular weight, leading to a decrease in properties. Typically, parts molded from degraded materials tend to become brittle and show a reduction in their mechanical properties. There can also be a loss in appearance ranging from a shiny finish to a dull finish. Other surface defects, such as splay, are also common. Hydrolytic degradation is a common problem with condensation polymers, such as polyesters, nylons, polycarbonates, and polyurethanes.

4.1.2 Presence of Surface Defects

A number of surface defects can be attributed to the presence of moisture during processing.

Splay

Any moisture that gets carried by the melt during injection of the melt into the mold stream gets released and then tends to form a film between the melt and the mold surface. The presence of this film prevents the melt from coming in contact with the mold surface and picking up the texture of the cavity surfaces. The film creates a smooth surface and once cooled down, the plastic shows shiny streaks in this area, see Figure 4.1. This phenomenon is called splay or silver streaks.

Figure 4.1 Splay on the surface of a part

Bubbles

If the moisture stays inside the melt and does not come out to the surface, internal defects, such as voids or bubbles, can form in the part. In some cases, if the bubble is too close to the surface, the moisture inside is still hot and pressurized when the part is ejected from the mold. This can show up as an external defect such as a bump or bubble on the part. These defects can either be microscopic or in some cases show up as large deformations common in large and thick parts. Figure 4.2 shows internal voids and external defects due to excessive moisture.

Figure 4.2 Internal voids and external defects on a part due to excessive moisture

Burn Marks

High injection speeds cause the plastic to be subjected to high shear rates. At these shear rates and in the presence of moisture some plastics can undergo degradation and burning of the molecules. This can show up as dark streaks or discoloration in the part. Sometimes, the burning can be seen at the end of fill. Vents are designed to help remove the displaced air in the cavity and runner system as the plastic fills the mold during the mold filling process. If moisture is present in the melt, the plastic can degrade, which causes an excess amount of gas that needs to be evacuated from the mold cavity. Once the moisture vapor gets to the front of the melt stream, it does not mix with the melt and always stays in front of the melt stream. If sufficient venting is not provided to allow the moisture vapor or gas to escape, it will get trapped and compressed under the high injection, pack, and hold pressures, resulting in a diesel effect that causes burning of the plastic. This burning can be seen as a black or a white mark at the end of fill location, depending on the type of plastic.

Plugging up of the Vents

The excessive moisture or the volatiles from the degraded polymer and its additives can overload the venting system of the mold. This degradation generates by-products that are low-molecular weight compounds. These compounds begin to collect in the vents, plugging them up. As production continues, the part quality starts deteriorating because of reduced venting capacity. Burn marks on the parts near the end of fill are common and frequent when this occurs. Cleaning of the vents is required to eliminate the burn marks. Figure 4.3 shows residue built up on the mold surface.

Figure 4.3
Residue built up on the secondary vents

Dimensional Variation

Plugging of the vents can cause dimensional changes in the part. Part dimensions are directly related to plastic pressure in the cavity. Therefore, to obtain consistent quality, the cavity pressure must be consistent during every injection molding cycle. However, if the vents are clogged, the air and gases cannot escape and the internal pressure of the cavity increases, producing a part of varying dimensions. Parts with tight tolerances can easily drift out of specification limits.

Loss in Properties at the Weld Line

When resins are not properly dried, moisture will be present at the flow front; when two flow fronts meet, the moisture interferes and does not allow weld strength to develop to its full potential. This causes reduced weld line strength. Sometimes the weld line is also more noticeable in such cases. This loss in weld line strength can also be a result of insufficient venting due to plugged vents as described earlier.

Nozzle Drool

This is a common phenomenon in crystalline plastics such as nylons. Under normal circumstances, plastic does not leak from the nozzle tip after the screw recovers and fills the barrel with resin. There are two main factors that prevent nozzle drool: First, at the end of the screw recovery process the screw is sucked back without rotation for a short distance. This reduces the internal plastic pressure and gets the molten plastic away from the nozzle tip. Second, the tip is controlled at a temperature that keeps the viscosity of the plastic high enough to prevent it from flowing out of the nozzle but low enough to allow plastic flow through the nozzle at the start of the next injection cycle. This control is extremely important and the presence of any moisture in the plastic can reduce the viscosity to an extent that the plastic will continually drool out of the nozzle tip. No amount of reduction of nozzle temperature or increase in suck-back can help because the plastic flow rate is high, preventing sufficient time for the plastic to cool down through the short nozzle tip. Drying of these resins to control the drool is therefore important. Since crystalline plastics have a narrow processing range, they tend to have lower viscosities compared to amorphous plastics. Drying of such resins to control the drool becomes important. When the material is first loaded into the dryer and is stagnant in the dryer for the required amount of time, the material near the outlet of the dryer is usually not exposed to the dry air. This is due to the design of the hopper where the cone that supplies the dry air sits above a certain level. Only plastic that is above this cone is subjected to the dry air and gets dried. It is advisable to drain out the plastic until dry plastic is encountered and then start the molding. In some cases, when the moisture content is very high, it is not uncommon to see sputtering of the melt out of the nozzle. Once the molding process is started, material is continuously flowing from the top to the bottom of the hopper and dry material is constantly supplied to the molding machine.

Inconsistency in Shot Control

The presence of moisture can mask the true volume of the plastic during the shot build up and therefore cause a reduction in the amount of the required plastic. Although the screw will always reach the shot size or build the required volume for injection, it may not always be equal to the volume of the shot that is essential to make a good quality part. Every shot will therefore be of a different weight, causing inconsistency from shot to shot.

Examples of hygroscopic materials that do not degrade in the presence of moisture are ABS, SAN, and acrylics. Nylons and polyesters are examples of materials that will degrade in the presence of moisture. Examples of nonhygroscopic materials are polyethylenes, polypropylenes, and polystyrenes. These materials do not absorb moisture; however, in humid environments moisture can settle on the surface of the unmolded pellet, resulting in cosmetic defects, such as splay if the material is not first dried to remove the surface moisture.

4.2 Hygroscopic Polymers

Polymers that absorb moisture are called hygroscopic. Whether a polymer absorbs moisture depends on its chemical structure, see Figure 4.4. The water molecule shown here is made up of two atoms of hydrogen and one atom of oxygen. Because the two hydrogen atoms and the oxygen atom tend to share electrons, the hydrogen atoms are usually on one side of the molecule, which results in the molecule being positively charged on one side and negatively charged on the other. A molecule with this kind of charge distribution is called a polar molecule.

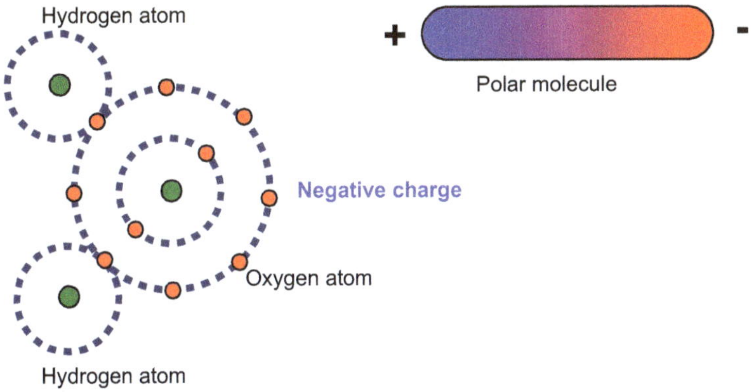

Figure 4.4 Chemical structure of the water molecule

Similar to water, certain groups on the main chain of a polymer can also form polar groups. For example, the carbonyl group (–C=O–) found in polyamides and polyesters is a polar group (see Figure 4.5). These polar groups on the polymer chains attract the polar water molecules like magnets to form weak secondary bonds resulting in the hygroscopic nature of the polymer (Figure 4.6). These secondary bonds are also called hydrogen bonds.

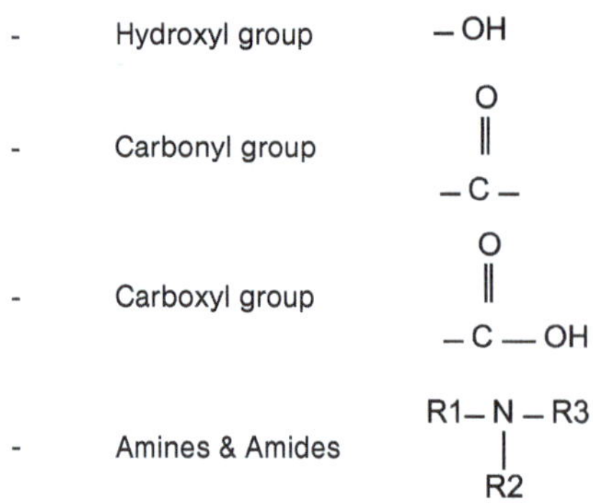

Figure 4.5 Examples of polar groups

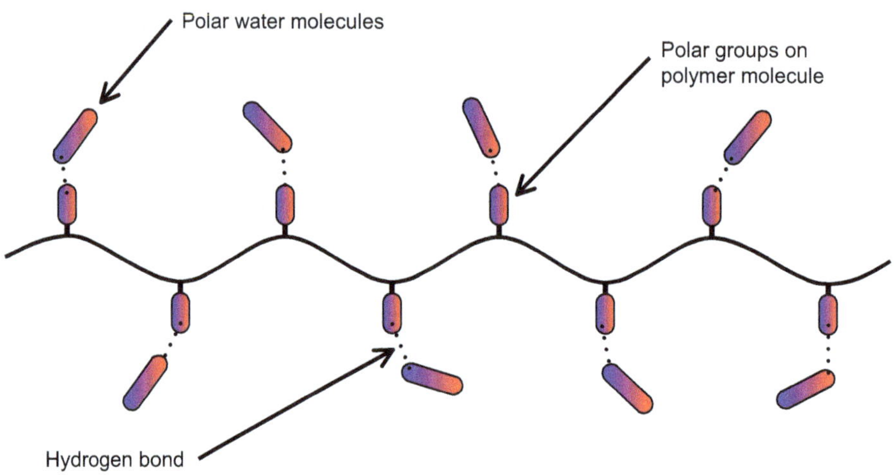

Figure 4.6 Polar water molecules forming hydrogen bonds with polar groups on the polymer molecule making the polymer hygroscopic

Some polymers, for example polyethylenes, are polymerized from nonpolar monomers such as ethylene. In an ethylene molecule, the charges are balanced around the carbon atoms and hence the polymer is nonpolar. The polar water molecule is therefore not attracted to the nonpolar polyethylene molecule. Polyethylene is therefore nonhygroscopic (see Figure 4.7).

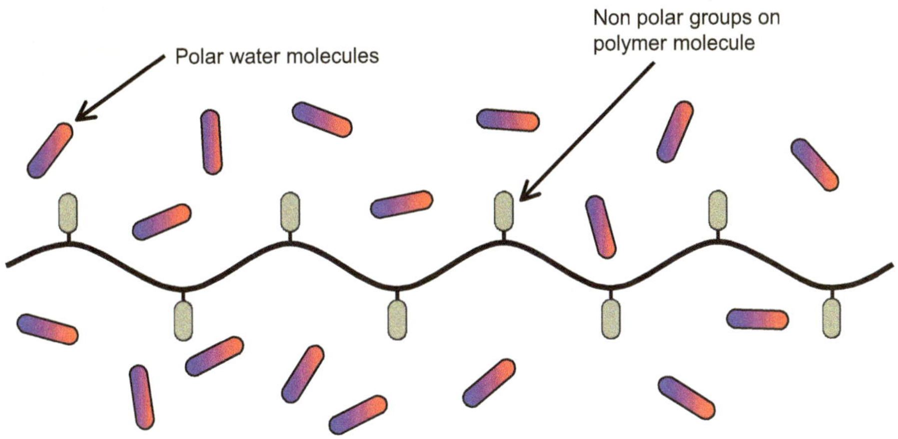

Figure 4.7 Absence of hydrogen bonding between water molecules and polymers with nonpolar groups making the polymer nonhygroscopic

4.3 Drying of Plastics

There are some important considerations that need to be taken into account for effective plastic drying before the melt processing stage.

4.3.1 Drying Temperatures and Times

The nature of each polar group for every plastic is different and the strength of the bond between the water molecule and the plastic varies. For example, in polyurethanes the bond strength is very high compared to the bond strength in ABS. For this reason, the drying times and temperatures for individual plastics are different. Depending on the specific grade, polyurethane is typically dried 4 to 6 hours at 132 °C (270 °F), whereas an ABS needs to be dried for 2 to 4 hours at around 75 °C (165 °F). The range in drying times is due to the fact that initial moisture content can vary depending on the humidity the polymer material is exposed to. If moisture content is higher, the drying time required to remove the moisture will be longer, while the rate per unit time of moisture removal stays the same. Only a given amount of moisture can leave the plastic in a given amount of time. The size and the shape of the pellet can also affect the drying times. The drying temperatures and times for some common plastics are listed in Table 4.2. Note that some of these materials may not be injection molding materials.

Table 4.2 Drying Temperatures and Times for Common Plastics (Courtesy: IDES.com)

Name	Drying temp (°C)	Drying time (h)
ABS	175–190	2.0–4.0
Acetal homopolymer	175–195	2.0–4.0
Acrylic (PMMA)	180	3.0–6.0
ASA	180–190	2.0–4.0
Nylon 6	160–180	2.0–4.0
Nylon 66	175	2.0–4.0
PBT	250–280	3.0–4.0
PC	250	3.0–4.0
PEEK	176	3
PEI	300	4.0–6.0
PLA	212	4
PPS	275	3.0–6.0
PS-GPPS	180	2.0–4.0
PS-HIPS	160–180	2.0–4.0
PSU	275	4
PUR	N/A	N/A
PVC	150	2.0–4.0
PVDF	302	1
SAN	160–180	2.0–4.0
SPS	176	2.0–4.0

With some plastics, various combinations of drying times and temperatures can be used to dry the material. For example, polyphthalamide (PPA) can be dried at 79 °C (175 °F) for 8 hours or at 212 °C (250 °F) for 2 hours. The choice of recommended combination depends on the optimal residence time of the plastic in the dryer. Sometimes the dryers can be large compared to the material used per hour. Care must be taken because certain resins can be very sensitive to higher drying temperatures, especially if subjected to longer times in the dryer than suggested by the manufacturer. Prolonged drying times can harm the plastic. In some cases discoloration can be seen in the plastic pellets. It is also possible that the plastic pellets may soften and stick together in the dryer. The pellets affected the most are those near the exit, or bottom, of the hopper, which take the weight of the pellets above them, causing bridging and making the situation worse. Higher temperatures can also lead to the loss of the low molecular weight additives in the base polymer.

In case of both hygroscopic and nonhygroscopic polymers, there can be water condensation on the pellet surface. If the pellet is transported from a relatively cold environment, such as a silo on the outside of the building, into a warmer and possibly more humid molding environment, water can condense on the surface of the pellets. This is similar to the water condensation we see on the outside of the glass of a cold drink on humid days. Condensation can also occur at the feed throat of the molding machine which is typically kept cool by cold water circulation in order to avoid bridging at the feed throat. If the area is not cooled, the plastic that is stationary between the end of screw recovery and before the mold opens melts to form a solid mass and prevents the plastic from being conveyed into the molding barrel. Nonhygroscopic materials are therefore surface dried at low temperatures to remove this moisture before processing. Exposing the plastic to room temperature or low temperature dry air for a short period of time is sufficient to dry off the moisture. The drying time can be as low as half an hour on low humidity days.

Additives can also be hygroscopic in nature. For example, certain calcium-based fillers can absorb moisture. Any plastics filled with these compounds must therefore be dried before processing regardless whether the base plastic is hygroscopic or not.

4.3.2 Relative Humidity and Dew Point

To extract the moisture out of the plastic, the air used for drying should be as dry as possible. When the plastic with moisture is exposed to the dry air, the system is looking to achieve equilibrium and therefore the moisture is continually extracted. The dryness of the air can be expressed with the help of two terms: the relative humidity and the dew point. Relative humidity is the percentage of moisture the sample of air holds compared to the amount of moisture it could hold when it is saturated. The saturation level changes with temperature. The lower the temperature, the lower is the maximum amount of moisture it can hold.

Dew point indicates the amount of moisture in the sample of air. The higher the dew point, the higher is the amount of moisture in the air. The dew point is defined as the temperature to which the given air sample will have to cool in order to reach 100% relative humidity or complete saturation. Therefore, a lower dew point indicates a lower amount of moisture in the air. A temperature of −40 °C (−40 °F) equates to very low amount of moisture (less then 0.4%) in the sample of the air. This temperature is therefore taken as the target dew point temperature to be achieved for the air that is supplied to the dryers.

Let us put this into perspective by considering an example. The dryer temperature is set to 100 °C. This means that the air being supplied to the dryer is at 100 °C. At this temperature, the air may still contain moisture, depending on its relative humidity. This moisture will prevent the drying of the plastic to the required levels. To reduce this moisture, the air must be dried. This is usually done by using desiccant beds that absorb the moisture from the air as it is passed through the desiccant. As the amount of moisture in the air is reduced, the temperature at which it will condense, or its dew point, also reduces. Lower dew point temperatures indicate that the amount of moisture being supplied to the dryer is also lower. Therefore, measuring the dew point will indicate the dryness of the air. If the dew point is around −40 °C, the quality of the air is acceptable to dry the plastic. The final moisture level in the air is a result of the temperature, relative humidity, and the dew point of the air.

4.3.3 Air Flow Rate

The hopper used to dry the material is supplied with air that has a low dew point, is dry, and has been heated to the recommended temperature. As the plastic dries, the moisture migrates to the surface and should be continuously carried out of the system. This requires a sufficient amount of flow of the air through the system. Therefore, the flow rate of the air through the system is important.

4.4 Equipment for Drying Plastics

Several types of dryers are available for drying plastics. The classification is based on the technology or other features, such as location of the dryer.

4.4.1 Oven Dryers

During the early years of molding, when processors realized the need for drying, ovens similar to baking ovens were used. The plastic was spread out on large trays and put into the ovens to dry. The large trays helped to spread out the plastic and increased the area of exposure to the heat. The plastic would not dry evenly if the layer of plastic was too thick. Handling of the hot trays and transporting the material to the hopper of the machine was not easy and often required two people. Smaller ovens are still in use in some R&D facilities and production facilities that process small amounts of material. One of the additions to these ovens is the use of a vacuum pump. The vacuum decreases the boiling point of any liquids and at the same time facilitates the removal of the moisture from the chamber speeding the drying process.

4.4.2 Hot Air Dryers

Hot air dryers supply hot air to the plastic at the bottom of the dryer and as the air passes through the plastic, it picks up the moisture from the surface of the pellets and transports it out of the system. The heat also helps to drive the moisture to the surface of the pellet. The supplied air is picked up from the atmosphere, heated and pumped into the dryer. These systems are best suited for surface drying of nonhygroscopic materials such as olefins or for materials that do not require moisture levels to be very low before processing.

4.4.3 Desiccant Dryers

These dryers are similar to hot air dryers except for the fact that the air first passes over a desiccant bed that adsorbs almost all of the moisture from the air. This dry air is then heated and supplied to the hopper. Since the relative humidity is now low, the air can pick up more of the moisture from the plastic, helping to achieve the lower moisture levels desired for materials such as nylons. As the moisture is adsorbed by the desiccant bed, the desiccant will eventually become saturated with the absorbed moisture. As the saturation level of the desiccant increases, its ability to absorb moisture decreases, making the drying process less efficient. To correct this, there is a regeneration cycle during which the desiccant is dried and then recycled back into the system. Desiccant dryers are widely used in most modern-day facilities because of their versatility.

4.4.4 Classifications Based on the Location of the Dryer

Dryers can be located next to the molding machine and a hose can be used to supply dried material from the dryer to the molding machine. These are conventional dryers that are most commonly used in small to medium size companies in which a variety of materials are processed and mold changes are frequent. The dryers are mobile and can be moved from one machine to another. In companies processing a mix of hygroscopic and nonhygroscopic materials it is not required to invest in the same number of dryers as the number of molding machines. The dryers can be moved where needed. Hopper dryers on the other hand are dryers that are mounted directly on the molding machine such that the dry material is fed straight into the feed throat of the molding machine. This is a good solution for extremely hygroscopic materials such as some polyurethanes that can pick up moisture during the transportation from the conventional dryer to the molding machine through the transport hose. The disadvantage is that these dryers are attached directly to the molding machine and are not easily moved to different molding machines on short notice. This type of dryer, for all practical purposes, is considered dedicated to the machine.

Central dryers are very large dryers that can be installed in companies that process large quantities of the same material. The material is dried at a central location and is then transported to the molding machine for processing.

4.5 Determination of the Amount of Moisture

It is critical to be able to accurately determine the amount of moisture in the plastic to be melt processed. Therefore, an accurate measurement system is necessary. There are various direct and indirect measurement methods, most of which suffer from a basic problem. When trying to extract the moisture for analysis and measurement, other volatiles, additives, and residual polymerization compounds including monomers are released. These can therefore mask the actual amount of moisture in the plastic and give false readings. There are ways to prevent this but they would not be practical in a production environment. Some moisture analysis methods are discussed in the following.

4.5.1 The Glass Slide Technique (TVI Test)

This test was developed by GE Plastics (now Sabic Innovative Plastics) and is called the Tomasetti Volatile Indicator test or the TVI test. It involves placing four to five plastic pellets on a glass slide that is heated on a hot plate. With the help of another slide, the pellets are pressed together and sandwiched between the two slides. Once the pellets are flattened out, the slides are removed and allowed to cool. Any moisture in the pellets shows up as bubbles in the slide. Sometimes it is helpful to use a microscope to view the bubbles. This is an easy way to determine the presence of moisture in the sample, although the slides and the hot plate must be handled very carefully to avoid safety issues. This test does not provide a numeric value of the amount of moisture present in the plastic and, as in other tests, volatiles from degradation or an additive cannot be identified. The TVI test is not a commonly used test in industry and is in fact outdated

4.5.2 The Karl Fischer Titration Method

This type of analysis will produce the most accurate results. However, because of the time required to conduct this test, the additional materials and equipment required, the test is also not commonly used. The basic underlying principle of this method is that a small amount of electricity will be generated with the chemicals involved in the test due to the reaction of the moisture in the plastic. The plastic is heated to a higher temperature to release the moisture. The amount of electricity generated is directly related to the amount of moisture present in the system. Accurate measurement of the electricity generated provides the amount of moisture present in the plastic. Figure 4.8 shows a setup for a Karl Fisher titration system. A disadvantage of this system is that at elevated temperatures, water can be produced in some plastics. This may be caused by degradation or by melt polymerization, a phenomenon explained in Section 4.1.1. Therefore, for these plastics the newly formed water can skew the results and provide false readings.

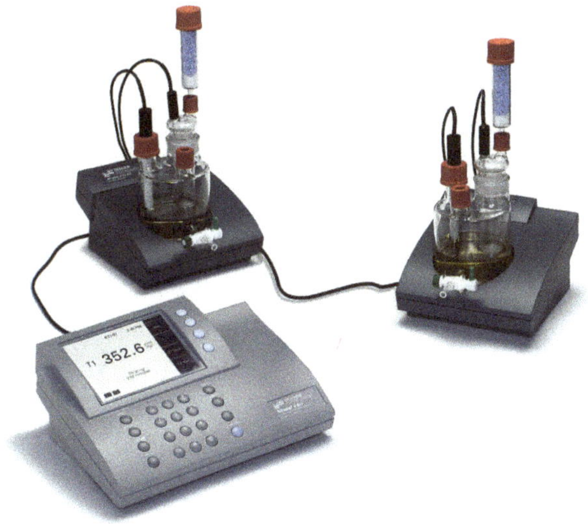

Figure 4.8 A Karl Fischer setup to measure moisture in a sample (Courtesy: Denver Instruments)

4.5.3 Electronic Moisture Analyzer

With the advances in technologies and with better understanding of the moisture absorption process, electronic moisture analyzers have become common in many production facilities. The biggest advantage they provide is their simplicity of use without any prior knowledge or experience in plastics or moisture determination. The test takes less than approximately ten minutes and provides a print out of the moisture level in the plastic. The instrument can also be hooked up to a PC to collect and record test data over time. These analyzers work on the principle of weight loss when the plastic is heated and the water leaves the system. In this test, a small amount of plastic is placed on a pan that is attached to a micro-scale. The weight of the sample is recorded and the sample is heated to a desired predetermined temperature for each plastic. As the water leaves the plastic, the weight drops and once the weight stays constant, it is assumed that the plastic has lost all its moisture. The final weight is recorded and the percentage weight loss is calculated, providing the moisture content of the plastic. The disadvantage of this system is that it does not take any other volatiles that may be lost during the heating process into consideration, which could influence the moisture content results. Figure 4.9 shows a picture of a moisture analyzer. These types of analyzers are widely used because of the simplicity of use and the sufficiently reliable data they produce.

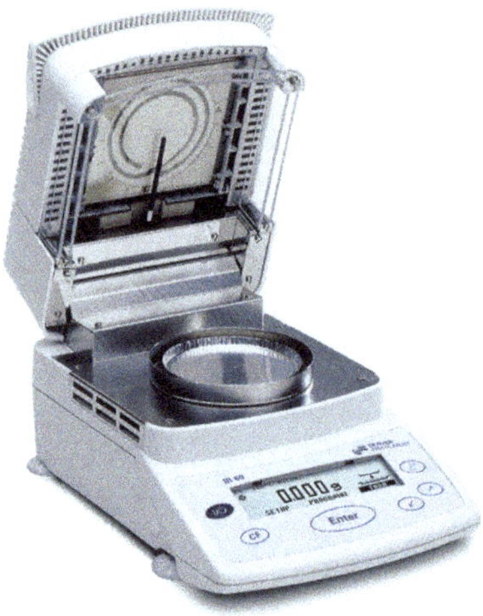

Figure 4.9
Electronic moisture analyzer (Courtesy: Denver Instruments)

4.5.4 Measurement of the Dew Point

A dew point meter helps to measure the dew point of the air that is supplied to the dryer. Although this method does not actually measure the dryness of the resin, it helps in assuring that the dryer is fed with dry air. A dew point of −40 °C (−40 °F) is a good indication that the amount of moisture in the air is at very low acceptable levels.

4.6 'Overdrying' or Overexposure to Drying Temperatures

Most polymers and plastics are blended with low molecular weight additives, such as heat stabilizers, processing aids, and other specialty additives. The additives are used to enhance the plastics' properties for specific applications and/or to reduce their cost. Such a blended polymer is called a resin. All plastics used for melt processing are considered resins because of these essential additives. Additives include plasticizers, lubricants, flame retardants, heat stabilizers, colorants, blowing agents, and biocides among others. Most of them are added to the polymers in small percentages and are low molecular weight compounds or oligomers.

The drying of the plastic must be controlled in order to not exceed the maximum recommended drying times. On the production floor this important rule is very easily overlooked. It is typical in the molding industry to first load the resin into the dryer and then to complete the mold change in order to ensure that the resin is being dried during the setup operation and no time is being wasted. This is a very efficient procedure. However, if the mold setup was not completed in time or the first shots from the mold were not acceptable and the mold had to be pulled out for maintenance, the plastic that is left in the dryer is now being subjected to additional drying time at elevated temperatures. The overexposure may lead to a potential loss of the low molecular weight additives in the plastic. These additives are usually not as heat resistant as the polymers, especially after prolonged drying times. Another scenario in which overdrying can be a potential problem is when the dryer is oversized for the mold in the machine. In an oversized dryer the residence time of the plastic is longer than the recommended maximum drying time of the plastic. For example, if the hourly usage of material is 10 lb and the maximum recommended drying time is 8 h, then, if the capacity of the dryer is more than 80 lb, the plastic that is on top or that is loaded last, will be subjected to a drying time of more than 8 hours before it is actually used. This must be avoided.

Case Study

A study [1] exemplifies this for two particular materials, a polyester and a nylon. Here, the polyester was a 30% glass filled polybutylene terephthalate (PBT) and the nylon was a 15% glass filled nylon 66. Both resins were dried for varying times and the impact of the overdrying was studied. Parts were molded from an existing production mold. For each material, the process was kept unchanged during the molding of the resins dried for the various times. The results were material specific and the effect of overdrying was different for both materials, as was confirmed by thermal analysis and mechanical tests. In a thermogravimetric analysis (TGA) the plastic sample is continuously heated, which causes the polymer to burn off at a given temperature which depends on the base plastic and its additives. Rheology studies are performed to study the viscosity of the plastic and to determine the effect of drying. The shear sweep study provided the viscosity versus shear rate plot and the thermal degradation study was done by plotting the viscosity versus time at a given shear rate. As the plastic stayed in the barrel, the plastic degraded and a change in the viscosity was recorded. The observations and the results are discussed in the following.

Results for PBT

The resin and the molded parts were tested using the procedures mentioned above. During the molding process it was seen that as the drying time increased, flash on the parts increased, indicating a drop in melt viscosity. However, there was no significant change in the actual characteristic values, such as fill time, cushion value, screw recovery time, and others. The TGA data is shown in Figure 4.10.

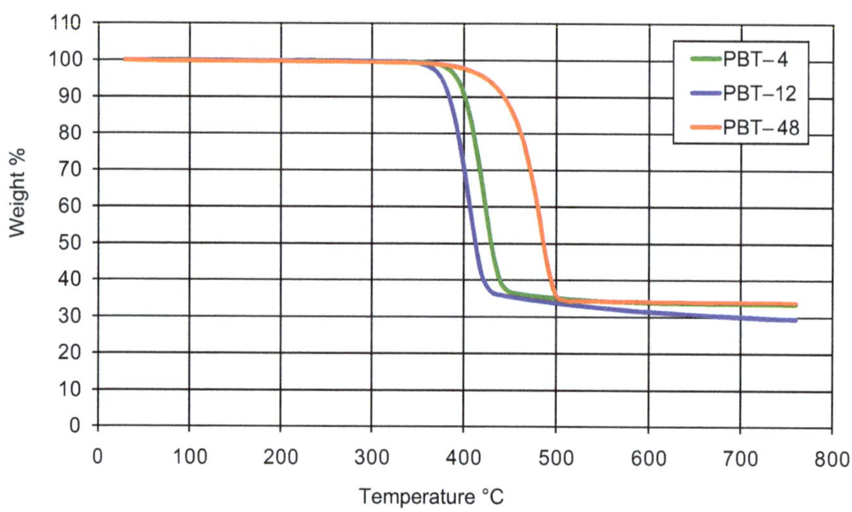

Figure 4.10 TGA graphs of PBT dried for 4, 12, and 48 hours [1]

The final residue left behind was about 33% of the initial weight. Because the resin was a 30% glass filled material, most of this residue must have been the glass. To better understand the results, $T_{1/2}$ is defined as the temperature at which 50% of the weight loss occurs. Considering the percentage of the residue remaining, 50% of the weight loss occurs at approximately 420 °C for PBT-4 and at approximately 475 °C for the PBT-48. This indicates that for the shorter drying times, a weight loss occurred at lower temperatures suggesting there must be a component in the resin that decomposed at lower temperature leading to an earlier loss in weight. This additive was probably decomposed and was taken out of the dryer during the lengthy drying process and therefore the PBT-48 had a $T_{1/2}$ that was almost 55 °C higher than the $T_{1/2}$ for

PBT-4. Looking closely at the numerical data and analyzing the initial weight loss in the PBT, it can be seen that the slope of the curve is higher for the PBT-4, indicating sudden weight loss similar to those seen in low molecular weight compounds and oligomers. After the initial differences in the slopes of the curves, the lines then seem to run parallel, suggesting the decomposition of the base resin.

Shear sweep data obtained by capillary rheometry is shown in Figure 4.11. The data did not show any difference in melt rheology. This could suggest that the additive lost during the drying process was not a processing aid intended to lower viscosity for ease of processing. No other conclusions could be drawn.

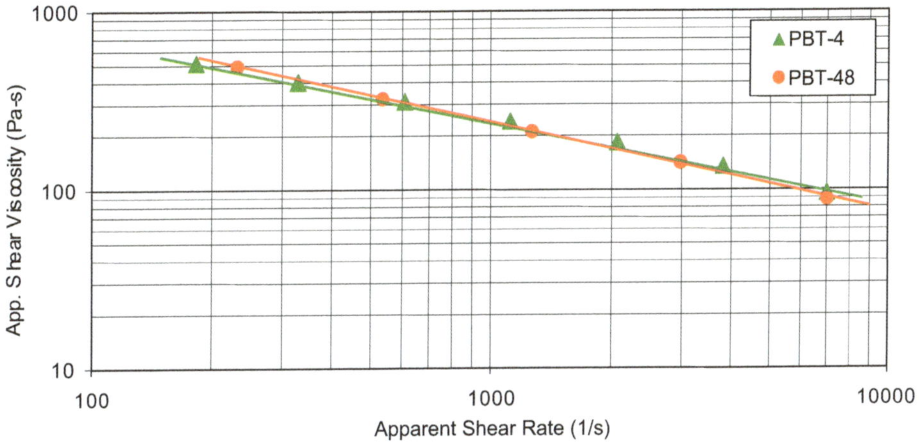

Figure 4.11 Capillary rheometry data (shear sweep) for PBT dried for 4 and 48 h [1]

Thermal degradation data also obtained using capillary rheometry is shown in Figure 4.12. The data showed a difference between the two resins only at residence times of 9 minutes and above. PBT-48 showed a lower viscosity compared to PBT-4. This could have been caused by the degradation of the base resin, lowering its molecular weight and thereby lowering its viscosity. It could also be an indication that the component lost during the excessive drying process was a heat stabilizer. It is also interesting to note that the viscosity curve almost flattens out past 9 minutes, possibly suggesting that the polymer is completely degraded and that it is the glass fibers now being carried by the degraded resin and contributing to the viscosity.

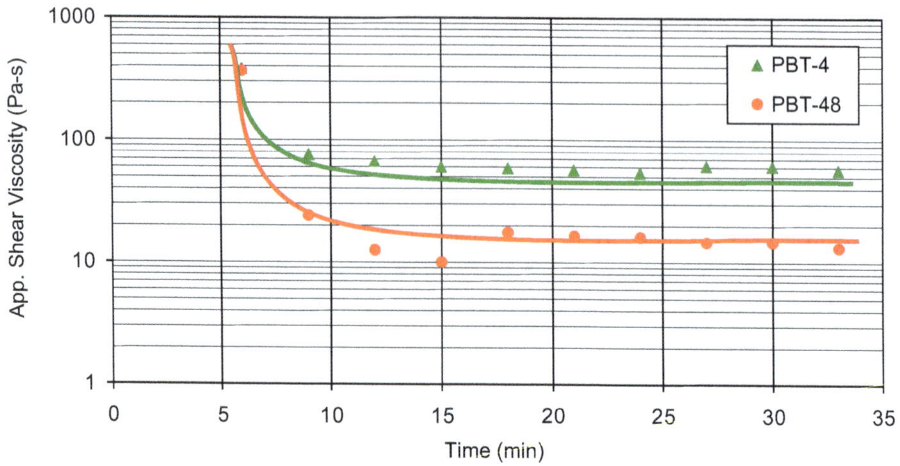

Figure 4.12 Capillary rheometry data (time sweep) for PBT dried for 4 and 48 h [1]

PBT parts molded from this resin after excessive residence times in the dryer crumbled into pieces when a small amount of force was applied. The drop impact testing of the parts demonstrated a clear difference between PBT-4 and PBT-48. This data is shown in Figure 4.13. As can be seen, the mean failure energy decreased with the increase in drying time. The parts tended to become more brittle. This could be the effect of either the loss of an additive, such as an impact modifier, and/or the degradation of the resin. There was a sharp drop in the mean failure energy between 12 and 36 hours of drying time. Before and after these times the curve stayed relatively flat. Looking at this data, drying times should be limited to 12 hours to retain the material properties.

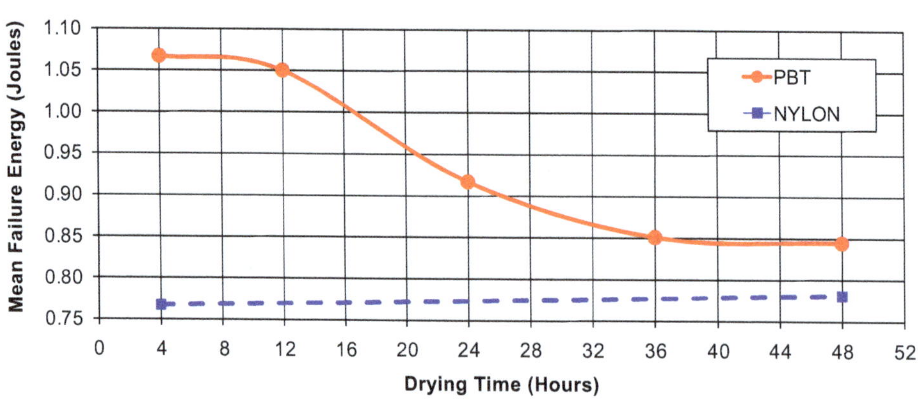

Figure 4.13 Mean failure energy of parts molded with PBT and nylon dried for varying drying times [1]

Results for Nylon

In the case of nylon, studies were conducted for drying times of 4 hours and 48 hours. The resin and the molded parts were tested with the procedures mentioned earlier and the TGA data is shown in Figure 4.14. Here, a 15% glass filled resin was used. The residue left behind was approximately 17% of the original weight. We can again be certain that most of the residue was the glass left behind. The $T_{1/2}$ values for both nylon-4 and nylon-48 seemed to be approximately 470 °C. The weight loss also started around the same time and imitated each other. We can therefore infer that there was no significant difference between the TGA curves for the regularly dried and the overdried resin. There was no noticeable loss of any additive.

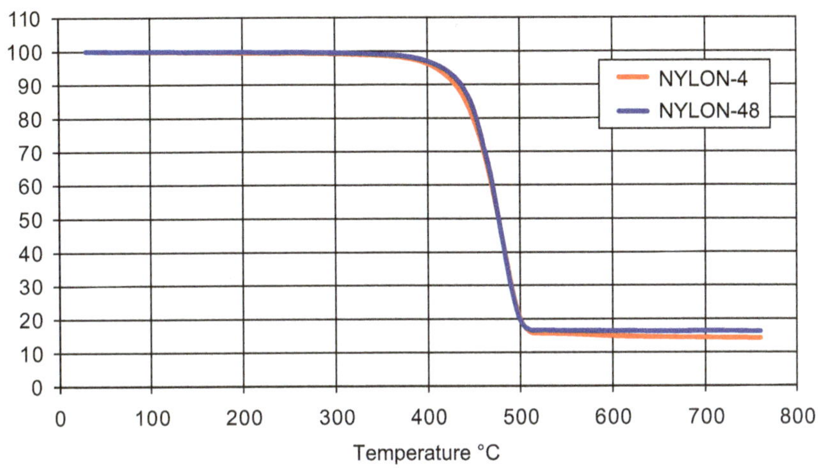

Figure 4.14 TGA graph for nylon dried for 4 and 48 h [1]

The capillary rheometry-shear sweep data is shown in Figure 4.15. It is interesting to note that the viscosity for the nylon-48 was higher than that for nylon-4. Khanna, et al. [2] have noticed a similar increase in the viscosity of nylons with increase in drying temperature and time held constant. Similar studies by Pezzin and Gechele [3] showed that the melt viscosity increased with time for lower moisture contents. Khanna hypothesizes that the increase in the viscosity could be explained by the following two reasons:

- The moisture in the nylon acts as a plasticizer for the melt, reducing its viscosity. With higher drying temperatures, loss of moisture leads to an increase in viscosity.
- The melt equilibrium is represented by a characteristic equilibrium constant, K_{cond}. If excess water is added to the melt in equilibrium, the reaction will go in the direction of the reactants (i.e., degradation), while the reverse (i.e., polycondensation) should occur if water is removed from the equilibrium melt.

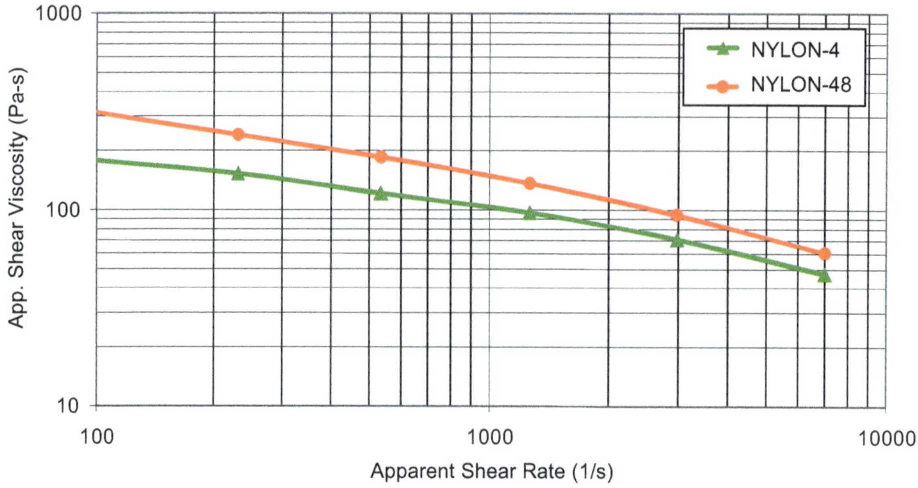

Figure 4.15 Capillary rheometry data (shear sweep) for nylon dried for 4 and 48 h [1]

In the present study, the drying temperature was held constant and the drying time was varied. Considering the theories put forth by Khanna, and the similar results obtained here, the net result of the increase in drying time must be the same as the increase in drying temperature. The loss of moisture between drying times of 4 hours and 48 hours must have contributed to the increase in viscosity of the melt. Evidence of this was also seen during the molding of the parts. A feature that was 48 mm long with an average height of 3.5 mm and a width of 6 mm was almost completely filled (99%) during the injection phase of the molding cycle when molding with nylon-4. When nylon-48 was used, this feature was only filled to approximately 62% of the original flow length. Photographs of this are shown in Figure 4.16.

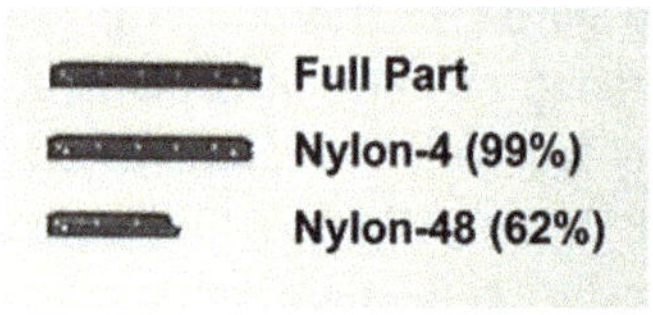

Figure 4.16 Complete parts and 'injection only' parts molded with nylon dried for 4 and 48 h [1]

Thermal degradation data using capillary rheometry are shown in Figure 4.17. The two resins do not seem to differ in viscosity for residence times of 12 minutes and beyond. With a residence time of 6 minutes, nylon-48 has a higher viscosity compared to nylon-4. The same reasoning as in the above section can be applied here. Once past the 6 minute interval, the resin probably begins to degrade and the final viscosity of the two resins matches.

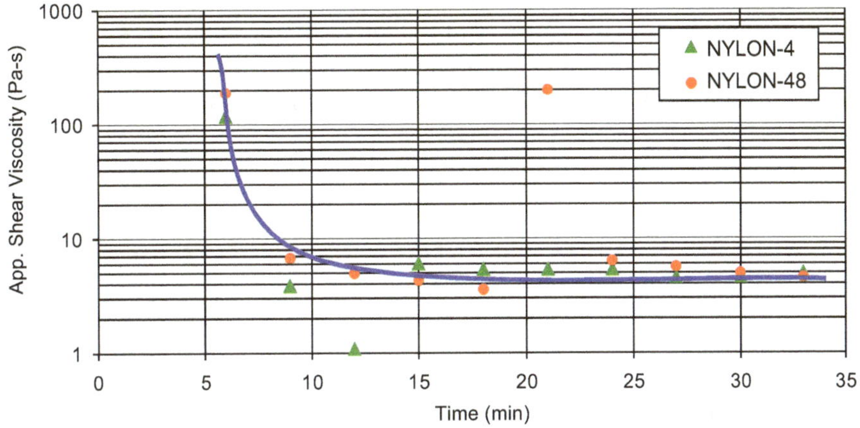

Figure 4.17 Capillary rheometry data (time sweep) for nylon dried for 4 and 48 h [1]

Drop impact testing of the parts demonstrated no significant difference between nylon-4 and nylon-48, as shown as a dotted line in Figure 4.13. The average mean failure energy for the two samples was 0.774 Joules. Long drying times did not seem to affect the drop impact strength of the resin. It is evident from the above discussion that it is important to control the drying process of the resin. Overdrying can result in a loss of physical properties as seen in the case of PBT, or lead to the increase in the viscosity as in the case of nylon. For nylons, because the water in the plastic has an effect on the viscosity of the plastic, the water can be thought of as a viscosity regulator. The viscosity changes will have an impact on the flow properties and the associated features, such as the weld line strength of parts. Other properties affected by flow, such as surface finish and polymer/filler ratio could also be impacted by long drying times. Secondary operations, such as ultrasonic welding and joining, require a certain amount of polymer to be present on the surface in order to achieve successful bonding. Therefore, a minimum level of moisture is required for nylons because water is a viscosity regulator. This level is usually around 0.015%. However, this is something each molder must determine and maintain individually.

4.7 Cautions

Drying of some resins such as PBT has a cumulative effect. The drying process removes the low molecular weight additives for good and they do not return if the drying is stopped and the resin returned to the shelf for the next run. However, in the case of nylons, the moisture can be absorbed back and the viscosity can be re-regulated. Since each resin and its additives is unique, customized experimentation is best suited to determine the type of control required for each individual process. The results above must not be taken as any sort of standard results. Experimentation is time consuming and expensive and therefore it is best to keep it simple and avoid overdrying.

4.8 Prevention of Overexposure to Longer Drying Times

On the shop floor there are a couple of efficient ways to prevent overdrying:

Turning down the dryer temperature: If molding is not ready to be started after the material is dry, turn down the dryer temperature to about 25 °C (80 °F), but keep the dryer running. Supplying the hopper with low temperature dry air will keep the moisture out and will not have detrimental effects on the resin.

Sizing the hopper: Size the hopper dryer such that the residence time of the plastic stays between the minimum and maximum recommended drying times of the plastic. When the hopper is larger than required this may be difficult. In this case, retrofit the dryers with a level sensor and adjust the level to maintain the required amount of material.

The material required for the run must be calculated and only the required amount of material should be dried. If the machine is going to be down for an extended period of time after the drying has taken place, the dryer must be shut off as soon as possible.

4.9 Overdrying Controller

Figure 4.18 shows this logic applied to a design of a controller, while Figure 4.19 depicts a concept for a programmable controller that would control the drying process of the resin.

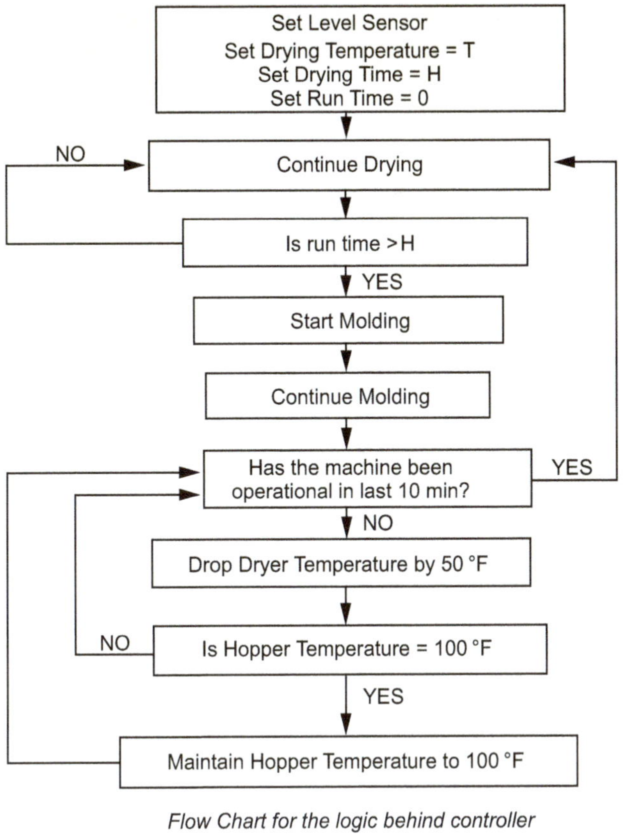

Figure 4.18 Controller logic to control residence time in the dryer [1]

Initially, the controller would set the drying time and temperature. Based on shot weight of the mold, a resin level sensor would be set such that the residence time of the resin in the hopper exceeds the recommended drying time by an hour to two hours. Once the set drying time has elapsed, the controller looks for a signal from the machine indicating that the machine is operational and molding parts. Such a signal can be picked up from screw rotation or mold open/close. If the controller receives this signal it will maintain the dryer temperature. If it does not receive the signal, indicating that the machine is not operational, the controller will start to drop the drying temperature by a preset value, for example 10 °C. If in a preset time, for example 15 minutes, the controller does not get the machine operation signal, it will further drop the temperature down by another step. It will do so until it reaches a temperature of 25 °C. At this time, dry air will be circulating in the hopper and keep the resin dry. As soon as the machine is ready for operation and the controller gets this signal, the temperature of the hopper will begin to rise to the desired drying temperature in preset steps of temperature and time. If normal molding operation is interrupted, the controller will follow the same logic described above and drop the dryer temperature. Such a control mechanism will ensure that the resin will never experience excessive drying times and prevent overdrying.

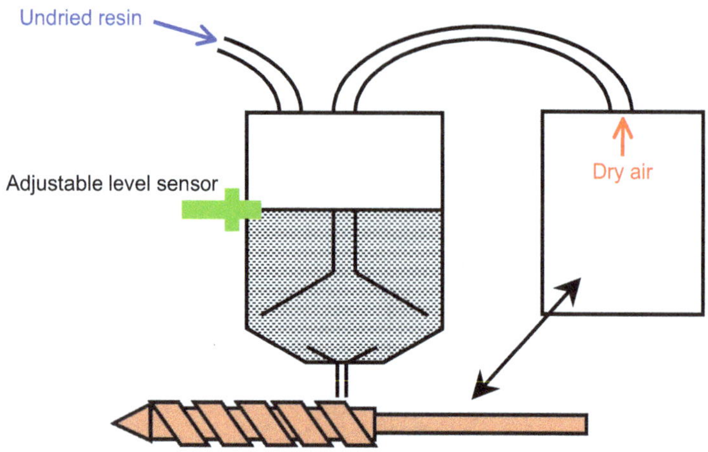

Figure 4.19 Flowchart for the logic behind controlling the drying process [1]

4.10 References

[1] Kulkarni, S. M., *SPE ANTEC Tech. Papers* (2003) p. 736

[2] Khanna, Y., et al., *Polymer Engineering and Science*, Vol. 36 (13), p. 1745, (July 15, 1996)

[3] Pezzin, G., Gechele G., *J. Appl. Polymer Sci.*, Vol. 8, p. 2195, (1964)

Suggested Reading

Brydson, J. A., *Plastics Materials* (1999) Butterworth Heinemann Ltd, Oxford

Harper, C. A., *Modern Plastics Handbook* (2000) McGraw Hill, New York, NY

Deanin, R. D., *Polymer Structure, Properties and Applications* (1972) Cahners, Boston, MA

Odian, G., *Principles of Polymerization* (1991) Wiley Interscience, NY

Shah, V., *Handbook of Plastics Testing and Failure Analysis* (2007) Wiley Interscience, NY

5 Common Plastic Materials and Additives

Injection molding processors are mainly concerned with the way the plastic flows and with the temperature at which the plastic can be processed. Flow characteristics of all plastics are very similar. They all follow non-Newtonian behavior and therefore exhibit shear thinning effects. A pattern from a flow simulation program for two different plastics will look very similar, because plastic flow is always laminar. The pressures and temperature distributions may be different, but the flow patterns are similar. All plastics have their unique processing temperature range, both for the mold temperature and the melt temperature. Even with these commonalities, it is still important to understand the different types of plastics and the additives that are incorporated into them. With this knowledge processors will be able to better understand the materials they work with and to take the necessary precautions during their processing. For example, while polyethylenes can be injected at very high speeds, one must be very careful with PVC, because it tends to degrade at higher speeds. This chapter will introduce the base materials, the additives, and the reason for their incorporation. Only the most common polymers by volume and the additives used in injection molding will be discussed.

5.1 Classification of Polymers

Polymers can be classified in a number of ways. In the field of molding the term plastic is most commonly used. We use the term polymer when we describe the basic nature of the molecules and their properties. Plastics are those polymers that can withstand a moderate to significant amount of force before showing any significant deformation. Polymers that are deformed under light loads are called elastomers. Technically, the difference can be seen clearly in the stress–strain graphs in Figure 5.1.

In the following, the term plastic will be used to describe all molding materials and the term polymer will be used when an intrinsic property needs to be referenced.

The following classifications of polymers are also used:

Thermoplastics: These can be repeatedly heated, melted, and processed into useful products. Example: ABS

Thermosets: Once these polymers are processed, they form a chemical network and all the molecules now are crosslinked. It is impossible to re-melt the polymer in this state because there is more energy required to separate the crosslinked bonds than the energy required to break the main chains. This results in a total destruction of the polymer. Example: liquid silicone rubbers (LSRs)

Organic Polymers: These contain carbon atoms in their backbone and are derived mainly from organic materials from nature. Example: polyethylene

Inorganic Polymers: These contain atoms other than carbon in their backbone. Example: polysilanes

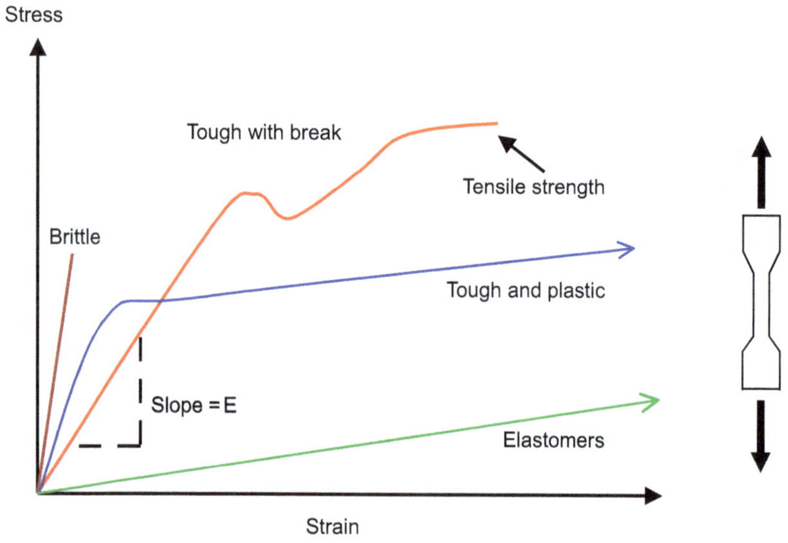

Figure 5.1 Tensile properties for different types of plastics

Elastomers (TPE): These are similar to thermoplastics except they are soft and rubbery at room temperature. Their glass transition temperatures are below room temperature.

Homopolymers: These are polymerized from only one type of monomer. Example: polyethylene

Copolymers: These are manufactured from two or more monomers. Example: ABS – polymerized from acrylonitrile, butadiene, and styrene

Alloys: These are polymers physically mixed with each other and there is no chemical interaction between the different polymers. Example: PC-ABS alloys are common in the appliance industry.

5.2 Commercially Important Plastics

Some of the commercially important plastics that are injection molded are described in the following. All descriptions relate to unfilled plastics without any additives.

5.2.1 Polyolefins

Polyolefins are the most basic hydrocarbons and contain carbon and hydrogen atoms only. Polyethylenes and polypropylenes are the most widely injection molded polyolefins.

Polyethylene: Polyethylene is one of the widest used materials because of its excellent electrical insulation properties, good chemical resistance, impact properties, and low cost. The properties of polyethylene (also known as polyethene) can be tailored to meet certain application requirements. The tailoring changes the density of the polymer which allows to classify the different polyethylenes. Low-density polyethylene (LDPE) has a density of 0.91–0.92 g/cm^3, medium density polyethylene (MDPE) has a density of 0.93–0.94 g/cm^3 and high-density polyethylene (HDPE) has a density of 0.95–0.96 g/cm^3. Linear low-density polyethylene (LLDPE) is technically a copolymer and is designed for better impact properties and improved flexibility. It is mainly used for films. The density of LLDPE is around 0.91 g/cm^3. To improve some of the mechanical properties and still retain the other properties of polyethylene, its molecular weight is increased. This product is called ultra-high molecular weight polyethylene (UHMWPE) with a density of 0.92–0.93 g/cm^3. UHMWPE is difficult to mold because of its high melt viscosity. The weight average molecular weight of UHMWPE can be in the range of 1–6 × 10^6, whereas for other polyethylenes it is around 50,000 to 300,000. The disadvantage of polyethylenes is that they exhibit high thermal expansion, poor weathering resistance, and low heat deflection temperatures. Although the mechanical properties of polyethylene are low, with the use of reinforcing fillers such as glass fibers these properties can be enhanced. Polyethylenes can also be cross-linked to improve their properties.

Polypropylene: Both polypropylene and polyethylene have similar structures and similar properties. Most commercially available polypropylenes are isotactic polypropylenes with densities of 0.90 g/cm^3. They exhibit good electrical properties, good environmental stress cracking resistance, and good heat resistance. Polypropylene dominates in the area of thin-wall molding which requires good flex properties. CD covers with living hinges are examples of these applications that can stand continued flexing. The biggest disadvantage of polypropylene is its low temperature flexibility or impact resistance. As the temperature reaches 0 °C, the polymer becomes brittle and often cracks. This issue is often resolved by either incorporating additives or copolymerizing polypropylene with a small percentage of polyethylene.

Other olefins such as cyclic olefins are used in molding but not to the extent as polyethylenes and polypropylenes.

5.2.2 Polymers from Acrylonitrile, Butadiene, Styrene, and Acrylate

Monomers and polymers of acrylonitrile, butadiene, styrene, and acrylates are compatible with each other and produce a variety of useful polymers and plastics. Each of these can be polymerized by themselves or with other monomers. Each of the monomers provides a unique property to the final polymer. For example, butadiene is a polymer with a low T_g value and therefore provides impact resistance; styrene adds luster to the product and helps to improve the processability; acrylonitrile adds heat and chemical resistance; acrylate adds weathering resistance. ABS is one of the most commonly used materials. Depending on the end use requirement, it can be produced with varying percentages of the monomers. An absence of styrene produces nitrile rubber, an absence of butadiene will produce styrene acrylate resins (SAN), and an absence of acrylate will produce styrene butadiene rubber (SBR) or impact modified styrene (HIPS). SBR is butadiene with a small percentage of polystyrene and HIPS is polystyrene with a small amount of butadiene. By themselves, the polymerization products are polyacrylonitrile, polybutadiene, and polystyrene. Similar combinations with acrylate monomer yield different polymers as was shown in Figure 5.2. Because of the endless possible combinations of the monomers, the properties of the end polymer can be tailored. Not only the percentages of the monomers play an important role, but the method of preparation can also affect the properties. The polymers can be copolymerized from the monomers, grafted onto a base polymer, or the individual polymers can be physically blended with each other. These polymers can be further blended with other polymers to create other customized polymers. A blend of polycarbonate (PC) and ABS called PC/ABS is widely used in thin-wall molding, such as cell phone cases.

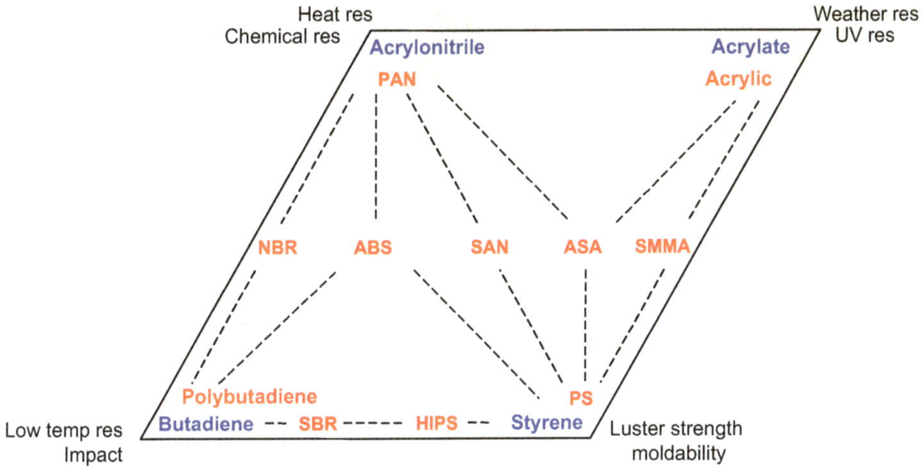

Figure 5.2 Polymers produced from acrylonitrile, butadiene, styrene, and acrylates

5.2.3 Polyamides (PA)

Most polyamides used for molding are commonly called nylons, although not all PAs are nylons. Nylons are crystalline materials; they are manufactured by polycondensation reactions usually producing water as the byproduct. Nylons produced from dibasic acids and diamines are identified by a numbering system that indicates the number of carbon atoms in the base monomers. A nylon 66 has six carbon atoms in the dibasic acid and six carbon atoms in the diamine. A nylon 412 has four carbon atoms in the dibasic acid and 12 carbon atoms in the diamine. For example, nylon 66 is made from hexamethylene diamine and adipic acid that each contain six carbon atoms. Nylons made from a single monomer are identified by just one number. For example nylon 6 is made by the ring opening polymerization of caprolactam that contains six carbon atoms. Commercially, nylon 6 and nylon 66 are widely used in injection molding. Nylons are tough and offer good impact and chemical resistance even at moderately elevated temperatures. This is the reason they are used in a lot of automotive applications, such as under the hood. Gears, cams, and bearings are also products made from nylons. The biggest disadvantage of nylons is due to the fact that they are strong hydrophilic materials, i. e., they absorb a lot of moisture. The presence of moisture drastically affects their properties, such as dimensional stability and electrical properties. Most data sheets for nylons are therefore available for 'dry' and 'conditioned' nylons. The properties of nylons are also affected by the crystallinity of the polymers. Table 5.1 shows the relative difference in properties between nylon 6 and nylon 66.

Table 5.1 Relative Differences in Properties of Nylon 6 and Nylon 66

Property	Nylon 6	Nylon 66
Crystallinity	Lower	Higher
Heat resistance	Lower	Higher
Chemical resistance	Same	Same
Impact strength	Higher	Lower
Wear resistance	Lower	Higher
Processibility	Lower melt and wider window	Higher melt and narrower window

Processing of nylon if fairly simple because of its favorable flow properties. Nylon must be dried to the right moisture level before processing. Some amount of moisture in the plastic acts as a viscosity reducer and regulator and therefore exposure of nylon to excessive drying times should be avoided. Drying of nylon is discussed on Chapter 4. If nylon is not dried prior to molding, it can degrade in the barrel due to hydrolysis causing a drop in properties. Surface defects such as splay is common with undried resin. Drooling from the nozzle tip between mold open and mold close is another issue because the water reduces the viscosity of the plastic and makes it drool out of the nozzle tip after the shot builds up. Because of the high T_m and T_g, cycle times can be short.

5.2.4 Polystyrenes (PS)

Polystyrene is polymerized from styrene and is also called poly(vinyl benzene). It can be either atactic or syndiotactic, as described on Chapter 2. Most commercially used polystyrene is atactic with good optical clarity because of its amorphous nature. The T_g of the polymer is high, resulting in good dimensional stability. Solvent resistance, weathering resistance, and impact strength of polystyrenes are low. Impact strength can be improved by incorporating a small amount of rubber, such as butadiene, creating a high impact polystyrene, commercially called HIPS; however, this affects the clarity of the final product. Another useful property of polystyrene is its ease to be processed into foam. Commercially available Styrofoam is polystyrene mixed with a blowing agent during the melt processing to produce the foamed product. It has extensive use and in fact dominates the packaging industry. Because of its low cost, ease of molding and availability in FDA approved grades, polystyrene finds its way into many household and kitchen products.

5.2.5 Acrylics

Commercial acrylic is mainly polymethyl methacrylate (PMMA). It is an amorphous polymer and has excellent clarity. Because of its high resistance to UV light and good weathering property, it is widely used to make lighted sign boards and outdoor lighting articles. It is a good substitute for glass, except for its low scratch resistance. The disadvantage of acrylics is their poor resistance to organic solvents and low stress cracking resistance.

5.2.6 Polycarbonates (PC)

PCs are almost considered engineering materials because of their high impact strength. Their high transparency and high impact strength make them an ideal choice for automotive exterior lighting parts, helmet visors, and safety goggles. Processing of PC is fairly easy, although the presence of moisture can cause a lot of problems in the molding of PC. Cosmetic issues, such as splay, are common. A variety of PCs are available on the market and care must be taken to study the data sheet before molding a new grade of PC. The processing may be so widely different that the processing temperatures of two PCs may not even overlap. PCs have low stress cracking resistance and their chemical resistance is fair.

5.2.7 Polyesters

Polyethylene terephthalate (PET) and polybutylene terephthalate (PBT) are the most common injection molded polyesters. They exhibit excellent mechanical properties and are easy to process. By volume, PET is more commonly used in blow molding to produce containers and bottles. Because of their high chemical resistance to automotive fluids, PBT is used in under-the-hood applications. PBT is prone to be attacked by certain solvents.

5.2.8 Polyvinyl Chloride (PVC)

PVC is one of the least stable polymers, but is processed successfully with the appropriate additives and stabilizers. It is probably the material that has had the worst reputation over the years because of safety concerns. However, PVC is one of the most versatile polymers. It can be easily modified to achieve a wide range of flexibility. PVC by itself is very rigid and is marketed under the name rigid PVC or UPVC. PVC has excellent chemical and weathering resistance and is therefore widely used in pool and garden equipment. PVC has excellent electrical properties and is also used as cable insulators. The addition of plasticizers softens rigid PVC. With the addition of a large amount of plasticizer PVC can be processed into films by nonmelt processing techniques. To make PVC more chemical resistant, it is chemically modified to incorporate chlorine in the molecule. This PVC is called chlorinated PVC. Heat resistance and weatherability decrease with the addition of the chlorine. Melt processing of PVC requires extreme care and caution. Decomposition of PVC produces hydrochloric acid, which is an irritant to humans. Therefore, the residence time in the molding barrel must be controlled and limited. Once the production run is over, all the PVC in the barrel must be thoroughly purged. All related equipment must also be cleaned to avoid any of the PVC from getting to the barrel with the next material. This is especially true if the next material to be molded is an acetal. Acetal and PVC are known to be an explosive combination in their melt form and therefore any chance of them being mixed must be avoided.

5.2.9 Polyoxymethylene (POM or Acetal)

Although acetal is the commonly used term (sometimes polyformaldehyde is also used), this material is actually polyoxymethylene. It is a crystalline polymer and is considered an engineering plastic. It has excellent stiffness, fatigue endurance, resistance to creep, and low coefficient of friction. POM also exhibits excellent chemical resistance below 70 °C to all organic solvents. Above this temperature, some phenolic materials can react with the polymer. Resistance to inorganic polymers is low. Gears, bearings, conveyor parts, or any moving parts in assemblies are good application candidates for acetals because of their unique properties. The cost of acetal is fairly high. The density of the unfilled polymer is 1.42 g/cm^3, which is very high compared to olefins or nylons. This makes it less attractive when it competes with other materials because the part will weigh more and therefore cost more, if molded in acetal. Melt processing of acetals is fairly easy. Overheating of processing barrels and large residence times must be avoided. Degradation produces formaldehyde gas which is an irritant. Acetals are non-hygroscopic and therefore do not require drying. Surface drying is helpful in humid environments. A typical problem occurring when molding acetal is that the parts show unusual shrinkage values, far different from the published values. The main reason is usually due to the gate freezing off before the part is packed out with the required amount of plastic. Enlarging the gate size helps.

5.2.10 Fluoropolymers

Fluoropolymers are polymers containing fluorine. Polytetrafluoroethylene (PTFE), also called Teflon, is the most commonly used fluoropolymer. However, because of its high melt viscosity it is almost impossible to be injection molded. It has excellent mechanical properties over a wide range of temperatures. It does not dissolve in any acids, alkalis, or organic solvents. It also has a low dielectric constant. Because PTFE cannot be melt processed, the sintering technique is used to produce PTFE components, such as pump valves. PTFE tape commonly used before installing a water pipe on to a mold is produced by the process of skiving. Polyvinylidene fluoride (PVDF) is a fluoropolymer that can be injection molded. PVDF also offers good chemical resistance although it is not as high as that of PTFE. PVDF has been successfully used in chemical and electrical equipment and even in the manufacture of internal battery components. Perfluoroalkoxy copolymer (PFA), fluorinated ethylene propylene (FEP), and ethylene-tetrafluoroethylene (ETFE) are other fluoropolymers that offer similar properties as PTFE, but they are modified to be injection moldable. Molds that need to be designed for molding fluoropolymers need to have wide runners and gates to reduce the pressure drop through the mold. Special screws with generous section dimensions, large feed sections, and smaller metering sections can be helpful in the processing of these polymers.

5.3 Additives

Today, all polymers that are supplied by the manufacturer have additives blended into them. Additives can change the properties of the polymers substantially. PVC is the best example of this. As mentioned earlier, PVC by itself is rigid but with the addition of a plasticizer it can become extremely flexible. All additives must be compatible with the base polymer and must not bleed out or plate out under extended service conditions. Some PVC films get tacky with increase in temperature. This is the bleeding out of some of the plasticizers contained in it. They must also be able to withstand the processing conditions such as melt temperature and shear rates. These additives can be solids, liquids, gases, or other polymers. Some of the important additives based on their use are described in the following.

5.3.1 Fillers

Fillers are added to polymers either to improve the properties or to reduce the cost. Glass fiber is an example of a filler used to increase the stiffness or creep resistance of the base plastic, it is also called a reinforcing filler. Other examples of reinforcing fillers are wood flour, carbon fibers, and nylon fibers used to increase the impact strength and rigidity of the base plastic. Glass beads are usually added to plastics to reduce the shrinkage or sometimes (depending on the base polymer density and the glass bead density) the weight of the polymer. These are called inert fillers. Talc and calcium carbonate are other examples of inert fillers. In theory, any material that is compatible and can be blended into the polymer matrix can be used as a filler. Over the years, experiments have been conducted with a large number of fillers, especially when a particular material was easily available in the local geographic area. For example, fibers from the coconut tree or from the coconut have been successfully used in some applications in some South Asian countries. In some cases, coupling agents are used to form a bond between the filler and the polymer since the two are not always compatible, typically with inorganic fillers.

5.3.2 Plasticizers

Plasticizers are added to provide flexibility and softness to a material. Because of their compatibility, they get in between the polymer molecules, increasing the intermolecular distance. They also help in reducing the melt viscosity for improved processing properties. PVC is the biggest consumer of plasticizer by volume. Plasticizers are usually low molecular weight compounds, usually oligomers. If the plasticizers are not compatible, they tend to migrate to the surface of the polymer. Phthalates such as di(isooctyl) phthalate (DIOP) are commonly used as plasticizers in PVC. Concerns regarding safety have led researchers to more natural products and in recent years epoxidized soybean oil (ESO) has gained popularity.

5.3.3 Flame Retardants

For most products that are in direct contact or in close proximity with humans, such as furniture, toys, televisions, and computers, flame retardants are added to prevent the spread of fire. Although the plastics used in these products will burn in the presence of a fire, they must not continue to do so when the source of the fire is removed. Most consumer materials, such as olefins and polystyrenes, will continue to burn even if the source is extinguished. Flame retardants are added to plastics to inhibit this continued burning. This is achieved by one of several mechanisms, such as forming an incombustible gas and cutting of the supply of oxygen required for the burning. Halogen-containing compounds, such as chlorinated paraffins, and phosphates, such as trixylyl phosphate, are used as flame retardants. Testing is done using various methods and the rate of burn, time of burn once the flame is removed, and the amount of burn of the sample are all recorded as vital data.

5.3.4 Anti-Aging Additives, UV Stabilizers

Over time, polymers tend to lose some of their properties because of a variety of changes in the polymer: The molecular weight can drop as a result of chain scission or certain additives can degrade or bleed out. Some of these can cause chain reactions because the reaction products can help in the propagation of the reaction. Antioxidants and UV stabilizers are commonly added to plastics as anti-aging additives. The antioxidants react with any free radicals that are formed and prevent the propagation of the chain scission. UV light can cause a change in the physical properties of the polymer, e. g., causing it to become hard leading to surface cracking. They may also discolor. Automotive head lamp covers tend to become yellow over time because of their extended exposure to the UV light present in sunlight. UV stabilizers absorb the UV light and prevent it from affecting the plastic. Certain amines are examples of UV stabilizers. All products that are exposed to sunlight must have a UV stabilizer incorporated.

5.3.5 Nucleating Agents

Nucleating agents are used in crystalline polymers to help speed up the nucleation process and lower molding cycle times. When the melt enters the mold, it is completely amorphous and as it starts to cool down, the crystallization process starts. Mold temperatures are maintained high enough to supply the energy for crystallization. Crystallization provides the required properties to the end product and therefore it must be completed before the plastic temperature falls below the glass transition temperature (T_g). Below the T_g, no meaningful crystallization can take place. Nucleating agents help to increase the onset and rate of nucleation. They are commonly incorporated in olefins, polyesters, and nylons that form the majority of the injection molded crystalline polymers. Talc is a good example of a nucleating agent for polypropylenes.

5.3.6 Lubricants

Lubricants are added to reduce the coefficient of friction in the final products. In many assemblies, moving parts slide against each other (cams) or drive another part by an applied mechanical force (gears). In such cases, a low coefficient of friction is desired for two reasons:

1. the energy required to move the parts is reduced, and
2. a lower coefficient of friction translates to less heat buildup in the part. Heat buildup can cause thermal expansion of the component changing its dimensions and leading to the failure of the assembly.

Graphite and PTFE are added to plastics in small quantities to act as lubricants.

5.3.7 Processing Aids

Some texts classify lubricants and processing aids in one category. Processing aids act as melt viscosity reducers that help in the melt processing of the plastic. They are also called flow promoters. Amine waxes are examples of these. Processing aids do not alter the final property of the plastic and their primary function is to ease the processing.

5.3.8 Colorants

Colorants are blended into the base polymer to impart the destined color. Compatibility is no doubt important, but in injection molding, where color pellets are mixed with the plastic, the carrier resin for the colorant also needs to be compatible. Colors can also be added as liquids directly into the barrel of the injection molding machine, although this can get messy. The advantage here is that the control of the amount of liquid can be very accurate. Mixing color powders with the resin must be done immediately prior to processing because the powder tends to separate during transport. The powder also tends to stick to the walls of the mixing equipment and hoppers.

5.3.9 Blowing Agents

Blowing agents are used to mold plastic parts with a cellular structure. In thick parts this helps reduce the overall weight of the part and it also improves the structural strength of the part. Defects related to thick sections, such as sink marks, are eliminated. The blowing agent mixes with the plastic and creates internal pressure to push the plastic towards the walls of the mold. The molded parts are therefore dimensionally more consistent as compared to conventional parts. The disadvantage of using a blowing agent is the additional cost and effort of mixing. Parts molded with a blowing agent do not have good surface finishes. Large hoppers or containers are examples of parts molded using blowing agents. When the process is used to improve the structural properties, it is also called structural foam molding. Styrofoam cups are molded using a blowing agent. Blowing agents can be chemicals that decompose in the injection barrel to form a gas and help foam the plastic (e.g., sodium bicarbonate) or they can also be gases, such as nitrogen and air.

5.3.10 Other Polymers

Technically, when polymer blends are produced, one polymer is added to another polymer. Therefore these polymers may also be considered as additives. For example, in a PC/ABS blend, PC can be considered an additive to ABS or vice versa. Thermal properties of a blend of polyhydroxybutyrate and polyvinyl acetate are shown in Figure 5.3.

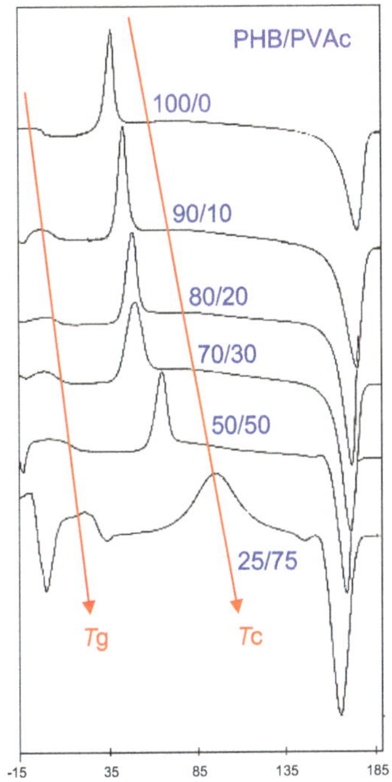

Figure 5.3 Thermal properties of blends of PHB/PVAc

5.4 Closing Remarks

In Chapters 2 and 3, the morphology and the rheology of the polymers were discussed. Additives can have a profound influence on both these properties. A polymer that is naturally crystalline can be made completely amorphous by incorporating additives. Figure 5.3 shows the thermal properties of a PHB/PVAc blend. The PVAc was blended in various proportions into the PHB starting with 10%. As the amount of PVAc increased, the glass transition temperature and the crystallization temperature of the PHB phase increased. Polymer properties can be tailored by blending techniques and by the addition of additives. However, there may not always be a win-win situation for all the properties. Some of them may have to be compromised.

5.5 Reference

[1] Alfrey, T., Gurnee, E. F., *Organic Polymers*, Prentice-Hall (1967) p. 51

6 Injection Molding and Molding Machines

6.1 The History of Injection Molding

The concept of injection molding for plastics was adopted from the metal die cast industry, which began to develop in the early to mid-1800s. Plastics were not known for their useful properties at this time. All polymers were naturally occurring materials and did not have much industrial value. In 1869, John Hyatt developed a concept to make billiard balls from cellulose nitrate, also known as celluloid. His machine consisted of a cylinder heated by steam to melt the celluloid and a hydraulic plunger to inject the plastic into the mold. Developments continued and several inventors produced several designs of molding machines and their components. Over the years, molding machines advanced from their manual operations to automatic operations with electrical controls. Almost 70 years after the first injection molding machine was developed, a major milestone was achieved. In 1948, for the first time, the two-stage screw was introduced in the injection molding industry. Until then melt homogeneity was achieved with the help of torpedoes in the injection barrel. The use of the two-stage screw provided the required homogeneity and allowed better control of the plastic shot size injected into the mold. As a result of the improved melt efficiency and better melt homogeneity provided by the two-stage screw, a larger volume of plastic could now be melted, leading to the possibility of molding larger parts. As molding machine control systems improved with the advent of advanced electronics engineering, the molding machines became more sophisticated, providing better control of the injection molding process. Multishot molding machines, better screw designs, clamp position versatility, and nonhydraulic all-electric machines were some of the many improved features and products being introduced to the molding industry. Today's machines have become highly complex and are capable of molding almost anything that a product designer would desire, from micromolded parts used in the watch industry to large parts used in the automotive industry and everything in between. The latest generation of machines can be connected to any computer in the world via the Internet and monitored remotely. This helps with quick debugging and problem solving of machine and molding problems without requiring an engineer on site.

6.2 Injection Molding Machines and Their Classifications

A conventional injection molding machine is shown in Figure 6.1(a). Plastic pellets are fed into the barrel of the injection molding machine where shear heat generated by the rotating screw and external heat provided by electric heaters around the barrel melt the plastic, making it ready to be processed. As the screw rotates, it augers the required amount of plastic for the shot to the front of the barrel. This plastic is then injected into the mold by the forward movement of the screw. The mold has a coolant flowing through it that helps cool and maintain its temperature. Once the part is cooled below its ejection temperature, it is ready to be ejected out of the mold. Although the process of injection molding seems fairly simple, there are multiple speeds, pressures, times, and temperatures that must be controlled in order to produce quality products. It is the optimization of these processing parameters necessary to run an efficient and successful molding operation that scientific processing is all about.

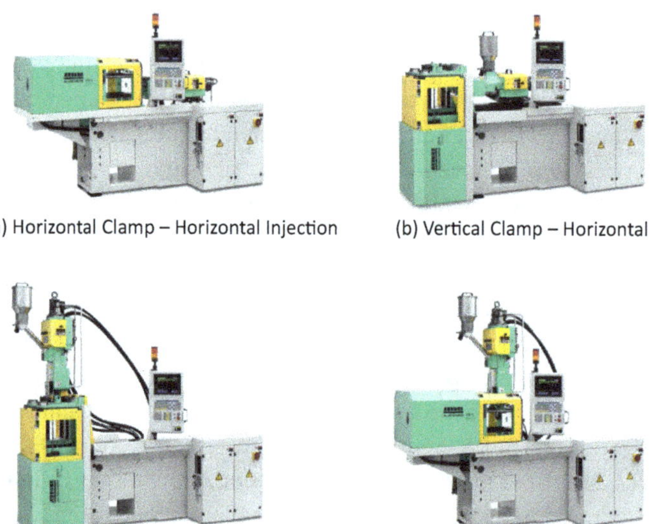

(a) Horizontal Clamp – Horizontal Injection (b) Vertical Clamp – Horizontal Injection

(c) Vertical Clamp – Vertical Injection (d) Horizontal Clamp – Vertical Injection

Figure 6.1 Machine classifications based on clamp and nozzle positions (Courtesy: Arburg Inc.)

Injection molding machines are classified in a number of ways. Based on the movement of their clamp, they are classified as horizontal clamp or vertical clamp molding machines (see Figure 6.1(a)–(d)). Horizontal clamp machines are suitable for most applications where the parts are ejected out of the mold and fall onto a conveyor or into a box due to gravity. Molded parts may then go through a secondary operation before being packed and sent to the customer. Horizontal clamp machines are the most versatile machine and are most common. Vertical clamp machines are suitable for producing insert molded parts, where a horizontal parting line is an advantage to seating inserts into the mold. It is the vertical clamp orientation of a vertical molding machine that allows the mold parting to remain parallel with the floor. This arrangement allows gravity to hold inserts or other components to be over-molded into position in the lower half of the mold. Once the inserts are positioned, the top half of the mold can be clamped into position over the inserts that have been placed into the bottom half. Vertical molding machines commonly have a rotary or a shuttle mechanism that moves the bottom half of the mold from underneath the clamping mechanism to an area easily accessible by either an operator or a robot for unloading the parts and loading the inserts. In most cases there are two or more moveable mold bases. While inserts are being loaded into one of the exposed mold bases, another mold base is clamped shut and being injected with plastic. This parallel activity saves time and increases output of the insert molding process. It also helps in reducing residence time of the plastic in the barrel.

Injection molding machines may also be classified based on the direction of plastic injection. Horizontal injection machines are most common. Here, the injection unit is mounted parallel to the ground and plastic is injected through a sprue bushing usually mounted in the center of the mold. Plastic flows into the mold via the sprue bushing, flowing through the center of the mold half, perpendicular to the parting line of the mold (i. e., the split line where both mold halves separate). The runner system delivers plastic to the cavities and is cut into one or both sides of the mold. The sprue intersects the path of the runner in the center of the mold, allowing plastic melt to be distributed to the cavities. Upon mold open, the molded parts and runners remain on the moving side of the mold half. In this scenario, the hardened plastic sprue is pulled out of the stationary mold half, leaving a long appendage of plastic that can be gripped by the jaws of a robotic sprue picker as part of the part ejection cycle. This is the standard horizontal molding arrangement. On the other hand, some molded components with part design and/or mold design restrictions, cannot use a sprue and runner configuration for delivery of plastic through the center of one mold half. Instead, the sprue is eliminated and the entire runner system has to be at the parting line. This is typically referred to as a parting line shot, since the plastic is injected into the mold where the mold halves meet, on the parting line. In such cases vertical injection of the plastic is required and therefore the injection barrel is rotated to a vertical position. There is a limitation on the size of the vertical injection unit because it is usually supported on tie bars. In some cases, a vertical injection unit will have its own independent support units, but these still have to be versatile in their horizontal movement to allow adjustment for the position of the parting line on the mold.

Figure 6.2 Two-shot (material) machine (Courtesy: Arburg Inc.)

Multishot machines have two or more injection barrels and are capable of molding parts that are comprised of two different materials. A toothbrush is a common example of a part produced by a two shot process. With the first shot, the plastic substrate is molded. The mold is then opened and the cavities with the first shot of plastic are indexed to align with the second injection unit. When the mold closes, the second shot is injected over the now hardened first shot or substrate, creating a two shot component that is ejected from the mold. Typically, the second shot is a soft elastomers-type material typical for the soft handle molded over the toothbrush substrate. A two-shot machine is shown in Figure 6.2.

6.3 Machine Specifications

Molding machines are most commonly specified by their tonnage and shot size. When selecting the molding machine, these two are the primary specifications that are considered. Once these requirements are satisfied, other specifications are considered.

6.3.1 Clamp Force (Tonnage)

The plastic is injected into the mold under high pressure. This pressure generates a force on the mold face that can very easily open the mold during injection. To prevent the mold from opening, the clamp is kept locked down by a counterforce. The maximum available force to keep the mold closed is called the clamp force of the machine. It is usually specified in tons and is therefore also called the tonnage of the machine. Appendix B lists the different units that are used to specify machine tonnage.

6.3.2 Shot Size

Shot size is another important parameter to consider when selecting a molding machine. The shot size of a molding machine is determined by the maximum weight of general purpose polystyrene (GPPS) that can be molded with a single stroke of the injection molding screw. A shot size of 100 g would indicate that a GPPS part weighing no more than 100 g can be molded on the machine. It is becoming more common to use the volume of the cylinder that comprises the barrel of the molding machine rather than the GPPS part weight because the latter can be misleading. The density of GPPS is 1.06 g/cm^3 and therefore, for the given volume of the cylinder/barrel of the molding machine, it can hold more or less weight of another material, depending on its density. To make matters simple, consider the shot size of the machine to be 106 g, which equates to 100 cm^3 of volume. Therefore, it can hold 106 g of GPPS. If this GPPS was replaced with low-density polyethylene (LDPE) with a density of 0.91 g/cm^3, the maximum shot size now is 91 g of LDPE. For a 30% glass-filled nylon (density of 1.33 g/cm^3), the shot size is 133 g. It is therefore now becoming common to specify the shot size in terms of volume rather than weight. The setting of the shot size on the machine is also taking this approach because it helps when moving the mold from one machine to another machine of a different screw diameter. It is easier to match the shot volume than calculate the new shot size in linear dimensions.

6.3.3 Screw Diameter and L/D Ratio

As the name suggests, the screw diameter is the diameter of the screw, either in millimeters or in inches. The barrel diameter is only slightly bigger than the screw. For example, for a 50 mm screw, typical barrel clearance is 0.1 mm (4 thousands of an inch). Over time, screw wear can cause material leaking over the screw flights, causing inconsistency from shot to shot.

The ratio of the length of the screw to the diameter of the screw is called the L/D ratio. The higher the L/D ratio, the better is the melt homogeneity.

6.3.4 Plasticating Capacity

The plastic has to be melted and heated to its processing temperature. The melting of the plastic is accomplished by the heater bands around the barrel and the shear heat generated by the rotating screw. The screw is also transporting the plastic to the front of the barrel, ready to be injected into the mold. Barrel temperature, exposure time, and screw speeds are all important to get a good homogeneous melt. If the material is moved too fast, the plastic pellets may not have enough time to melt, and the melt can contain plastic pellets that are not or only partially melted. During the purging process, it is common to see unmelted pellets of polyethylene coming out of the nozzle tip if high screw speeds were used. Plasticating capacity is the maximum weight of GPPS that can be raised to the molding temperature and metered in front of the screw. It is usually expressed in kilograms or pounds per hour.

6.3.5 Maximum Plastic Pressure

Thin-walled parts require more pressure to fill than do thick-walled parts. Therefore, depending on the application, the maximum available pressure needs to be specified. For hydraulic machines, the maximum available plastic pressure depends on the maximum available hydraulic pressure and the intensification ratio of the screw. Plastic pressure is specified in pounds per square inch or bar or other pressure units.

There are other parameters that are specified on machine data sheet. Discussions of these are beyond the scope of this book. Suggested books on this topic are mentioned at the end of this chapter.

6.4 The Injection Molding Screw

The injection molding screw and barrel assembly is responsible to help deliver the right quality of melt to the mold. Electric heater bands are installed around the barrel and supply radiant heat energy to melt the plastic. However, it is the screw that plays the vital role in the process of achieving a homogeneous melt. The screw provides shear heat to assist in the melting process along with the required mixing and homogenizing of the melt. It also helps in accurately measuring the volume of the shot to be injected into the mold. Several designs of screws have been developed based on different materials and their needs. The most common screw is called a general purpose screw or the GP screw. The design of a GP screw is shown in Figure 6.3.

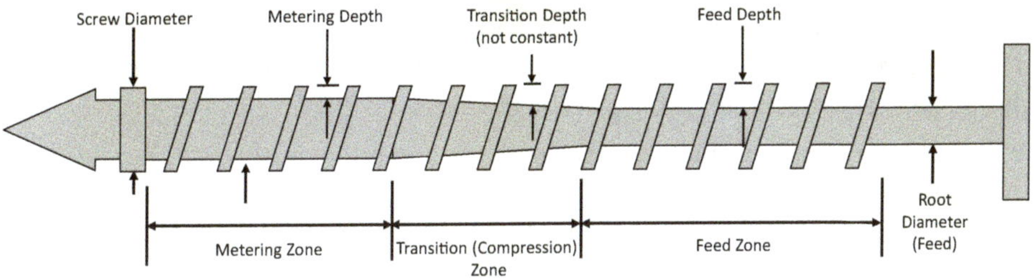

Figure 6.3 The general purpose screw

The GP screw has three main sections (or zones), each of which serves a special purpose. They are described in the following together with some other terms. The screw construction can be described as a shank with flights that are wrapped around it. In most cases, there is only one ribbon that creates the flights.

Outside Diameter: This is the diameter of an imaginary cylinder that is created by joining the outside area of the flights. The outside diameter is constant and is slightly smaller than the internal diameter of the barrel.

Root Diameter: This is the diameter of the shank. The root diameter changes from the back to the front of the screw, depending on the section of the screw.

Channel Depth: The difference between the outside diameter and the root diameter is the channel depth. Since the root diameter changes, the channel depth also changes from the back to the front of the screw.

Feed Zone: This is the section of the screw that picks up the material from the feed opening (base of the hopper) and begins to soften the material as it is being conveyed. The root diameter is the smallest here and is constant. Since the root diameter is constant, the channel depths are also constant; they are also called the feed depth. In the feed zone, the material is picked up and is softened as it is conveyed by the rotation of the screw. The material must never be completely molten in this section because that would prohibit the picking up of additional material. A term commonly used to describe this phenomenon is *screw slipping*, where the melt rotates with the screw and prohibits the screw from moving back to pick up more material and build the next injection shot.

Transition Zone (Compression Zone): In this section, the root diameter increases gradually, resulting in the decrease of the channel depth. At the start of this section, the root diameter is the same as the root diameter of the feed section, where it gradually increases until the section ends. This causes the feed depth to steadily decrease. As the screw rotates, the softened pellets begin to get compressed and the air and any other volatiles are forced out from between them because the feed depth is decreasing and the plastic is being conveyed. With the help of heat from the external heater bands and the shear from the rotation of the screw, the plastic begins to melt. As the feed depth reduces because of dispersive and distributive mixing, the plastic ends up as a homogeneous melt by the time it reaches the end of the transition zone. Distributive mixing occurs when the melt streams distribute and reconvene. On the other hand, dispersive mixing is similar to a smearing action. There is a combination of distributive and dispersive mixing taking place in the barrel.

Metering Zone: The metering zone is the last zone and is the closest to the nozzle of the machine. The depth of the channels in this section is minimal compared to the other two sections. The root diameter stays constant and therefore the channel depth is also constant. Since the shot is built by moving the screw back until it reaches a set linear position (shot size), the metering depth must be as minimum as possible to reduce the variation in the amount of melt for each consecutive shot. With a larger metering depth, the amount of material that is fed in front of the screw can vary, leading to inconsistencies. However, as the depth reduces, the shear increases and, therefore, the risk of material degradation is also increasing, especially for shear sensitive materials such as PVC. A compromise must be found and special screw designs are therefore necessary for certain types of materials.

Figure 6.4 shows the melting progression of the plastic as it travels through each of these sections. In the feed zone, the pellets have softened and begin to adhere to each other. When they travel to the transition zone, there is a combination of melted and unmelted plastic. There is still evidence of plastic pellets that have been compressed together. The metering and transition zones are the same, and the feed zone is usually twice the length of any of one these sections. In custom designed screws, these lengths can be altered. Longer feed zones increase the throughputs, longer transitions decrease shear, and a longer metering section will output a more homogeneous melt but will create more shear.

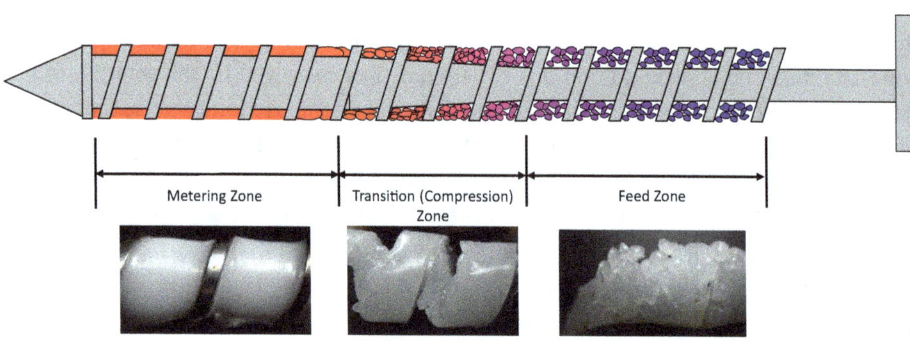

Figure 6.4 Melting progression of the plastic as it travels through the sections of the screw

Compression Ratio: This is the ratio of the feed section channel depth to the metering section channel depth. It defines the amount of compression to which the material has been subjected. The higher the compression ratio, the better is the melt homogeneity, but also the higher is the shear. The depth of the channels also contributes to the amount of shear heat, melt homogeneity, and the throughput.

Typical compression ratios are mentioned below:

- Low compression ratio: 1.5:1 to 2.5:1 used for shear sensitive materials such as PVC
- Medium compression ratio: 2.5:1 to 3.0:1 used in general purpose materials
- High compression ratios: 3.0:1 to 5.0:1 used for crystalline materials such as nylons

Helix Angle: This is the angle of the flight in relation with the plane perpendicular to the screw axis.

L/D Ratio: The L/D ratio is the working length of the screw flight to the outside diameter of the screw. Most injection molding screws have L/D ratios of 20:1. Greater L/D ratios allow more exposure of the plastic to heat and shear, improving the melt homogeneity and therefore increasing throughput at the desired processing temperature.

6.5 Screw Designs

The screw described in Figure 6.3 was a general purpose design that can be used for most materials and in most situations. Screws have been designed for specific requirements, such as for shear sensitive materials or to improve the throughput to reduce cycle times. Screw designs also depend on the degree of crystallinity, the viscosity, and the additives present in the plastic. In injection molding, mixing screw designs and barrier screw designs are most common. Mixing screws, as the name suggests, help to mix the additives such as colorants and also help to improve melt homogeneity. There are certain sections incorporated into the screw that create the mixing effect. Barrier screw designs have two screw channels in the transition section of the screw separated by a barrier flight. The unmelted plastic stays in the first channel until it is completely molten and then moves on to the second channel. This ensures that the plastic is completely molten before it reaches the metering section. Examples of screw designs are shown in Figure 6.5.

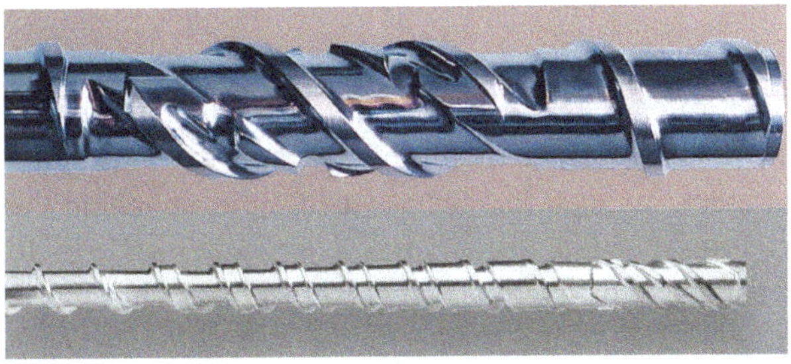

Figure 6.5 A mixing screw and a barrier screw (Courtesy: Westland Corporation)

6.6 The Check Ring Assembly

The check ring assembly is essentially a nonreturn valve. The most popular design is shown in Figure 6.6. During the shot build-up process, the screw is rotating to pick up the material. During this time, the check ring is in the forward position and allows the passage of the material to the front of the screw. During injection, the check ring seats itself on the body and stops the plastic from being pushed back over the flights, acting as a one-way nonreturn valve. Check rings wear over time, and the plastic will begin to leak into the flights. This causes inconsistency in the shot and therefore shot-to-shot variations. Check rings must be checked periodically and changed at the slightest sign of leakage. Various designs of check rings are available on the market for specific materials and applications.

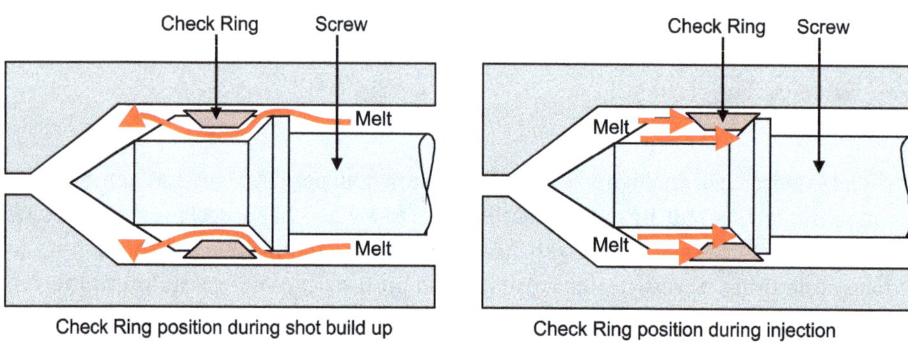

Figure 6.6 Working of a check ring

Other types of check rings include ball check valves that are suitable for unfilled, non-shear sensitive materials such as polyolefins. Smear valves are used to process highly viscous materials such as rigid PVC. There is no shut-off mechanism in a smear valve. The high viscosity helps to prevent the back flow of the plastic during injection.

6.7 Intensification Ratio (IR)

To understand IR (intensification ratio), consider a hydraulic piston: a cylinder arrangement as shown in the schematic (Figure 6.7) A pressure of 800 psi is applied on the larger cross section of 6 in^2. Therefore, the total force on the hydraulic oil behind the piston is 800 × 6 = 4800 lb. There is a reduction of the cross section area from 6 in^2 to 2 in^2. The force of 4800 lb is therefore applied on the smaller cross section of 2 in^2. Therefore, the calculated pressure on other side of the smaller piston is 4800/2 = 2400 psi. The reduction of the diameter from 6 in^2 to 2 in^2 increases the pressure from 800 psi to

2400 psi, so the pressure is intensified 3 times. This number is the ratio of the area of the larger piston to the area of the smaller piston and is called the intensification ratio.

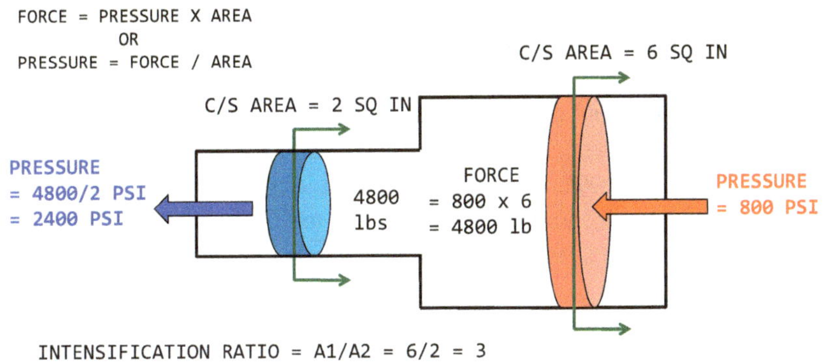

Figure 6.7 Schematic for intensification ratio

To drive a nail into a wall, the force is applied to the head of the nail. The cross section of the area at the head of the nail is comparatively larger than the point. Therefore, the applied force of the hammer at the head of the nail is intensified at its point and this is what pushes the nail into the wall. This is only an analogy because in injection molding, we deal with fluids and not solids. The laws of physics are different for fluids and solids. In injection molding, there is a hydraulic ram at the back on the molding screw where hydraulic pressure is applied. This force is intensified at the screw tip and converted to plastic pressure. The ratio of the cross sectional area of the ram to the cross sectional area of the screw is called the intensification ratio (IR), see Figure 6.8. Plastic pressure at the nozzle is equal to the product of the applied hydraulic pressure and the IR. Intensification ratios range from about 6:1 to as high as 23:1. Higher IRs are necessary to mold thin parts with long flow lengths. In electric machines, there is no hydraulic ram and the pressure applied to the back of the screw is what the plastic experiences so that the IR for an electric machine is equal to 1.

Plastic Pressure = Hydraulic Pressure × Intensification Ratio

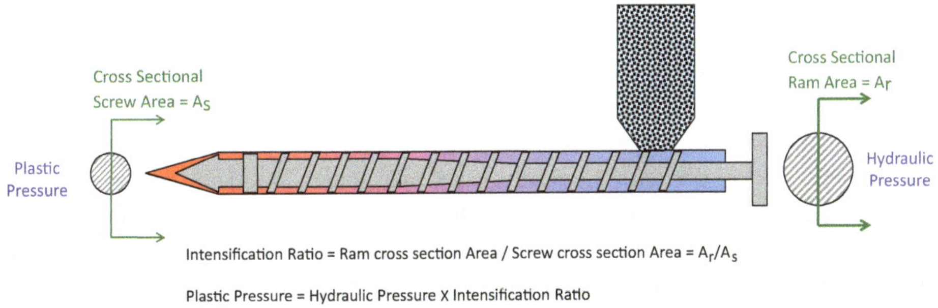

Figure 6.8 Intensification ratio for hydraulic machines

When a mold is moved from one molding machine to another, in order to mold identical quality parts, the plastic molding process on both machines must be identical. One of the factors that need to be matched is the plastic pressure. The plastic pressure needs to be calculated with the formula above and then the hydraulic pressure on the second machine needs to be back calculated from the same formula. An example of this is shown in Table 6.1. If the packing pressure used on Machine 1 is 800 psi and the intensification ratio is 10, the applied plastic pressure is 8000 psi. If the mold is moved to Machine 2 with an intensification ratio of 13.5 then to get a plastic pressure of 8000 on Machine 1 a pressure of 8000/13.5 = 593 psi must be used. If the hydraulic pressures are matched then the plastic pressure will equal to 800 × 13.5 = 10,800 plastic psi which could result in flashed parts and/or dimensionally nonidentical parts.

Table 6.1 Process Transfer between Machines, Matching Plastic Pressures

Parameter	Machine 1	Machine 2
Intensification Ratio	10	13.5
Case 1:		
Holding Pressure – Hydraulic ↓	800 psi	8000/13.5 = **593 psi** ↑
Holding Pressure – Plastic	800 × 10 = 8000 psi →	Required = 8000 psi
Case 2:		
Holding Pressure – Hydraulic	800 →	800 psi ↓
Holding Pressure – Plastic	800 × 10 = 8000 psi	800 × 13.5 = 10,800 psi → This could result in flash and/or dimensionally unacceptable parts

6.8 Obtaining Intensification Ratios

Getting the IR number directly from the machine manufacturer can sometimes be a challenge. However, there are a couple ways that these can be obtained.

Some machines provide a graph like the one shown in Figure 6.9. The different screws sizes that are compatible with the machine are mentioned. Based on the screw size of the machine, one can calculate the IR by dividing the plastic pressure value by the hydraulic pressure value. The graph can also be used to find the plastic pressure at any given hydraulic pressure setting on the molding machine.

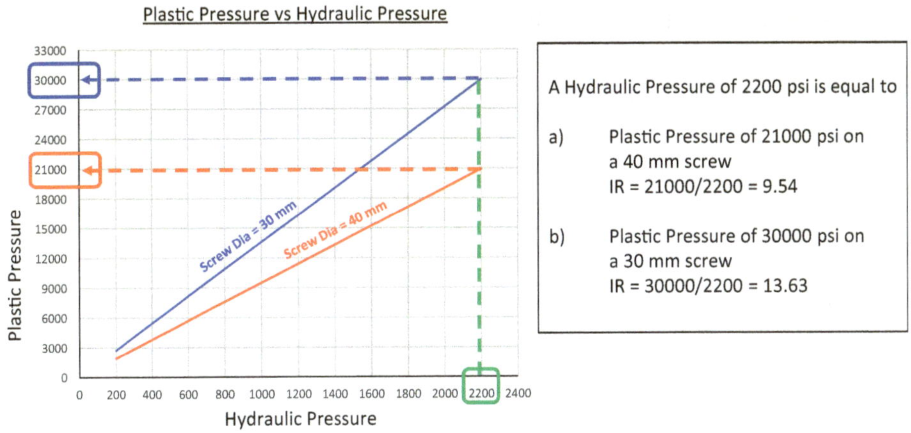

Figure 6.9 Calculation the IR from the graph provided by the machine manufacturer

If such a graph is not available, the machine specification sheet can be helpful. The machine specification sheet should list the maximum plastic pressure. On the molding machine, if the injection pressure setting is selected the machine usually displays a range of pressures. The higher number is the maximum hydraulic pressure of the machine. If the maximum plastic pressure from the specification sheet is divided by the maximum hydraulic pressure of the machine, the resulting value is the IR. If the machine does not show the range of pressures, an input of a very high number in the field will cause an alarm and provide the user with the maximum hydraulic pressure value.

IR = Maximum plastic pressure seen on the machine datasheet/Maximum machine pressure seen on the molding machine screen

6.9 Selecting the Right Machine for the Mold

The machine selection is one of the five important factors that will contribute to the quality of the part. The mold and the machine must be compatible, and this is often overlooked. Usually, only two factors are taken into consideration: whether the mold physically fits in the machine and whether the clamp tonnage is sufficient. However, one of the most important factors is what percentage of the machine shot size will be used. The residence time of the material in the barrel and the lower limiting size of the mold are other factors. Both are described in the following.

6.9.1 Physical Size of the Mold

The mold size is defined by three variables (see Figure 6.10).

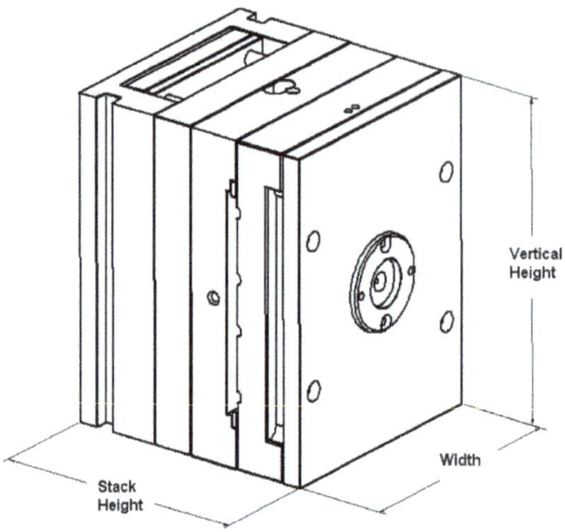

Figure 6.10 Mold size specifications

Mold Stack Height (H): This is the distance between the sides of the mold in the direction of the mold open and close when the mold is fully shut.

Mold Width (W): This is the distance between the vertical sides of the mold, looking in the direction of injection of the plastic. This applies to molds mounted in horizontal machines, but the definitions are extended to other molds also.

Mold Length (L): This is the distance between the top and bottom side of the mold, looking in the direction of injection of the plastic. This again refers to horizontal molds.

Naturally, the mold must fit in the machine such that at least two sides can be bolted to the platen of the machine. The mold can hang off the platens on the other two sides, but the molding area must not be outside the platen. The molding area must always be supported by the platen, see also Figure 6.11. If the cavity is not supported, the injection pressure can easily deflect the plates and cause flash in the part. Injection pressures can be very high, applying tremendous force on the mold base, and over time, can damage the mold components if the mold is not properly supported by the mold platens. On the other hand, the mold must not be considerably smaller than the platen. It must cover at least 70 to 75% of the area between the tie bars. This is especially true for toggle machines, where the clamping force is applied on the outside and not in the center of the platens. There is a possibility of platen deflection, if the mold is too small, causing platen damage over time. The toggle system also provides reduced support in the center of the mold where the main injection pressure is applied. Even with adequate support pillars in the mold, there still could be deflection because of the lack of support, causing part defects.

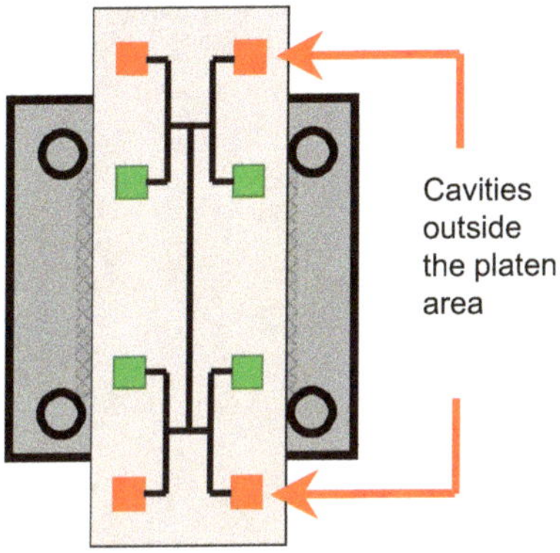

Figure 6.11
Cavities outside the molding area

In regards to the mold height, every machine has a minimum and a maximum mold height that it can accept. The moving platen closes the mold and applies the set tonnage to the mold. Because of the limit on the travel distance during closing, the mold height needs to be greater than this limit. If the mold is less than the minimum mold height, the platen can never let the two halves of the mold touch and apply the tonnage. Therefore, minimum mold height is important. On the other hand, the mold must be smaller than the maximum mold height in order to fit in the machine.

The required mold open stroke will depend on the part, usually the part dimension in the direction of ejection. The open stroke should be such that the mold halves are far enough apart when fully opened so that the part falls out of the mold after ejection. The mold open stroke must be greater than the longest dimension of the part in the direction of ejection. For example, on a rectangular part as shown in Figure 6.12, it should be the diagonal of the part. Even with this distance, there is still a danger of damaging the part because it may still hit the side of the mold as it falls off the ejectors. For this reason, the mold opening stroke must be set as wide as possible to avoid damage, but not such that cycle time is lost due to unnecessary movement of the mold.

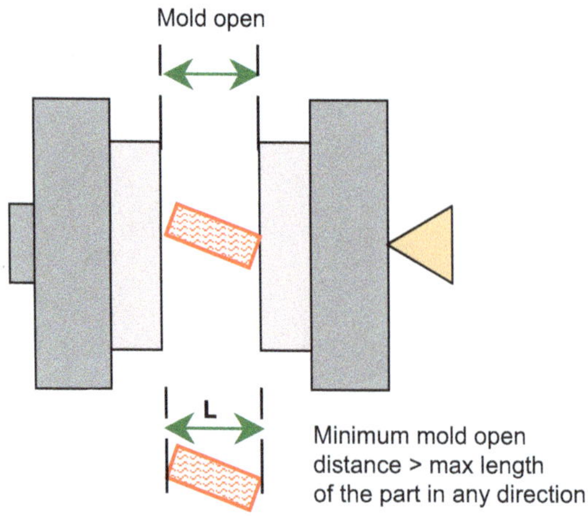

Figure 6.12 Minimum mold open distance

6.9.2 Calculating the Required Machine Tonnage for a Mold

The injection pressure of the plastic applies an outward force on the mold cavities, which works to separate the mold halves. This force must be counter balanced by the machine in order to keep the mold halves closed. If the plastic pressure is higher than the clamp force applied to keep the mold closed, the mold will open, and the plastic will escape from the mold at the parting line where the mold splits open, causing part defects, typically flash. The force that keeps the mold closed is called the clamp tonnage of the machine. The applied clamp force is measured in tons. Further, to mold an acceptable part, the part must be packed out with the right amount of plastic. This packing will result in an increase in the cavity pressure and will also apply a force that can open the mold. The packing pressure for a given part will depend mainly on the part design. The plastic will need to the fill the cavity and then pack the cavity with additional plastic to compensate for the shrinkage that is taking place as the plastic cools. As the plastic undergoes the fountain flow effect to fill the cavity there is a loss in the pressure and the pressure inside the cavity typically reaches on an average between about 4000 and 8000 psi (275 to 550 bar). Depending on the part design, it could also be outside these numbers. One US ton is equal to 2000 pounds. Therefore, the average cavity pressure is between 2 and 4 tons per square inch. This is how material manufacturers recommend the tonnage for their materials and, as an example, will mention that their material needs 3 to 4 tons per square inch.

For the sake of understanding the calculation of the required tonnage, let us consider a pressure inside the cavity of 5000 psi, which equals 2.5 tons per square inch. This force can open the mold if there is no counter force being applied on the mold halves to keep them closed. If the area on which the pressure is applied is 1 square inch, then the force needed to keep the mold halves closed will be slightly more than 2.5 tons to counter the force. Projected area is defined as the 2D area on which the plastic pressure is applied looking in the direction of injection as shown in Figure 6.13. Note that it is not the surface area of a part. A simple way to understand this is to visualize opening the mold and looking at it in the direction of injection. A photograph with a schematic representation is shown in Figure 6.14, where the molded part was left in the mold to have the photograph taken. The projected area is shown as the shaded area in the schematic.

Molders use a simple formula based on the above discussion for the calculation of the required tonnage. The rule of thumb for calculating the tonnage required for the part is given in the following:

Required tonnage = (Projected area of the part × Number of cavities + Projected area of the runner) × (Tons/in^2 required for the resin)

Molders use a simple formula based on the above discussion for the calculation of the required tonnage. Referring to Figure 6.13, if the projected area of one cavity is 6 sq. inches and this is a 4-cavity mold, the projected area for the four cavities is 6 × 4 = 24 sq. inches. If the projected area for the runner is 4 sq. inches, then the total projected area for the mold is 28 sq. inches. If the suggested tonnage requirement by the material supplier is 4 tons per sq. inch then the tonnage required for this mold is 28 × 4 equals 112 tons. The formula above is a good estimation of the required tonnage for a given mold.

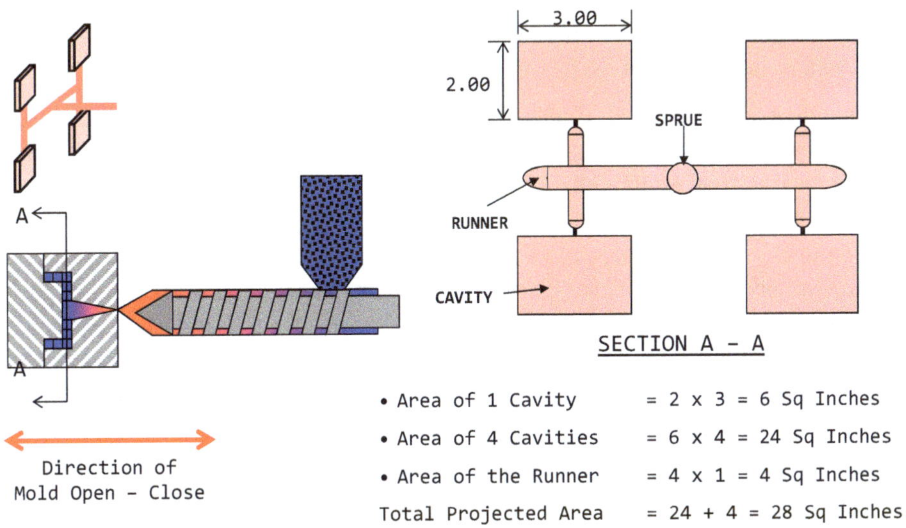

Figure 6.13 Projected area of a mold

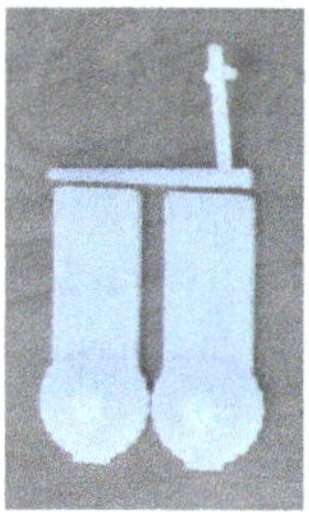
Molded parts and runner

Ejector side of the mold with no parts

The shaded area is the area on which the plastic pressure is applied.

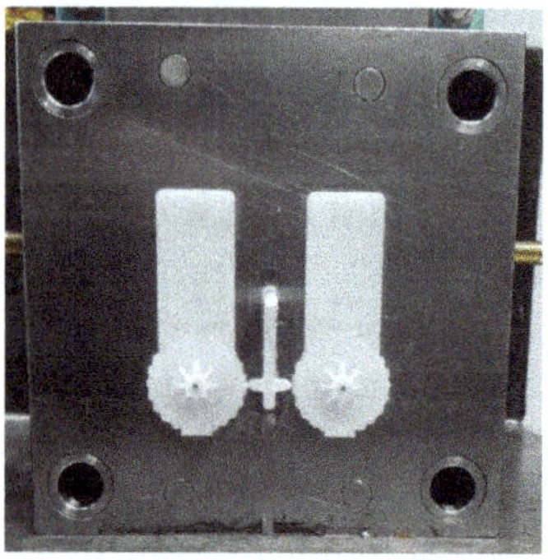
Ejector side of the mold with parts

Figure 6.14 Example of a projected area of a mold

However, there are several limitations to this type of a calculation. These are discussed below. Refer to Figure 6.15 and Figure 6.16.

- Injection pressure is used to fill the cavity with plastic, pack pressure to compensate for the shrinkage and hold pressure to hold the plastic inside the cavity until the gate area freezes off. Consider a large round trash bin used to collect the yard waste. The flow length to thickness ratio (L/T ratio) of this bin is large and therefore the injection pressures required to fill the cavity is high. Most of the part is a thin wall part except for maybe the areas of the handles and therefore the pack and hold pressures are relatively smaller. Now consider a part design that has a very thick wall. In such a case, the injection pressure will be low but the pack and hold pressures will be high. In some cases, they can be much higher than the injection pressure. Therefore, a thin wall part and a thick wall part with the same projected areas can end up requiring different tonnages.
- Visualize a plastic desk ruler that is about 300 mm long (about 12 inches) with one side gate. Now visualize the same part with two gates, one at each end. With our knowledge of molding we know that the part with two gates will require lower injection pressures and lower pack and hold pressures, thereby reducing the required tonnage for the mold. In this case, the plastic does not have to travel a lon-

ger distance compared to the part with one gate. The number of gates including the gate design will have an effect on the required tonnage. An edge gate will fill the part more easily than a sub gate or a cashew gate.

- Extending the discussion with the same part above, if the gate is moved to the center of the part from the end of the part, one can follow the same theory and conclude that we will need less pressure in the center gated part and therefore less tonnage. The location of the gate will also affect the required tonnage.
- Large parts such as car bumpers and automotive side panels use sequential valve gating for the mold filling process. With sequential valve gating the individual gates are opened and closed on demand. The biggest advantage of using sequential valve gates is to get rid of any weld lines. Weld line strength is always of concern since the two flow fronts that meet may not fuse and become homogeneous. Sequential valve gates solve this problem by opening the gate only once the first flow front has reached the second gate position. The melt flow front progresses all in one direction rather than flowing in opposite directions. In the case of parts with the sizes of car bumpers, the projected area is very large but since each valve gate opens one at a time or in a certain sequence, the plastic material is still not beginning to cool down and therefore lower pressures can be used to fill and pack the parts out. Most of these parts can be molded on 1500 to 2000 ton machines even though the projected area calculation would mandate maybe a 5000 or 6000 ton machine. Figure 6.15 summarizes the above discussions.
- Further, the orientation of the part in the mold will influence the projected area. See Figure 6.16. It is the same part where the shaded area is the projected area. In the second case the projected area is much lower than in the first case but this does not mean we can get away with lower tonnage since the flow length in the direction of mold open is much longer.
- There are different types of side mechanisms to mold side features on the part. A slide can be on angle pins or they can be activated by other means such as by hydraulics, pneumatics, or electrical systems. If they are on angle pins then the load in the direction of mold open must be calculated since this force will open the mold and therefore affects tonnage. If they are activated by other means then the force of activation need not be taken into consideration.
- Lack of or insufficient venting will also affect the required tonnage. Consider an extreme case where a mold has no vents. In this case as the injection starts, the air in the mold will get compressed to a certain amount since it has nowhere to go but there will be a point where the cavity pressure will increase and separate the mold halves, causing flash and other problems. This will also happen with inadequate venting, which often is the case in several molds. In theory, the flow rate of the plastic into the mold should be the same as the flow rate of the air out of the mold. Not having adequate venting will increase the required tonnage.

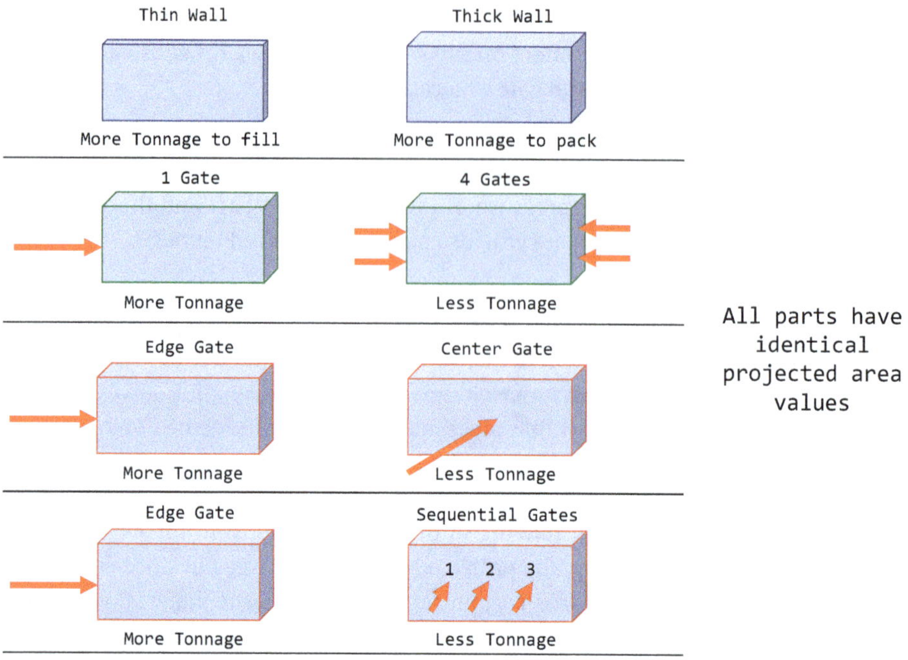

Figure 6.15 Factors affecting required tonnage on parts with identical projected areas

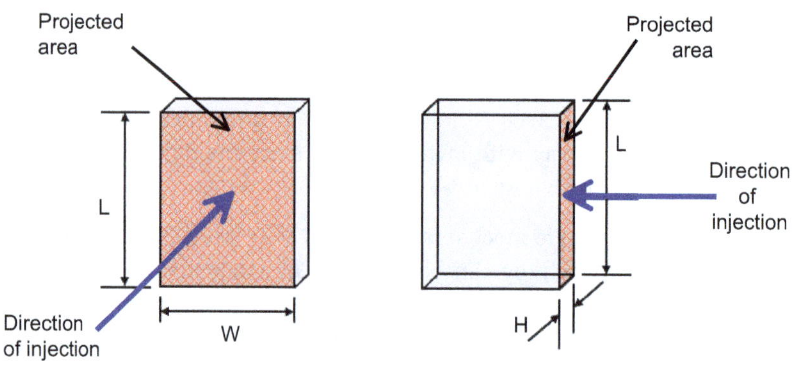

Figure 6.16 Part projected area in the direction of injection

From the discussion above, it is clear that the tonnage calculation should not be based on projected area alone. There are several other factors that must be considered. These are part thickness, flow lengths, position of gates, number of gates, type of gates, gate size, gating mechanisms, orientation of the part in the mold, presence of slides, the type of slide mechanism used for the mold, and the extent of venting. The formula for required tonnage based on the multiplication of projected area and material tonnage specification is therefore not accurate by any means.

Most molders will select a machine based on the projected area formula for lack of a better resource. When the mold is delivered to the molding floor, most processors will set the tonnage to the maximum available on the machine. Using maximum available tonnage will result is several defects and therefore must be optimized. A procedure and an explanation for process optimization are given in Chapter 9.

Depending on the plastic material properties, every material requires a certain amount of force during the mold fill and then requires a certain amount of pressure to pack the part out. Typically, crystalline materials require 3.5 to 4.5 tons of clamp force per square inch of projected area, while amorphous materials require anywhere between 2.5 to 4.0 tons of clamp tonnage per square inch of projected area.

Tonnage calculation is very complex and not easy to predict. Computer simulation programs do an acceptable job in this calculation, but caution is warranted when applying the results. This book will explain a procedure to set the clamp tonnage for a given mold. This will be provided in the chapters on process development.

6.9.3 Percentage Shot Size Used and Number of Shots in the Barrel

The percentage of the shot size used is one of the most important factor for molding consistency and is often overlooked. The percentage of the shot size gives an idea of the amount of plastic injected into the mold with respect to the maximum amount of plastic the barrel can hold or the fraction of the shot size that is injected into the mold. The formula for this calculation is given as:

% Shot size used = ((((Part weight × Number of cavities) + Runner weight)) × (1.06/(Density of the plastic)))/(Shot size of the machine) × 100

An example is shown in Figure 6.17, where the % usage is 25%.

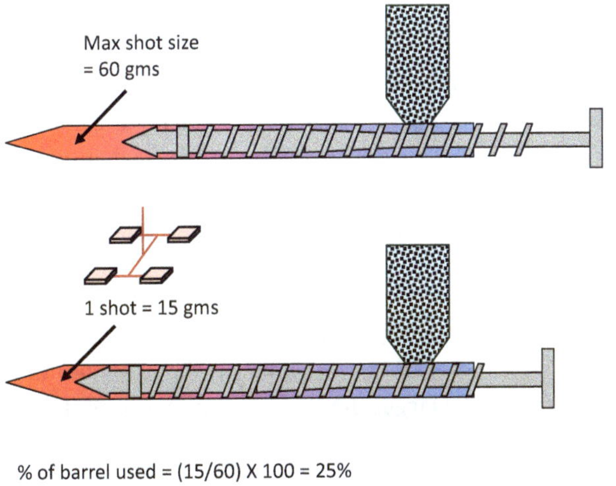

Figure 6.17 Percentage of barrel shot size used

An acceptable percentage shot size is 20 to 80% of the maximum available shot capacity of the machine. For example, if the machine shot capacity is 100 grams, then the weight of the molded shot should not be less than 20 grams and not more than 80 grams. A shot is the weight of the all the parts and the runners. There can be exceptions, and these will be discussed in the following paragraphs.

A number below 20% could result in the following problems:

- The machine needs a finite amount of time to build and achieve the set pressures and speeds. It is similar to listening to a commercial for a car that claims "0 to 60 mph in 5 seconds." This tells us that the car will require 5 seconds to reach 60 mph and so at the end of 2 seconds, the speed is not yet what the driver has set to achieve. Therefore, if the molding shot is very small, the injection phase can be inconsistent because not enough time is given for pressure and velocity to reach required levels. Moreover, the plastic that has now built up the pressure is suddenly stopped, and the momentum is unpredictable, leading to large variations in the fill.
- In the case of crystalline materials, the shear heat from the turning screw is an important factor for melting the crystallites and achieving a homogeneous melt. If the shot is very small, then the screw may make very few revolutions to reach the set shot size limit. This results in the loss of shear heat, therefore affecting the melt homogeneity.
- With smaller shots, the residence time of the plastic in the barrel increases. This can lead to material degradation.
- Plastic melt is compressible. When pressure is applied to a small shot, some of the applied pressure is lost in compressing the melt leading to the inconsistency in fill. On larger shots sizes, there is a compression of the melt but the percentage compression is much smaller as compared to a smaller shot.

A number above 80% could result in the following problems:

- As described in Section 6.1, the material needs to spend the required amount of time in order going through the melting and homogenization process. A large shot size will transport the material quickly, and the material will not have a chance to form a homogeneous melt for the injection shot. For example, during the startup of a machine, when the machine is being purged using high screw speed and back pressures, sometimes unmelted pellets can be seen coming out of the nozzle tip. This is because the pellets did not have enough time to melt and homogenize because of the higher percentage shot sizes used in purging (Figure 6.18). In the case of hot runner molds, there is an added challenge of transferring the pressure applied to the screw through the hot runner. When the shot is large, the pressure applied has to first compress the plastic in the barrel, then in the hot manifold, and then inject the plastic in the mold. The larger the shot, the more inconsistency.

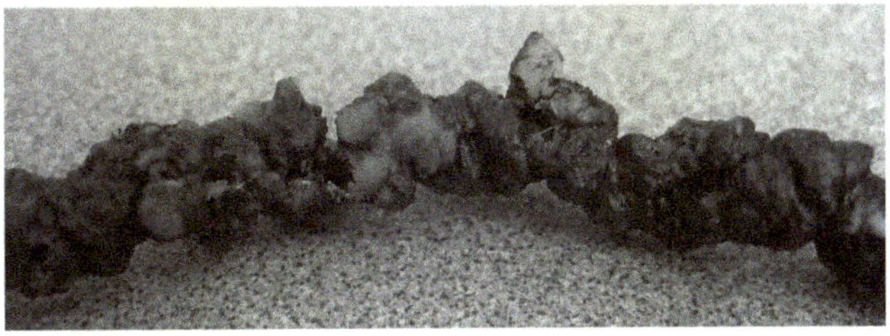

Figure 6.18 Unmelted plastic pellets in the purge as a result of moving too fast through the molding barrel

Deviation from the 20 to 80% rule:

It is not always possible to have a machine that fits the 20 to 80% rule for the shot size of a given mold. Almost every molder has such a mold and they can mold perfectly acceptable parts.

On the low end, the preferred number is 20%. Following are the situations where numbers less than 20% could be acceptable:

- If the molded part has wide dimensional tolerances. In such cases, any variations from the inconsistent fill that cause a variation in the dimensions can be absorbed by the tolerances.
- If the molded part is being molded out of a heat stable material, such as a polyethylene or a polypropylene. The chances of material degradation are low.
- Recent advances in machinery controls have improved the accuracy and control of the injection phases of the molding cycle. In such cases, deviations from the rule can be considered.

On the high end, the preferred number is 80%. Unless the screw design is a specially optimized screw for efficient melting with large shot sizes, the rule should never be deviated from. Larger L/D ratio screws help in efficient melting.

As the term suggests, the number of shots in the barrel would mean how many shots are present in the barrel. As an example, if the machine shot size is 60 grams and the injected shot weight is 15 grams, then there would be 4 shots in the barrel. The barrel usage would be 25%.

Cautionary Notes: Hot runner manufacturers can provide information on the volume of the hot manifold. This volume should not be used in the calculation of the percentage of the shot size. The formula mentioned above is a good estimate because to get the exact percentage shot size number, one needs to add on the amount of melt that is on the molding screw, and at the same time, subtract the material that is not being used for the shot. Only the amount of plastic required for the shot is melted and collected in front of the screw. The machine does not collect the complete maximum shot and then inject what is required.

6.9.4 Residence Time of the Material in the Barrel

Residence time is related to the percentage of the shot size used and can be defined as the total amount of time the plastic resides in the injection unit. The total time from entering the feed throat to exiting the nozzle is called the *residence time* of the plastic in the barrel. The calculation is shown in the following equation:

$$\text{Residence time} = \frac{(\text{Shot capacity of the machine})}{((\text{Part weight} \times \text{Number of cavities}) + \text{Runner weight}) \times (1.06/(\text{Density of the plastic}))} \times \text{Cycle time}$$

In simple terms, residence time can also be calculated as shown here:

Residence time = Number of shots in the barrel × Cycle time

An example of the residence time is shown in Figure 6.19.

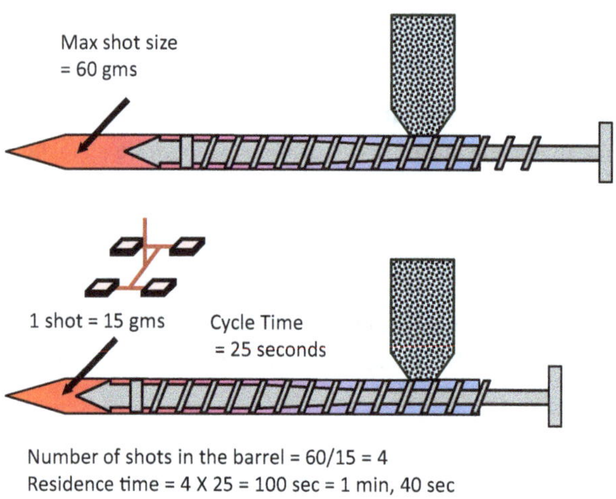

Figure 6.19 Residence time in the barrel

Every material has a maximum recommended residence time in the barrel. Because plastics are heat sensitive, they will degrade when overexposed to molding temperatures. PVC is a prime example of a material that cannot be left in the barrel for long periods of time. It degrades quickly and releases hydrochloric acid, an irritant and hazard. Residence time numbers, typically in minutes, are provided by the material manufacturers. Depending on the grade of PVC, the maximum residence time is between 4 and 7 minutes. Maximum acceptable residence time for the same material can be different based on the molding temperature used. The higher the molding temperature, the shorter is the recommended residence time. In the case of Ultem, a polyetherimide from the Sabic company, the maximum residence time at 365 °C is 14 minutes whereas at 410 °C it drops to 6 minutes.

Residence time depends upon the cycle time, the L/D ratio of the screw, and the lengths of the metering and compression sections of the screw. In the case of a hot runner mold, the residence time in the hot manifold must also be added to the time in the barrel. Hot runner manufacturers can provide the volume of the hot manifold and this can be used in the calculation.

In the injection molding process the screw turns only to pick up the material needed. In the example shown the shot weight is 15 grams and the maximum machine shot weight is 60 grams. The number of shots in the barrel is calculated as 4. The screw will stop rotating when it has picked up the 15 grams. For the calculations of percentage shot sizes and residence times it is assumed that the other 3 shots present on the screw flights as shown in Figure 6.20. This is a good assumption but may not always work. It is best to find the actual number of shots as shown in the next section.

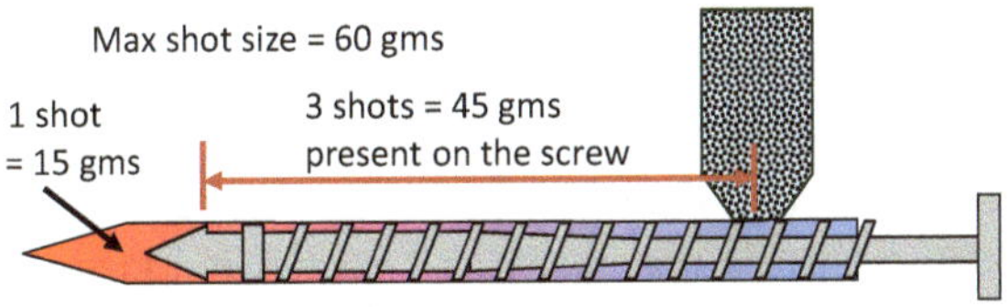

Figure 6.20 Assumption of the number of shots on the screw

6.9.5 Practical Methods to Find Percentage Shot Size, Shots in a Barrel, and Residence Time

Please note that extreme caution needs to be exercised during any tests at the molding machine. Only those with sufficient experience with molding machines should perform these tests. A simple and practical method can provide the number of shots in the barrel: As the molding machine is in operation, the hopper system can be pulled away, and a color pellet is dropped at the bottom of the feed throat when the molding screw is visible. Next, push the hopper system back in place, and starting from the next injection, count the number of shots until the color shows up in the part. This will be the number of shots in the barrel.

Calculate the residence time by multiplying the number of shots in the barrel by the cycle time. Calculate the percentage of the shot size used by taking the reciprocal of the number of shots in the barrel and multiplying it by 100.

For example: Cycle time = 15 sec

One shot weight = 5 grams

Number of shots until color showed up in the parts = 6

Calculated residence time = 6 × 15 = 90 minutes = 1 min 30 sec

Calculated % shot size used = (1/6) × 100 = 16.7%

6.9.6 Residence Time Distribution

The screw is designed to mix the plastic and additives, such as color pellets. Therefore, the system is not truly a first in, first out system. There is no buildup of the plastic in any part of the barrel or the screw, but because of the mixing, the molecules could spend more or less time in the barrel. The more effective the screw mixes in, the higher the distribution. Needless to say, the higher the number of shots in the barrel, the larger the distribution seems to be, but this can get misleading. For example, if there are 3 shots in the barrel, then the added color pellet may get mixed in only one of the three shots where as if there are 12 shots in the barrel then the added color will get distributed in 4 shots. If any change is made in the processing parameters involving times and temperatures, then the number of shots in the barrel should be molded and discarded before collecting parts for a quality check. See Figure 6.21, which shows the residence time distribution. The color showed up in the 3rd shot suggesting there are between 2 and 3 shots in the barrel. The color can be seen even in the 7th and some in the 8th shot. The pellet of added color mixed over almost 5 shots.

This is a very good visual to understand contamination by a single pellet of another material or color. It is therefore very important that machine hoppers and loading systems are cleaned thoroughly before the start of each job.

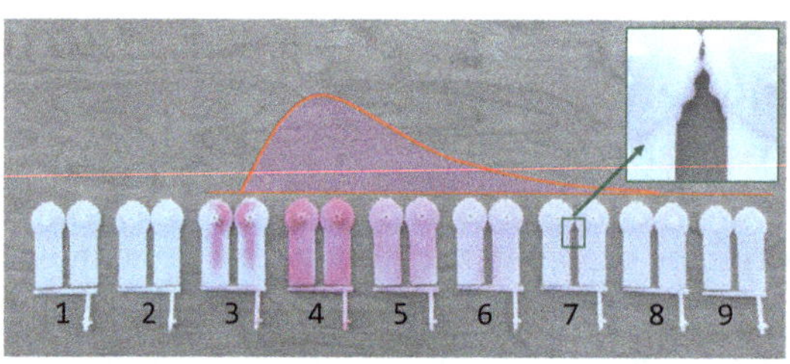

Figure 6.21 Residence time distribution

Suggested Reading

Chabot, J., *The development of plastic processing machinery and methods*, John Wiley and Sons, Inc. (1992) New York

Osswald, T. A., Turng, L., Gramann, P. J., *Injection Molding Handbook* (2007) Hanser Publishers, Munich

Beaumont, J. P., Nagel, R., Sherman, R., *Successful Injection Molding* (2002) Hanser Publishers, Munich

Rosato, D. V., Rosato, D. V, *Injection Molding Handbook* (2000) CBS, New Delhi, India

Dray, R., How to compare barrier screws, *Plastics Technology* (Dec. 2002), p. 46

Dealy, J., Wissbun, K., *Melt Rheology and its Role in Plastic Processing Theory and Applications* (1990) Van Nostrand Reinhold

7 Scientific Processing, Scientific Molding, and Molding Parameters

7.1 Introduction

Several parameters determine a successful molding process. There are various speeds, pressures, times, and temperatures to be considered. Scientific processing encompasses an understanding of the underlying scientific principles of each parameter and the application of these principles to achieve a robust process and consistency in part quality. Scientific processing covers the complete molding process, from the time the plastic enters the facility to when it leaves as a finished product. A robust process is one that can accept reasonable natural variations or a small purposeful change in an input but still delivers consistent output. The term consistency means molding parts with the least variation in the quality of the part. The quality of the part can mean its dimensions, appearance, part weight, or any other aspect that is important to the form, fit, or function of the part. The variation should be from special cause variations and not from any natural cause variations. Special cause variations are variations that are caused by an external factor. For example, if the chiller unit shuts down, the mold temperature will change causing a change in the quality of the part. Natural cause variations are inherent to the process. Their effect can be minimized but not eliminated. For example, if the plastic used to mold the parts has 30% of glass fiber mixed in it, in every molded shot the amount of glass will not be exactly 30%. It will be slightly more or less, for example, between 29.7 and 30.3%. If one weighs 100 consecutive parts from the molding process, each part will weigh differently although the process was not changed. This variation cannot be eliminated, but the mixing process can be improved, and the variation can be reduced.

Robustness and consistency should not be confused with parts being molded within the required specifications. Parts can be out of specifications but the process can be robust and the quality can be consistent. The goal of scientific processing is to achieve a robust process at each stage of the molding process the pellet is subjected to.

The term scientific molding was coined by two pioneers in the field of injection molding, John Bozzelli and Rod Groleau. Their principles and procedures are widely used today and are industry standards. Scientific molding deals with the actual plastic that enters the mold during the molding operation at the molding machine. The term introduced here is scientific processing, which is defined as the complete activity the plastic is subjected to from the storage of the plastic as pellets to the shipping of the plastic as molded parts. Scientific processing is applying scientific principles to each of the steps involved in the conversion of the plastic to the final product, see Figure 7.1. Chapters 8 and 9 will focus on the understanding and the application of the theories to each of these steps and then optimizing them. Successful process development results in a process that is robust, repeatable, and reproducible.

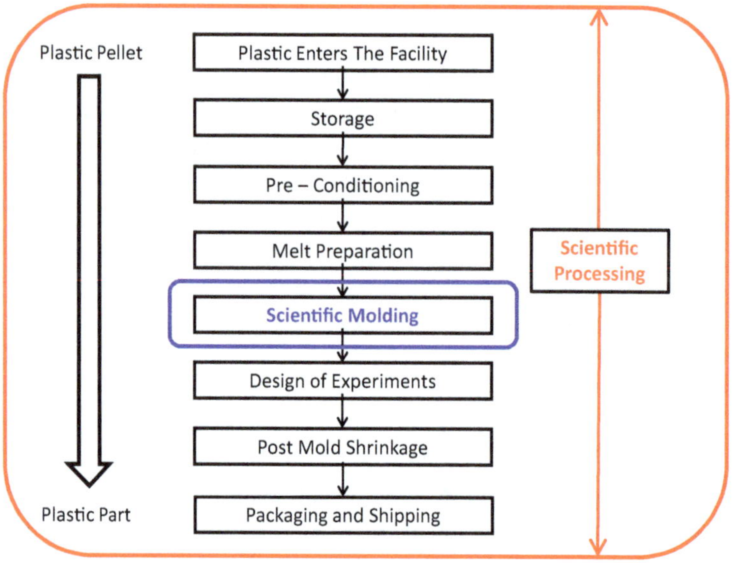

Figure 7.1 Scientific molding and scientific processing

7.2 What Scientific Molding and Scientific Processing Are Not!

Good molding practices start with a robust process development strategy. Molders gather information from a seminar, a book, and/or online resources about the concepts of scientific molding and scientific processing. It is very important to recognize that these techniques and procedures should not be followed blindly. There are exceptions to every technique and procedure. After understanding the science behind these techniques, the molder needs to know when and how to use these techniques. This is the importance of the word 'scientific'. Understanding the science behind molding is essential. For example, the in-mold rheology study (viscosity study) is the first study that is taught in molding seminars. There is a lot of science behind this, and a molder gets fascinated by generating these graphs; a eureka moment for some! However, there are several exceptions to the use of this technique. The same follows for all the other techniques that are used in process optimization. One must do what is necessary to understand and optimize the process rather than follow procedures because they were a requirement.

7.3 Process Robustness

A process is considered robust when the natural variations in the input parameters do not cause an unacceptable variation in the product quality. For example, if the heater bands on the molding machine fluctuate by +/- 2 °C or applied packing pressure on the molding screw varies by +/- 5 psi, this is considered natural variation. The molded product should still be within tolerances with these variations. The molding process outputs such as actual injection pressure or cushion values should also show a least amount of variation. The word 'natural' in natural variation is very important. A special cause variation would mean the presence of an external cause. For example, if a thermocouple is malfunctioning and the temperature of the incoming water is fluctuating by 20 °C, this is not considered natural variation.

A word on the set tolerances is also necessary. If the product designer has set very tight tolerances that a process cannot hold, this does not mean that the process is not robust. Tolerances must be set realistically understanding the capabilities and limitations of the molding process. Most plastic part designers come from the metal part design industry, where it is easier to hold tighter tolerances. The concept should not be applied to plastics.

A robust setting of an input would mean a setting such that the corresponding output is not adversely affected. For example, after a certain injection speed is reached on the molding machine, the viscosity of the plastic remains constant. The viscosity curve is in a robust area where variations in injection speed have little effect on the viscosity and therefore the amount of fill into the mold. At low injection speeds, a slight change in the injection speed causes a large change in the viscosity, resulting in shot-to-shot fill inconsistency. Therefore, a lower injection speed setting is not in a robust area of the process and should therefore be avoided. In addition, it must be understood that natural variations can never be eliminated. Taking these conditions into consideration will help ensure building a robust and consistent process.

7.4 Process Consistency

A process is considered consistent when it meets the following two requirements:

1. All variations in the outputs of the process are a result of only natural cause variation.
2. The standard deviation of the variation is at a minimum value.

For example, the cushion value is an output of the injection, pack, and hold phases. If the cushion value shows minimum variation, and a distribution curve of the cushion value over time is normal, then the process is consistent. In this case, the process under consideration would include only the injection, pack, and hold phases.

A robust process will always produce parts of consistent quality because there is little variation in the output. It also goes without saying that for the quality to be consistent, the process must be robust. For injection molding, whenever there is an inconsistency in part quality, the robustness of the process is usually suspect because the process is reflected in the part quality. In general, based on how robust the process is and the required tolerance limits, we consider four possible resulting production process scenarios, as shown in Figure 7.2, which shows a representation of a run chart for a particular dimension.

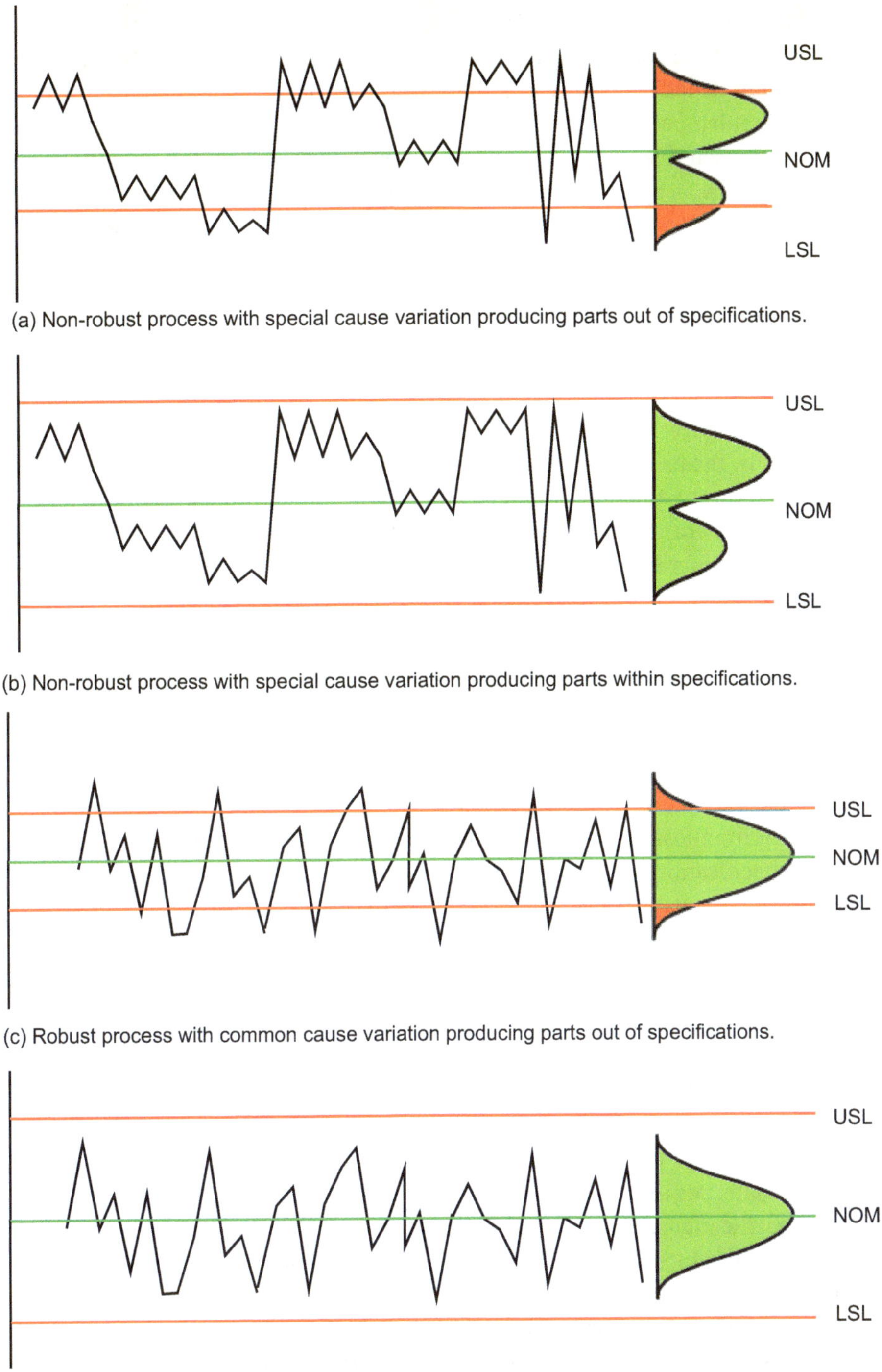

(a) Non-robust process with special cause variation producing parts out of specifications.

(b) Non-robust process with special cause variation producing parts within specifications.

(c) Robust process with common cause variation producing parts out of specifications.

(d) Robust process with common cause variation producing parts within specifications.

Figure 7.2 Types of processes based on variations and tolerances

Figure 7.2(a) shows a process that is not robust because of dimensional changes in the part. The first four data points are closer to the upper specification limit, but the next data point drops down towards the lower specification limit. There are some parts being molded out of specifications. Figure 7.2(b) shows the same process, however with increased tolerances; this process is producing parts within specifications. In both these cases, a special cause variation seems to be contributing to the inconsistent part quality. An attempt must be made to eliminate this variation even though in the second case, the parts are within the required specifications. Figures 7.2(c) and (d) both represent a robust process, because the quality distribution is normal. In Figure 7.2(c), the tolerance limits are such that this process although robust, produces parts out of specifications. In Figure 7.2(d), the tolerances are wider than in Figure 7.2(c) and, therefore, the same process now produces acceptable parts. Clearly, the process in Figure 7.2(d) is the most desirable process. Setting of the tolerance limits is done by the product engineer. In some cases, the engineer does have the flexibility to open up the tolerances based on the form, fit, and function of the part. If opening of the tolerances is not acceptable to the product engineer and if it is not possible with the current process setup to reduce the variation, alternative solutions such as selecting another plastic, changing the amount of filler, or using cavity pressure control must be considered. As mentioned earlier, an attempt must always be made to eliminate the special cause variation, even if the parts are within the required specifications. This makes part quality more predictable and the manufacturing process less vulnerable to molding defective parts. Another benefit of implementing the discipline of developing robust processes is a reduction in part inspection frequency and sample sizes. The goal of process development, using scientific principles and techniques, must be to establish a process that is consistent and well within the specification limits similar to the one shown in Figure 7.2(d). Simply molding parts within the specification limits does not necessarily mean that the process is robust and stable.

There are ways to improve the robustness of a process, reduce variation, and improve consistency. The aim of process development should be to develop a stable and robust process.

There are systematic steps that must be followed in order to achieve this goal. Unfortunately, these steps are often ignored because they are time consuming and can increase the number of mold trial iterations. Often overlooked is the amount of time, energy, and materials that are wasted and scrapped due to production of parts that do not meet specifications. The mold needs constant attention of a technician to adjust the process to produce the parts within specification and the parts molded out of specification must be scrapped and rerun. This is a waste of material, machine time, and human resources. The time lost cannot be regained. Often, parts that are out of specifications are shipped to the customer where they will be discovered. This causes a loss of reputation, return of parts for rework, time consuming corrective action investigations, and the added cost to scrap and remold the parts that cannot be reworked. With all the competition in a truly global market, efficiency in every area becomes an essential requirement. Consistency in cavity-to-cavity, shot-to-shot, and run-to-run are results of a robust process.

7.5 The Injection Molding Cycle

The basic phases of an injection molding cycle are shown in Figure 7.3. The cycle starts with the mold closing and the buildup of clamp tonnage. Once the mold is clamped under the tonnage necessary to keep it closed under injection pressure, the injection of the plastic starts. This is followed by the packing phase and then the holding phase. Once the cavity filling is complete, the part is cooled down to or below its ejection temperature, the mold is opened, and the part is ejected out of the mold. During the cooling phase of the cavity, the reciprocating screw rotates to pick up new material for the next shot. This is referred to as the screw recovery phase.

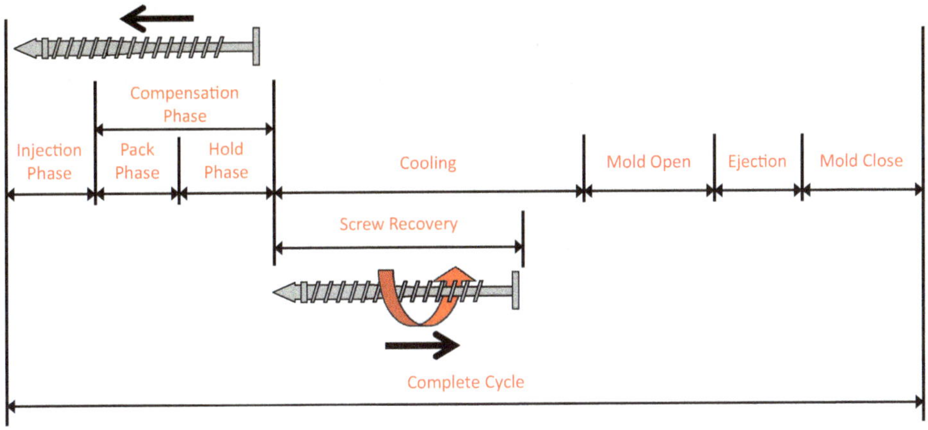

Figure 7.3 The complete injection molding cycle

7.6 The Injection, Pack, and Hold Phases of Injection Molding

The process of filling the cavity is divided into three phases: injection, pack, and hold. During the injection phase, the screw begins the injection process and starts to fill the cavity with molten plastic. Because of the fountain flow phenomenon, the plastic begins to cool down as soon as the plastic hits the walls of the mold and forms an outer frozen skin. As the injection continues, the plastic flows in between the frozen layers. At the end of the injection phase the cavity is completely filled with molten plastic. As the plastic begins to cool, the molecules start to get closer to each other resulting in shrinkage. At this point, if the addition of plastic were stopped, the parts would be under-packed and would exhibit a defect known as sink. Sink shows up as depressions on the surface of the part. To avoid the under-packing and to compensate for the shrinkage, the injection phase is followed by the packing phase, where the required amount of plastic is packed into the mold. The required amount corresponds to the volumetric shrinkage. At the end of the packing phase, the amount of plastic in the cavity must equal to the theoretical weight of the part. Any less plastic in the cavity will result in an under-packed part and any more will result in an over-packed part. Please note that in practice there is an acceptable weight range and not necessarily an exact weight that needs to be achieved. At the end of the packing phase the cavity is packed with the required amount of plastic, but the plastic is still molten and therefore under pressure. At this time the screw is still in the forward position and the gate is still not closed or frozen. If the pressure on the screw is released, because of the pressure inside the cavity, the plastic will flow back out of the gate resulting again in sink in the part. Therefore, a certain amount of pressure must continue to be applied in order to hold the melt inside the cavity. This is called the holding phase, where the plastic is not entering the cavity nor is the plastic coming back out of the cavity. More than the required pressure will pack the part out and less than the required pressure will cause the plastic to come back out. The holding pressure must be applied until the gate is frozen. The time required for the gate to freeze is referred to as holding time. In each of the three phases, pressure, velocity, and time are important. However, the injection phase is dominated by velocity of fill, the packing phase by the pressure of pack, and the holding phase by the time of hold.

One can further understand the pack and hold phases by visualizing a balloon being filled with air. The air inside the balloon is under pressure and if a knot is not tied and the balloon is released, all the air will come back out of the balloon. Tying a knot is analogous to adding the holding phase. Refer to Figure 7.4, which describes the three phases.

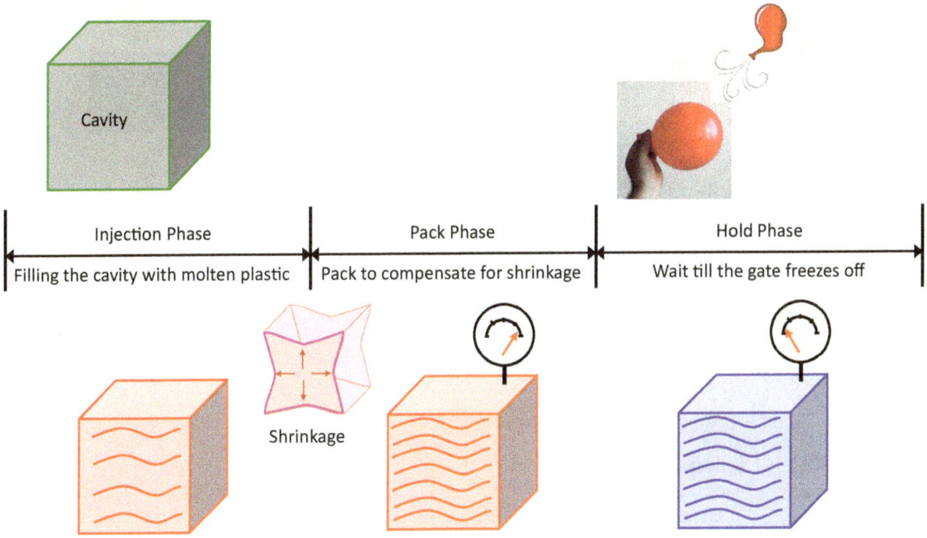

Figure 7.4 The injection, pack, and hold phases

Figure 7.5 shows the effect of the presence and absence of the holding phase. For this particular part, because of the part design, the packing pressure required to pack in the required amount plastic inside the cavity was close to 1400 psi. This number is considered to be on the higher side but is not uncommon. The cavity pressure was therefore high. The processor was packing the part out but did not give any time to hold the plastic in there until the gate froze off. This resulted in sink near the gate. However, when the holding phase was added with a holding pressure of 500 psi, the plastic flow stopped and the gate froze off, so the part was molded free of sink.

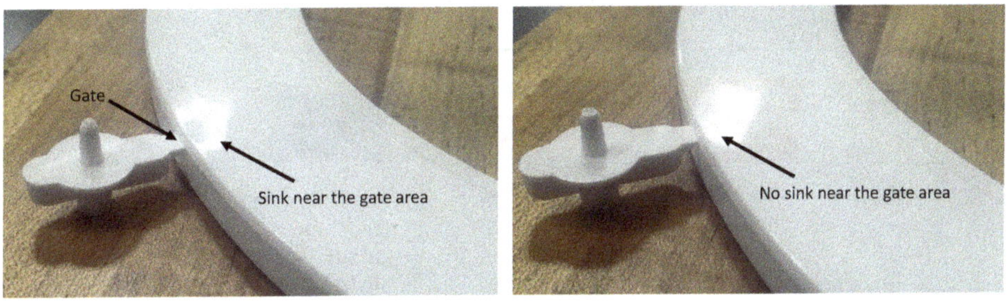

Figure 7.5 Effect of lack of the holding phase

Figure 7.6 shows what happens when each of the phases is not optimized correctly. Sink is the most common defect when there is lack of pressure and time in the pack and hold phases.

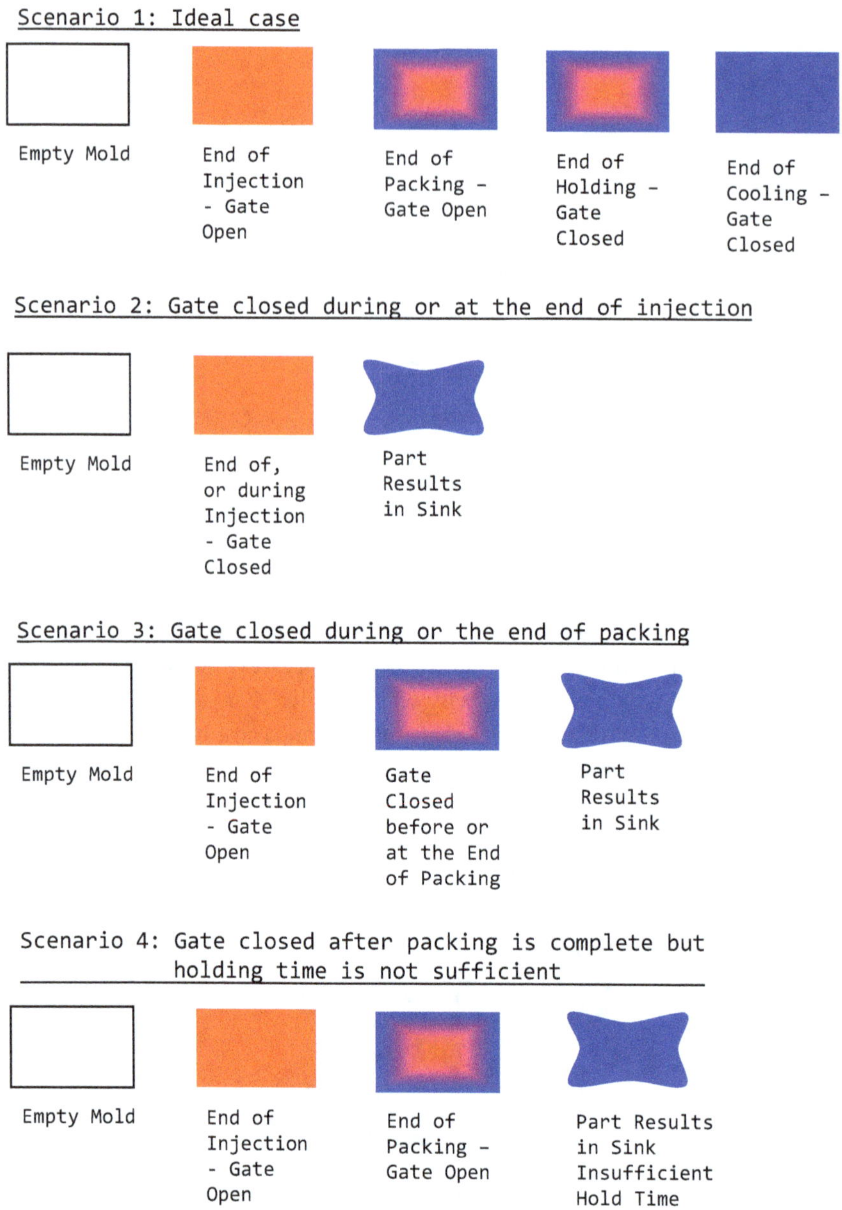

Figure 7.6 Effect of pack and hold on part filling

The pack and hold phases are used for compensating the shrinkage during the filling process and are collectively called the compensation phase. It is difficult to determine the switchover point between pack and hold without employing some advanced methods, such as using cavity pressure measurement equipment. For the most part, pack and hold are taken as one phase and are usually called the holding phase. In many cases, the processes are set with profiles on the holding pressure to yield acceptable parts via trial and error. In doing so, the processor has actually set up a process that has a pack and a hold phase but is not aware of it or has not been able to distinguish the two phases from each other. Under-packing results in defects such as sinks and internal voids in the parts. Such parts usually exhibit some amount of post-molding shrinkage as well. Over-packed parts can have molded-in stresses that usually get relieved after the parts are ejected, resulting in defects such as warpage or premature failure.

A systematic procedure to develop a process with a clear differentiation between pack and hold is described in the next chapter.

7.7 Speed and Pressure

The molding machines have settings for speeds and pressures, for example injection speed and injection pressure. Processors often find these confusing and interchangeable since in some processes increasing both will have the same effect of filling the parts further. There is a clear difference between the two that must be understood.

To illustrate the phenomenon at hand, consider a gaylord filled with a light packing material such as Styrofoam that needs to be pushed across the hall at 100 meters per minute. The task requires one person, so the force required to move the gaylord is one person. Now, if the gaylord is filled with plastic pellets, and the speed still needs to be 100 meters per minute, the force required to perform this action may now require four people, see Figure 7.7. The force has changed, but the speed has remained the same. In other words, as the resistance to movement changes, the required force changes. Viscosity is the resistance to flow. Therefore, in molding, the weight of the gaylord is analogous to the viscosity of the plastic. The speed of the gaylord is analogous to the injection speed. The force is analogous to the required injection force needed to move the screw at the set speed. Refer to Figure 7.8. To move the screw at 200 mm/sec, higher pressures are required for a high viscosity material such as a 40% glass filled PC compare to an unfilled LDPE.

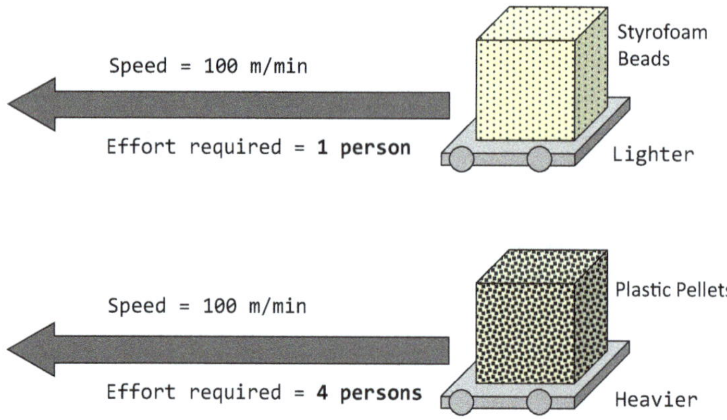

Figure 7.7 Effort required to move a gaylord of Styrofoam and plastic at the same speed

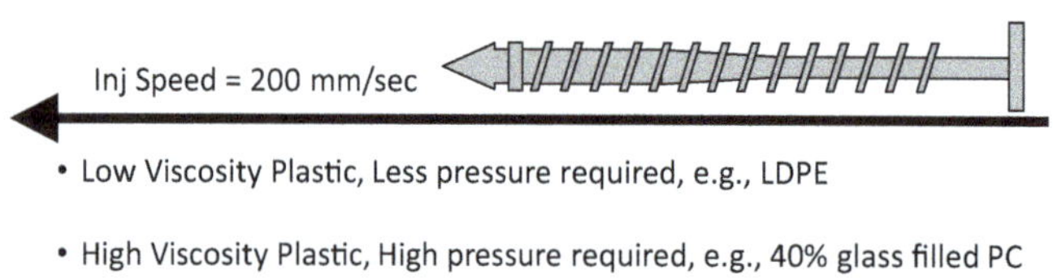

Figure 7.8 Effort to inject a low viscosity and a high viscosity plastic

In the same example just discussed, if there were only three people instead of four to move the gaylord of plastic pellets at 1000 mm per minute, the required speed cannot be reached. In injection molding this means that if the screw needs 2500 psi of injection pressure to move at 5 in/s, but the machine is only capable of producing 2200 psi of the 2500 required, then the screw will never reach the required speed. The limited supply of pressure will limit the injection speed to the speed attainable at 2200 psi. When this situation exists, the process is said to be pressure limited. This is discussed further in the Chapters 8–11, covering process development.

7.8 The 11 + 2 Plastic Injection Molding Machine Input Parameters

There are 11 primary plastic molding parameters on the molding machine that have a direct influence on the quality of the molded product. These do not include parameters such as mold open or close speeds and so on. In most molding processes, the compensation phase, also known as the 2nd stage, consists of only one subphase known as the holding phase. In some cases, molders will differentiate the compensation phase into two phases known as the pack phase and the hold phase. In such cases, two more parameters are added to the 11 primary parameters, so the total number of parameters is 13. These 13 parameters are shown in Figure 7.9. Optimization of each of these parameters will be discussed in the following chapters.

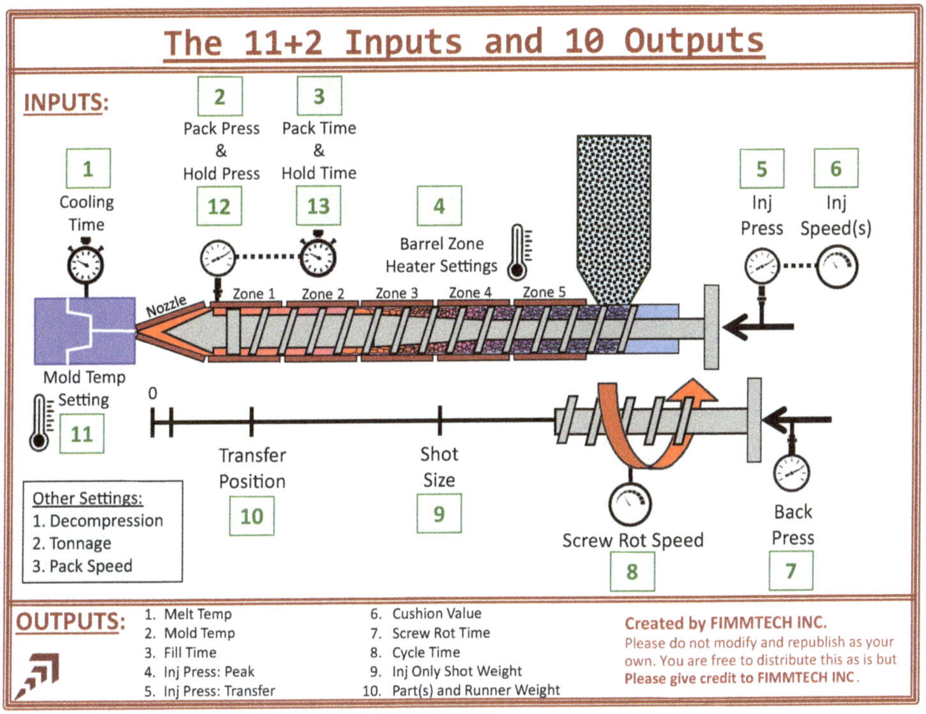

Figure 7.9 The 11 + 2 plastic injection molding parameters

Following are the 13 molding parameters that are machine inputs:

7.8.1 Barrel Temperatures

The plastic needs to be in a molten state to be injected into the mold. The heaters around the barrel heat the barrel, which conducts the heat to the plastic on the inside to aid the melting. Depending on the length of the barrel, there can be several heater bands, and the temperature for each of these will have to be individually set depending on the material that needs to be processed. The result of these settings is reflected in the melt temperature. The shear from the screw rotation also contributes to the melt temperature. The recommendations for melt temperatures are available from the material manufacturer. The tip of the barrel has a nozzle that also much be heated. The back zone is the feed zone, which is kept cold in order to avoid the melting of the plastic in this section, which would prevent the plastic from being moved forward by the turning screw. Figure 7.10 shows an example of the nozzle and the barrel heater bands.

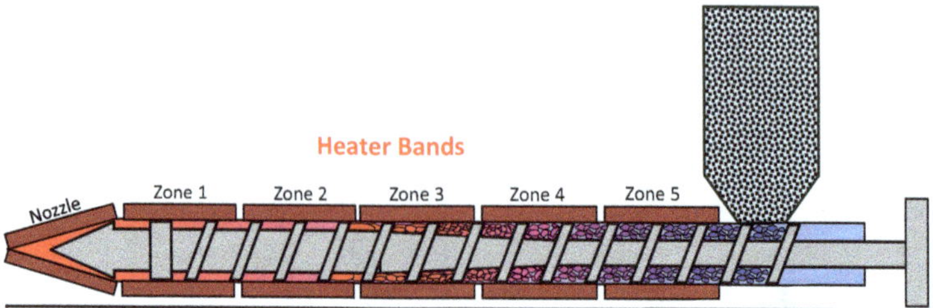

Nozzle	Zone 1	Zone 2	Zone 3	Zone 4	Zone 5	Feed
245 C	240 C	240 C	235 C	235 C	230 C	90 C

Figure 7.10 Nozzle and barrel heaters

7.8.2 Mold Temperatures

The word coolant is always used for the fluid that is used to control the temperature of the molds since the melt must be cooled down in the molds. For molds that require temperatures above room temperature the mold temperature controllers have a heating unit that will heat the coolant to the required temperature and for those that need the mold temperatures below the room temperature chillers are used. Some production facilities will cool the molds with water directly from a water tower heat exchanger usually installed outside the building. The issue with such a system is that the water temperature cannot be maintained at the desired level and will fluctuate based on the ambient temperature and the seasons. The water temperature in the winter can be much cooler than that in the summer. Needless to say, these systems are not used in locations where the ambient temperatures can go below the freezing point of water. Tower water systems are used in molding facilities to cool molds that mold products where the dimensions and other quality requirements are not very tightly controlled. Good examples of these are commodity items such as forks, spoons, buckets, and pails. Every molding facility has some sort of the heat exchange system since the water exiting the mold will have picked up the heat and will have to be cooled back down to be recirculated in the mold. Images of a chiller, a mold temperature control unit, and a water cooling tower are shown in Figure 7.11.

Figure 7.11 Chiller, mold temperature controller, and water tower (Courtsey: Conair)

In some molds where it is difficult to get a coolant line into a section of the mold, cartridge heaters are used. Refer to Figure 7.12. The part that was to be molded was a cylinder made from a polyetherimide (PEI) material. The cylinder had an internal diameter of about 12 mm with a length of 40 mm. The mold temperature requirements for PEI range around 175 °C (350 °F). It was impossible to get a coolant line into a core that was about 12 mm in diameter. A cartridge heater was used to get the temperature of the core up to the desired mold temperature. When using cartridge heaters it is important that there still be a coolant line that will act as a heat sink to remove the added heat from the injected plastic melt.

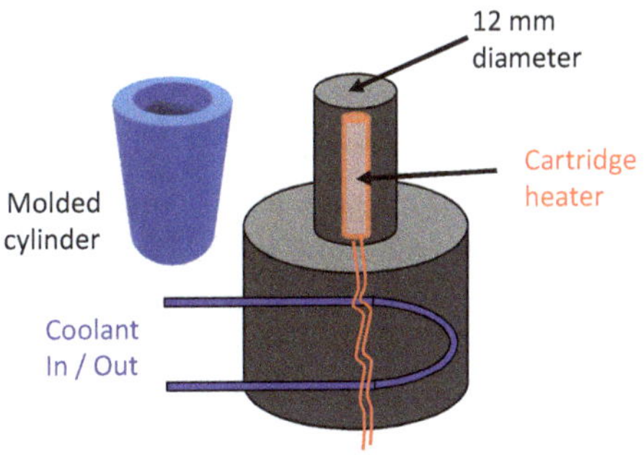

Figure 7.12 Cartridge heater

7.8.3 Shot Size

Refer to Figure 7.13, which depicts the shot size, which is also known as the dosage or screw back position. The zero position of the screw is when the screw is all the way in the forward position. The screw rotates and moves back to pick up the molten material. The set distance that the screw moves back to is called the shot size. This is measured in linear distance or in terms of volume. The shot size can be calculated based on the total shot weight and the melt density of the plastic. However, because melt densities are temperature dependent and it is difficult to estimate the exact temperature, shot size calculations are usually estimations. Appendix H provides the calculation for the calculation of the shot size and Appendix I provides the calculation of the melt density. Note that these would be approximate numbers.

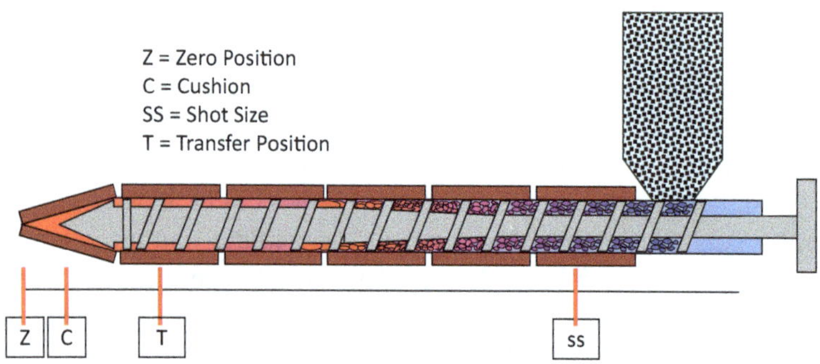

Figure 7.13 Screw positions in a molding cycle

7.8.4 Transfer Position

The point of transfer from the injection phase to the compensation phase is called the transfer point. Refer to Figure 7.13. This can be set in units of linear distance or volume similar to the shot size. This transfer can also be done via time or hydraulic pressure, or by external means such as a cavity pressure sensor. In theory, at the transfer position the mold should be filled exactly 100% with the molten plastic. In practice, however, the mold is filled less than the 100%. This is explained in detail in Section 7.1.

7.8.5 Injection Speed (Velocity)

The terms injection speed and injection velocity are used interchangeably. This is the speed at which the screw moves to inject the molten plastic into the mold. The injection speeds must be set fast enough so that the plastic remains in the molten state until the mold is completely filled to the optimum requirement. Once the mold is filled with the required amount of molten plastic, the injection of the plastic in the injection phase is stopped, though continues at a slower pace in the compensation phase. The technique of in-mold rheology, also known as an in-mold viscosity curve, is used to optimize the injection speed of the machine.

It is always preferred to have the screw move at one injection speed from the start of injection to the transfer position. Visualize driving a car at 60 km/h and the next traffic signal turns red. One must start slowing down the car before coming to the stop position. Applying the brakes at the traffic light will slide the car into the traffic intersection because of the momentum. When the car slows down there is a change in velocity from fast to slow. Similarly, if high injection speeds are used then the screw must be slowed down for better consistency from shot to shot. This is called velocity profiling and will be discussed in Chapter 10. For velocity profiling, injection speed switchover positions will get added.

7.8.6 Injection Pressure

This is the pressure that is applied to the molding screw to maintain the set injection speed. If the viscosity of the plastic increases, the effort or the pressure required to maintain the set speed will also increase. It is important that the injection speed is kept constant during injection. The machine should always have a surplus of pressure available to compensate for material viscosity variations and achieve a robust process. The actual injection pressure should always be less than the set injection pressure. A pressure drop study is used to optimize the injection pressure.

7.8.7 Pack Pressure

This is the first subphase of the compensation phase. Once the mold is completely filled in the injection phase, to compensate for the shrinkage that will take place, additional plastic is packed in. This pressure is known as the pack pressure. Pack pressure is one of the most important parameters that determines the shrinkage and hence the dimensions of the part. Data from a Design of Experiments (DOE) is used to optimize the pack pressure.

7.8.8 Pack Time

This is the time for which the pack pressure is applied. Data from the gate seal study is used to optimize the pack time.

7.8.9 Hold Pressure

This is the second phase of the compensation phase. This is the pressure that is applied to plastic to ensure that the excess plastic does not go into the cavity nor does the molten plastic come back out from the cavity. In theory, during this stage, there is no movement of the plastic until the gate is frozen. Data from the gate seal study is used to optimize the hold pressure.

7.8.10 Hold Time

This is the time for which the hold pressure is applied. Data from the gate seal study is used to optimize the hold time.

7.8.11 Screw Rotation Speed

This is the speed of the screw rotation during the picking up of the plastic during the shot buildup. Crystalline materials need higher screw speeds than amorphous materials. It is mainly based on measuring the melt temperature and examining the melt for burning and unmelted particles.

The setting of the screw speed is another critical parameter, especially when it comes to crystalline materials. The shear from the rotating screw contributes a significant amount of energy to help melt the plastic. During the machine cycle, when the material is being processed through the barrel, the heat from the heater bands alone is not sufficient to melt the crystallites. The additional heat required is supplied by the shear energy created by the rotation of the screw. High screw speeds generate high shear and help in melting the crystallites. Recommended screw speeds are provided by the material suppliers. However, the values are typically given in revolutions per minute (rpm). This is very misleading because, depending on the diameter of the screw, a given value of rpm will result in high shear for large diameter screws while the same value will result in low shear for a smaller diameter screw. The screw speeds should therefore be specified in distance per time units, such as inches per second or meters per second. An example is given in Figure 7.14, which shows that with the same rpm of 30 on two screws of 10 and 20 mm diameters the shear applied on the plastic will be different. The 20 mm screw will have a higher shear.

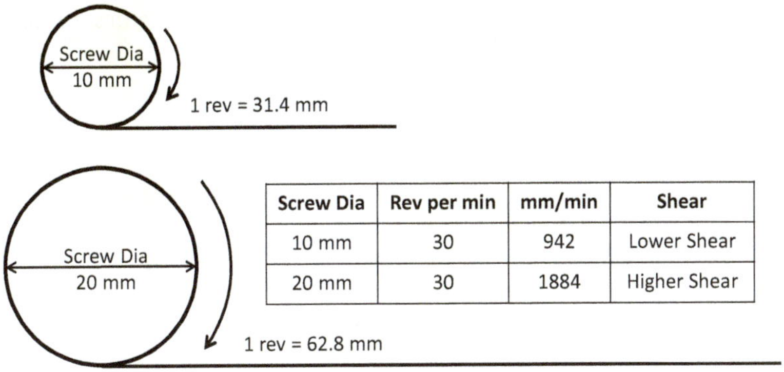

Screw Dia	Rev per min	mm/min	Shear
10 mm	30	942	Lower Shear
20 mm	30	1884	Higher Shear

Figure 7.14 Speeds in rpm versus linear units

Highly filled plastics or plastics filled with long glass fibers typically require slower screw speeds to avoid a breakdown of the fibers. In these cases the screw design plays an important role in processing. Screw design also plays an important role in determining the melt homogeneity for different types of materials, depending on their shear sensitivity. In some cases, the cycle times can get longer because the available screw on the molding machine cannot generate enough shear heat to melt the plastic fast enough. Again, specialty screws can make a difference here.

7.8.12 Back Pressure

During the screw recovery process the screw is turning to auger the plastic to the front of the screw tip. The back of the screw is the feed section of the screw and the plastic here is unmelted and in the form of pellets. The plastic melts as it travels further because of the heat from the heater band and the shear of the rotating screw. Visualize that the molding process has already started, and plastic has already been injected into the mold. As the screw turns and the plastic is collected in the front of the screw, it pushes the screw back towards the shot size position. During this stage there is no mechanism in the machine that is pulling the screw back. Back pressure is the pressure applied to the back of the screw during the screw recovery process. There are three main reasons to apply the back pressure.

Imagine going down an incline with a load as shown in Figure 7.15. The person handling the load puts a force in the direction opposite to travel in order to be able to control the speed of going down the incline and accurately reaching the stop sign. The force applied is just enough for control of the load and not high enough to either slow down, stop, or pull the load back. If no such control was present then the load would simply go crashing down the incline in an uncontrollable manner. Similarly, as the screw is turning to pick up the plastic a small amount of pressure is applied to the back of the screw to get the screw back in a controlled manner to consistently reach the shot size position of the screw. This is shown in Figure 7.16. This is the first reason to apply back pressure.

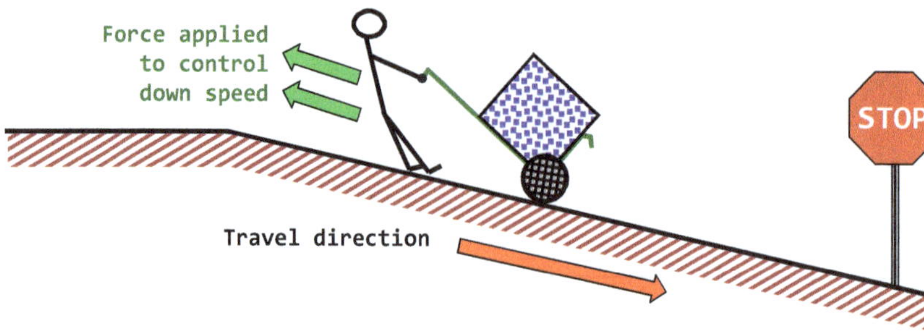

Figure 7.15 Applying a force in the direction opposite to the direction of travel for better control

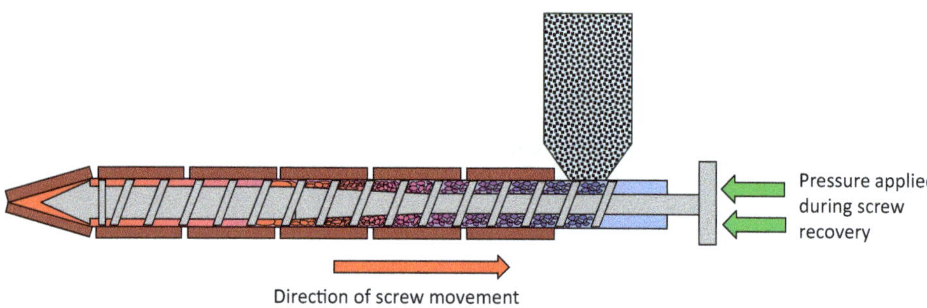

Figure 7.16 Pressure applied at the back of the screw during shot buildup

The molding screw consists of three zones. They are the feed zone, the compression zone, also called the transition zone, and the metering zone. Refer to Figure 7.17, which shows the plastic in the compression zone. The job of the compression zone is to compress the soft plastic pellets formed at the end of the feed zone and squeeze out all the air and any other volatiles that are present between the pellets. Some volatiles may be formed during the melting process inside the barrel. The application of the back pressure helps in this compression action and pushes the volatiles out. This is the reason why processors can often get rid of splay on a molded part by increasing the back pressure. Figure 7.18 shows the presence of splay with no back pressure being applied. This is the second reason to apply back pressure.

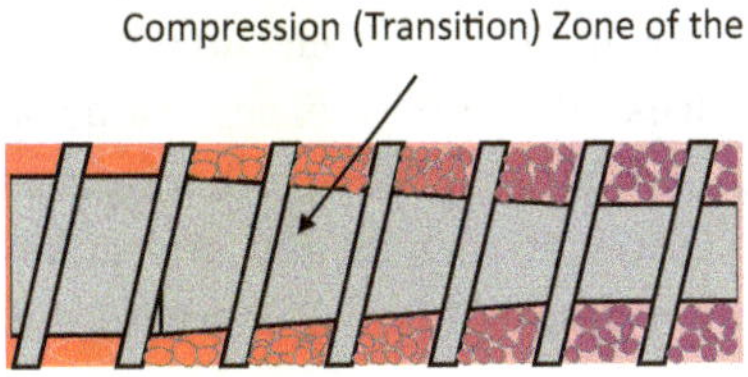

Figure 7.17 Plastic in the compression (transition) zone of the screw

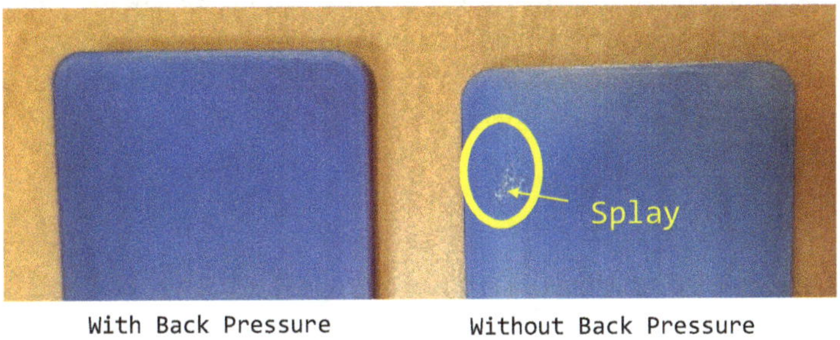

Figure 7.18 Influence of back pressure on splay

The goal of process development should be to achieve a robust process that will provide cavity-to-cavity, shot-to-shot, and run-to-run consistency. Consistency here would mean dimensional as well as cosmetic consistency. Dimensional consistency translates to shrinkage consistency, which translates to having the same amount of molecules and therefore to a consistent weight of plastic in every shot. Weight equals volume multiplied by density. Since the shot size is fixed and assuming the cushion is not varying, the volume of the melt being injected into the mold is therefore fixed. This means that to get a repeatable shot weight, we need to have a repeatable melt density. The back pressure does the job of compressing this melt and enough back pressure must be applied to achieve consistent melt density. This is the third reason to apply the back pressure. Refer to Figure 7.19, which explains this concept.

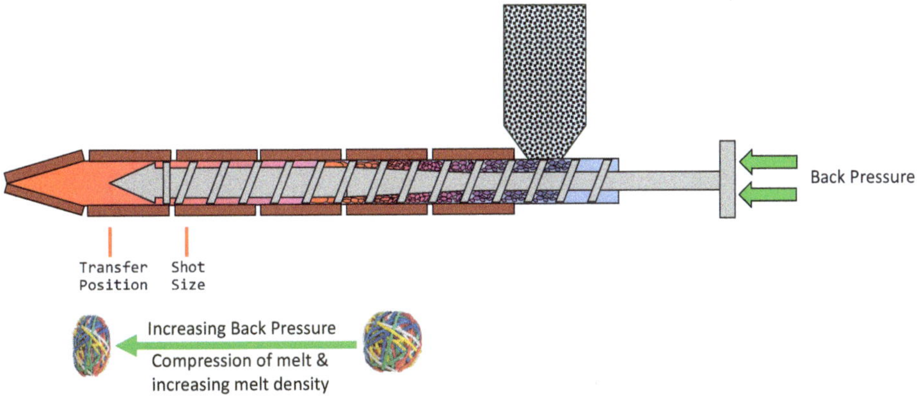

Figure 7.19 Compression of melt with increasing back pressure

Too much back pressure can cause excessive shear in the material, which can degrade the material. Some fillers, such as glass fibers, can break down with excessive back pressure resulting in a loss of properties in the final product. Excessive back pressures and screw rotation speeds can cause degradation of certain low molecular weight additives, which can result in the formation of gases. This will increase the number of defects in the final parts and over time cause the vents in the mold to build up with residue and become clogged. Vents on the parting lines are easy to clean; however, the internal vent components, such as vent pins, are difficult to clean without pulling the mold from the machine and dismantling it. Excessive back pressure and screw rotation speeds can also result in increased wear of barrel and screw components.

7.8.13 Cooling Time

There are two types of cooling times. The first one is the set cooling time and the other is the actual cooling time. Set cooling time is the one that is set on the machine settings' screen. For example, it can be input to 5 seconds. Visualize fountain flow. The plastic starts to cool off as soon as the plastic touches the mold and therefore the actual cooling time is equal to the addition of the injection time (fill time), set pack time, set hold time, and set cooling time. In Figure 7.20 the set cooling time is 5.0 seconds and the actual cooling time is 11.0 seconds. Actual cooling time is the total time the mold is kept closed to let the plastic reach the ejection temperature and/or to achieve the required dimensions. Cooling times are mainly optimized via the technique of Design of Experiments.

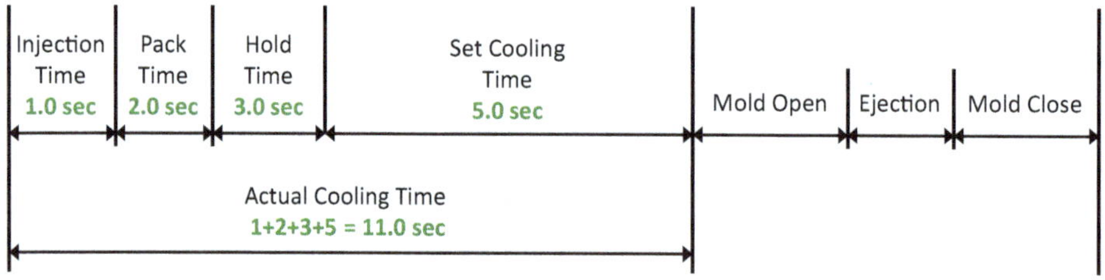

Figure 7.20 Set and actual cooling time

7.9 Additional Settings

There are three other parameters that do not and should not get adjusted as often as the 13 just mentioned. They also do not have a substantial effect on the part quality.

7.9.1 Decompression or Suck Back Position

Because the shot is built up in the presence of back pressure, the plastic that is in front of the screw is compressed and is under pressure. At this time there is an injected shot in the mold. As soon as the mold opens and the molded sprue moves back with the mold, the melt pressure is released. This can cause the plastic to flow out of the machine nozzle. For this reason, at the end of the screw rotation, the screw is pulled back in a linear movement without any rotation and the pressure is released to decompress the melt. This distance through which the screw is pulled back is called decompression or suck back. Refer to Figure 7.21, which shows the effect of decompression. These distances are usually very small (less than 1 cm) and can affect the cosmetics of the part if not set correctly. In case of hot runner molds, lack of decompression can cause some drooling of the melt from the hot tips as the mold opens. In case of valve gated systems, there will be no drool but with no decompression, the pressurized melt can cause a very small amount of drool as the valve gates open and cause some blemish or cosmetic defect around the gate.

The decompression distance is added after the screw reaches the shot size. Therefore, the start position of injection is the shot size value plus the decompression value.

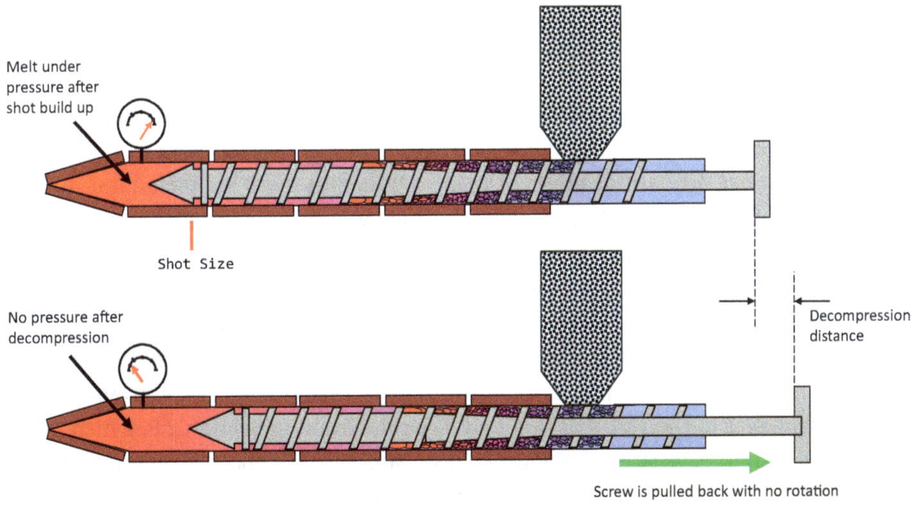

Figure 7.21 Decompression of melt

7.9.2 Pack Speed

This is the speed at which the screw moves linearly during the pack phase. Some machines may not have this setting. The pack phase is a pressure-controlled phase but the screw must have some velocity to move the plastic into the cavity. The setting of this speed will be discussed in the process development section.

7.9.3 Tonnage

Tonnage is the force used to keep the mold halves closed during the injection cycle. It should be optimized during the process development process but then should not be changed during production to alter the quality of the part. This will be discussed in the process development section.

If one looks at the screens of the modern day molding machines, one can get overwhelmed with all the different settings for the parameters and options. There are several process monitoring screens that are available to track part quality and the cycle efficiency. The machine manufacturers sometimes tend to provide a number of redundant and unnecessary controls. For example, some machines will have an option of 10 second stage pressures for the pack and hold phase. In most systematically developed molding processes, two pressures should be more than sufficient. In some cases, to compensate for an unoptimized part and/or mold design, the processor may need to add another profile. The optimization of the settings of the 13 parameters listed in this chapter should be the focus of process development.

7.10 Process Outputs

The molding parameters discussed in the previous section are the machine inputs. These are what a processor will set on the molding machine control panel. There are 11 outputs that are a result of these 13 main inputs and the 3 supplemental inputs. The quality of the part, the shot-to-shot consistency and the run-to-run consistency are a function of the outputs and not the inputs. If a mold is moved from one machine to another machine, matching the outputs will provide the parts with identical quality. Machine screw diameter sizes, screw L/D ratios, tonnage, heater band wattages, the condition of the molding machine, and several other factors can be different between two machines. Setting identical inputs on both machines will not provide identical outputs. Refer to the PVT diagram and its explanation in Section 2.8, which mentions it is the cavity pressure, the volume of plastic, and the cooling rate that decide the part quality. These should be duplicated from machine to machine, which translates to matching machine outputs. Following is the description of the 11 main process related machine outputs:

1. Actual melt temperature
2. Actual mold temperature
3. Injection (fill) time
4. Peak injection pressure
5. Pressure at transfer
6. Cushion value
7. Screw rotation time
8. Cycle time
9. Injection only shot weight (including runners)
10. Full part(s) weight
11. Runner weights

The following descriptions will help the reader understand more about them and the importance of each. Each of these outputs must be clearly documented on the process sheet and through the process development process.

7.10.1 Actual Melt Temperature

The heat from the heater bands and shear from the turning screw provide the energy to melt the plastic. These are the inputs. The output will be the quality of the melt and the actual melt temperature. The melt temperature should be measured with the help of an insertion pyrometer as shown in Figure 7.22. The measurement procedure is described in Chapter 9.

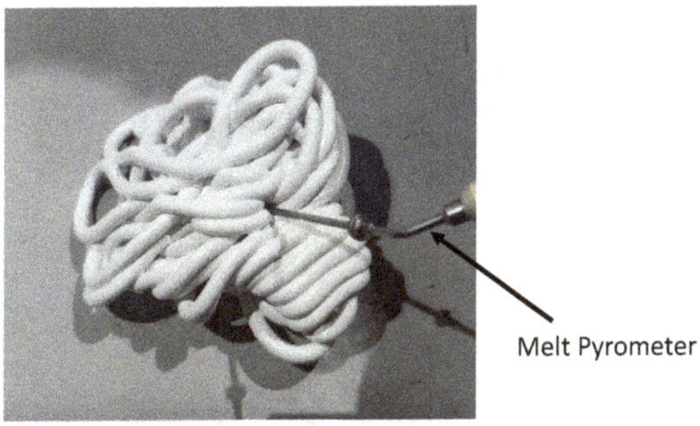

Figure 7.22
Measuring the melt temperature with the help of an insertion pyrometer

7.10.2 Actual Mold Temperature

Depending on the mold temperature requirement of the plastic to be molded, either a chiller, a mold temperature controller, or tower water is used to control the mold temperature. There are settings on these units that will set the exit temperature of the water from the units. However, in the case of a tower water, there is no such setting and the actual water temperature leaving the tower will vary based on the environmental conditions. The water leaves these units and circulates in the mold. There are heat losses from the unit to the mold and then again in the mold. At the same time during the injection molding cycle, heat is constantly being added. The system should be in a state of thermal equilibrium. The actual temperature of the cavities is therefore not necessarily the same as what is set on the units. The actual temperature is what plays a role in the thermal transfer rate or cooling of the plastic and should therefore be recorded as an output. The temperature can be measured with the help of a surface pyrometer or a thermal imaging camera. Refer to Figure 7.23 and Figure 7.24, which show these two techniques and the results. The most accurate technique is by the use of the surface pyrometer. However, when there are several cavities such as a 16- or 32-cavity mold, measuring the temperature of each one before the mold steel cools down can be an issue. A thermal imaging camera works best for such large molds. The disadvantages of thermal imaging cameras is that the reflections on the mold surfaces can sometime mislead the measurements.

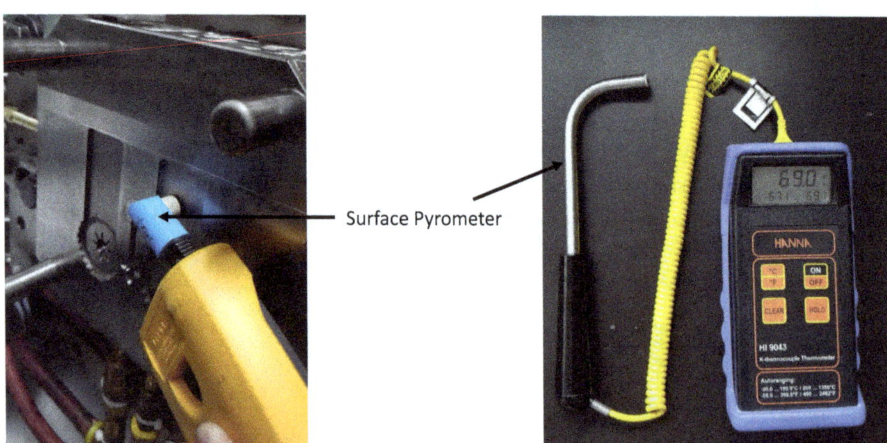

Figure 7.23 Measuring mold temperatures with a surface temperatures

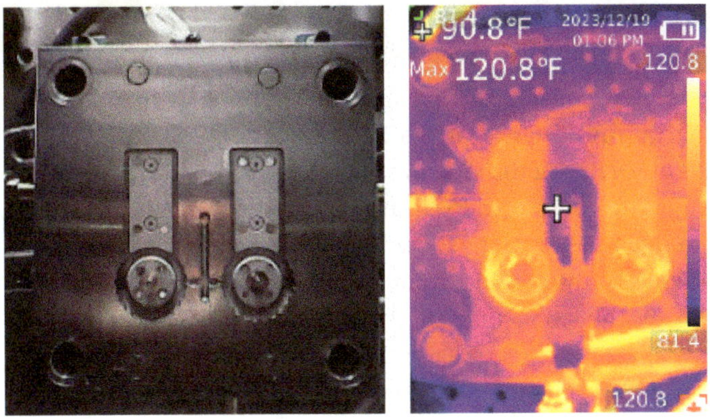

Figure 7.24 Measuring mold temperatures with the help of a thermal imaging camera

A thermal imaging camera also gives a quick indication if one or more of the water lines are not turned on or have a different temperature. See Figure 7.25, which illustrates how a thermal image shows a blocked water line with a temperature of the line of 74.1 °F; when the water was opened the temperature rose to the required temperature of 80 °F.

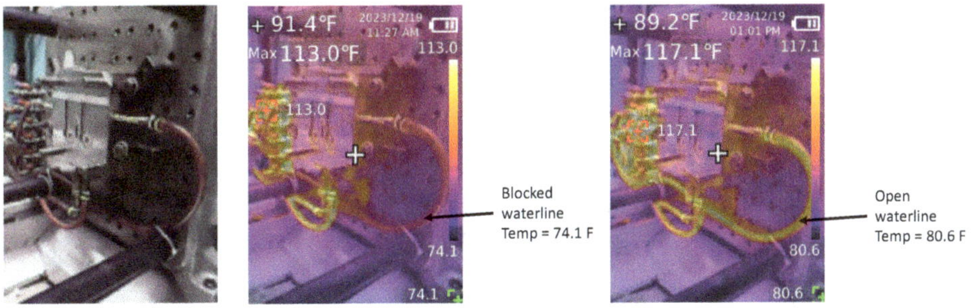

Mold image showing waterlines

Thermal image showing a blocked waterline

Thermal image showing a blocked waterline

Figure 7.25 Thermal Imaging helps to show a waterline that did not have water flow

7.10.3 Injection (Fill) Time

After injection, the screw rotates to pick up the plastic, reaches the shot size, and then moves linearly back equal to the distance of the decompression setting. The final screw position is therefore at a value of the shot size plus the decompression position. The screw will start the injection cycle from here and this position is called the injection start position. Refer to Figure 7.26, which shows the injection start position of 204 mm. The actual time the screw takes to move from the injection start position to the transfer position is called the injection time or the fill time. This is measured in seconds and is output on the screen of the molding machine. The screw moves from 204 mm to 30 mm in the injection phase. If the screw took 0.92 seconds to travel this distance, then the fill time will be 0.92 seconds. The value is output on the screen of the molding machine.

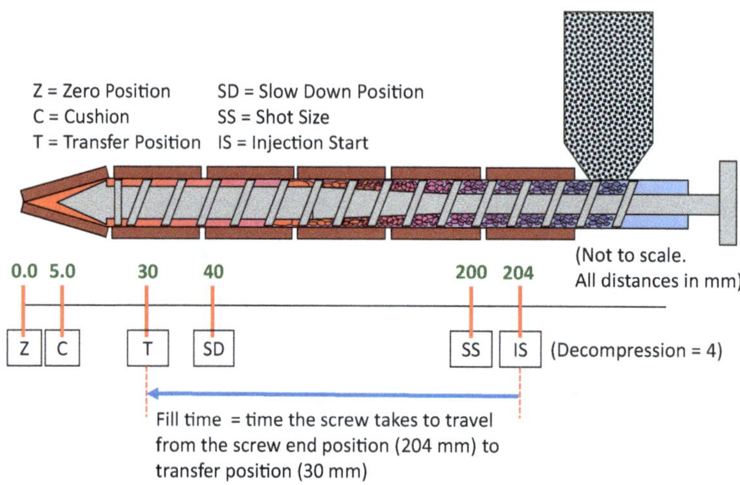

Figure 7.26 Screw positions and the fill time

7.10.4 Peak Injection Pressure

This is the maximum injection pressure value seen during the entire injection phase of the molding cycle. The set injection pressure should always be more than this peak pressure setting. The peak injection pressure and set injection pressure are shown in Figure 7.27.

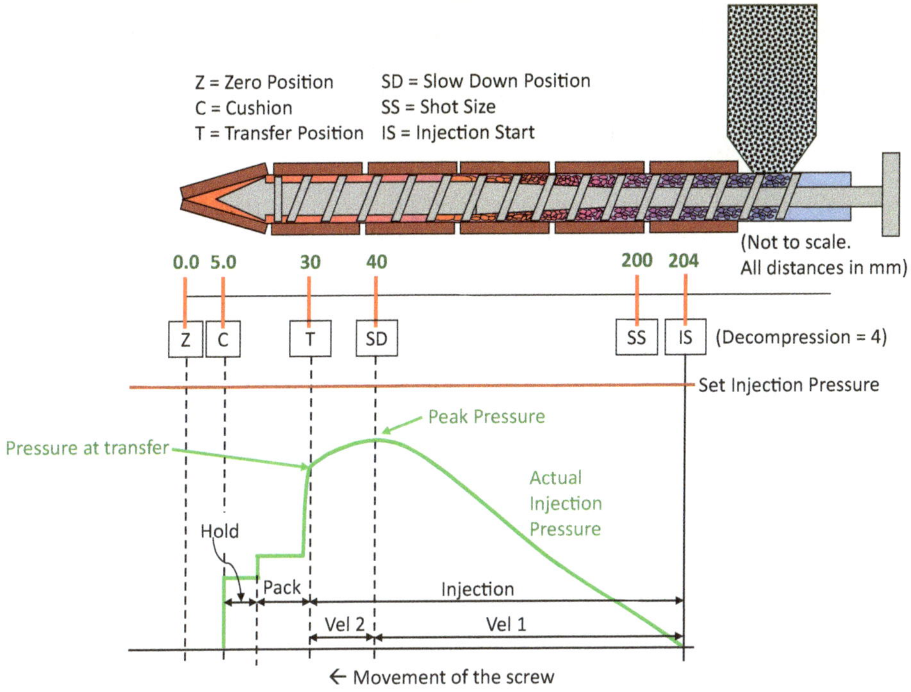

Figure 7.27 Peak pressure and pressure at transfer

7.10.5 Pressure at Transfer

In some processes the injection velocities are profiled and injection screw is slowed down before it reaches the transfer position. The pressure required to move the screw at the lower speeds will be lower than the peak pressure. This is called the pressure at transfer and is shown in Figure 7.27.

7.10.6 Cushion Value

Refer to Figure 7.28, which shows two chambers, A and B, connected with a connector. In the first scenario there is oil in chamber B, in the connector, and some in chamber A. When a force is applied on the piston, the force is applied on the oil in chamber A. This gets transferred to chamber B and the pressure in chamber B increases. In the second scenario there is no oil in chamber A and therefore no amount of pressure applied on the piston can be transferred to the oil in chamber B. It is the presence of the oil in chamber A that helps apply pressure in chamber B. The pack and hold phase are responsible for adding the required amount of plastic to compensate for the shrinkage that is happening inside the mold. Therefore, at all times, similar to the oil in chamber A, there must exist some molten plastic in front of the screw tip that will help in apply pressure and pack the cavity with the required amount of plastic. This plastic is called the cushion and the position of the screw at the end of the pack and hold is called the cushion value as shown in Figure 7.29. The cushion value in the example is 5 mm.

Screw values should be kept between 5 and 10 mm for screw sizes below about 60 mm. For larger screws higher numbers should be used but care must be taken that cushion values are not large. Refer to Figure 7.30. Larger cushions will give more inconsistency from shot to shot since the plastic melt is compressible and some of the pressure is lost in compressing the plastic in the large cushion.

The front heater zones of the barrel are usually set to higher temperatures. If the plastic is close to being degraded or is getting degraded because of higher residence times, it will be the plastic in the cushion that will be degrading or close to being degraded. Visualize fountain flow. If this is a cold runner mold then this degraded plastic will form the skin of the runner and can be discarded. A large cushion may result in the degraded plastic ending up in the part causing failures in quality.

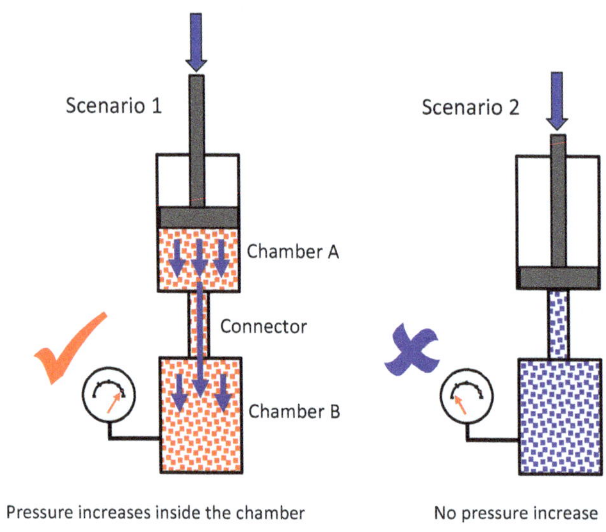

Figure 7.28 Conditions for transfer of pressure in the lower chamber

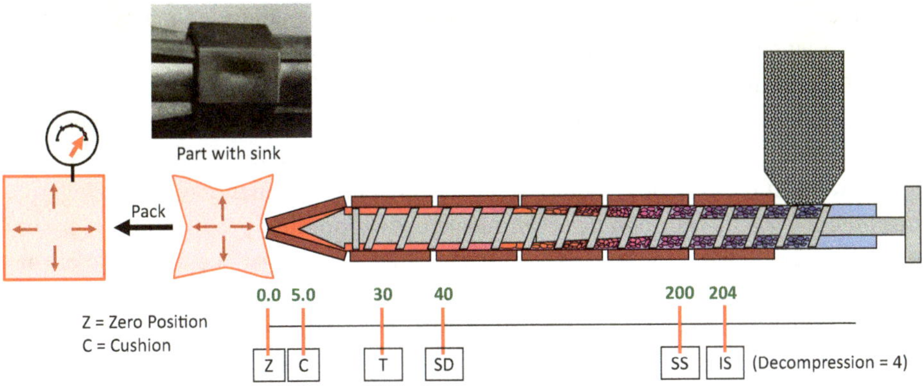

Figure 7.29 Cushion value and the importance of cushion

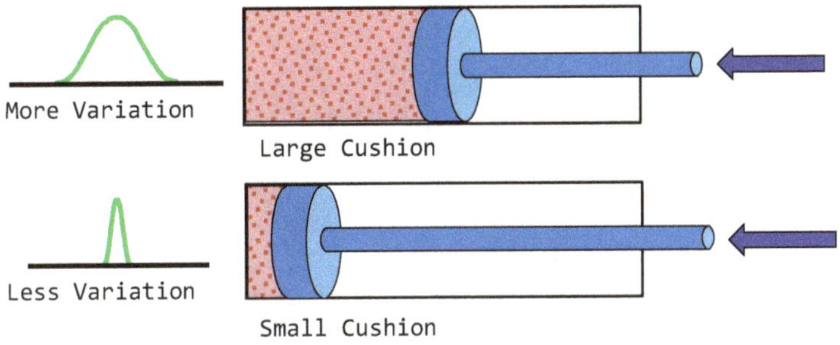

Figure 7.30 Importance of small cushion values

7.10.7 Screw Rotation Time

Once the pack and hold phases are complete, the screw starts to rotate and auger the plastic to for the next shot. It turns until it reaches the shot size position and then moves linearly back to decompress the melt. The total time it takes to reach the injection start position is called the screw rotation time. In some geographical locations it is also called screw refilling time.

7.10.8 Cycle Time

The time it takes to complete one complete molding cycle is the cycle time. The cycle ends when the ejection is complete and the ejectors go back to their zero position. The cycle time clock starts as soon the mold starts to close and ends at the end of the ejection sequence. In reality, one could start and end the timer at the start or end of any function of the molding cycle. For example, injection start of one cycle to injection start of the next cycle as shown in Figure 7.31.

For a molding business to be successful it must be able to mold as many parts as possible in a given shift or day. Therefore, cycle times must be optimized and the least possible cycle time should be selected. A simple calculation will show that if the cycle time is reduced from 30 seconds to 29 seconds, the saving of just one second per cycle equals 4 hours in a 5 day period or 208 hours in a year with a 5 day work week. This can also mean tremendous savings in terms of energy and manpower. Freeing up a machine for 208 hours could possibly mean running another production run or simply being able to mold more of the same parts in a given time. The downside of running fast cycle times is usually seen when processes have not been optimized. In such cases it is easy to mold parts out of specifications even when the inputs on the machine have not changed and with fast cycle times scrap or rework parts can accumulate very quickly. These processes are called processes that run at the edge of a cliff and cannot cushion in any natural variations. This book deals mainly with process development and will discuss the concept of robustness of processes in detail.

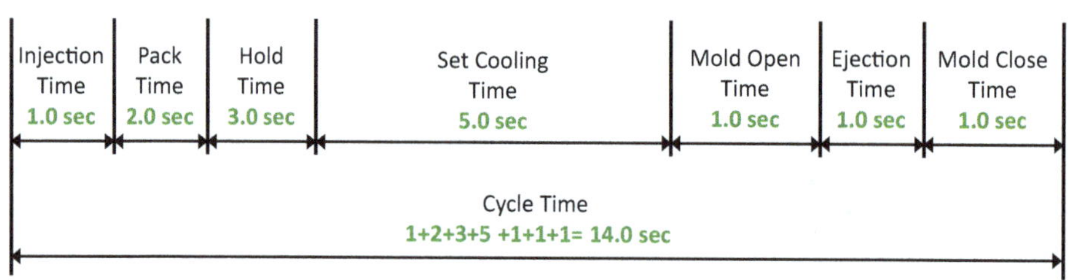

Figure 7.31 Cycle time

7.10.9 Injection Only Shot Weight (Including Runners)

In the injection phase the cavity is completely filled with molten plastic, and as the plastic starts to shrink the pack phase kicks in to add plastic to compensate for the shrinkage. This point of transfer is a critical point that plays a big role in part quality and therefore the weight of the plastic injected in the injection phase should be recorded. The weight of parts and the runner is recorded as the injection only shot weight. Figure 7.32 shows such as example.

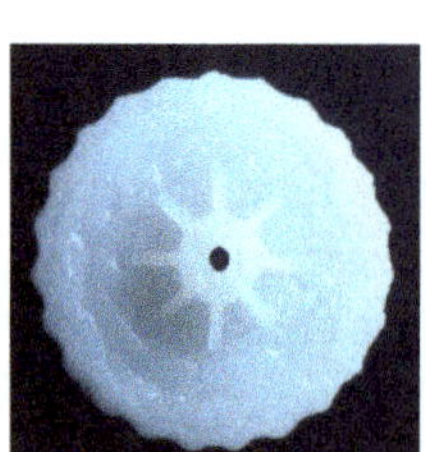

Figure 7.32 Part molded with the injection phase only

7.10.10 Full Part(s) Weight

Once the part quality is achieved the weight of the individual cavities should be recorded. Refer to Figure 7.33. The final part weight will help in determining the amount of plastic that will be needed for future production runs. This can also be a quality control requirement in some molded products. For example, automotive companies try to keep the weight of the vehicle down to improve energy usage efficiencies and this trickles down to the weight of the molded components.

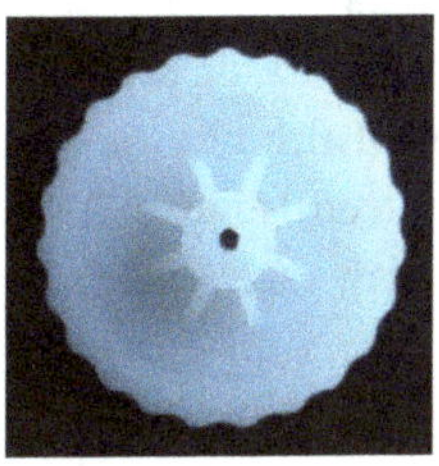

Fully packed out part

Figure 7.33
Part molded with the injection, pack, and hold phases

7.10.11 Runner Weights

Similar to the part weights, the runner weight should also be recorded. Refer to Figure 7.34. The fully packed out part and runner weights also help to determine the amount of plastic that is required for future production runs. The weight or the parts and runners from one shot is called the shot weight. The runner weight is also required in case the molder wants to regrind the runners and reuse the plastic. The weight ratio of the runner to the parts will determine the maximum percentage regrind that is available but it will be up to the product designer to decide how much regrind can be allowed in the parts.

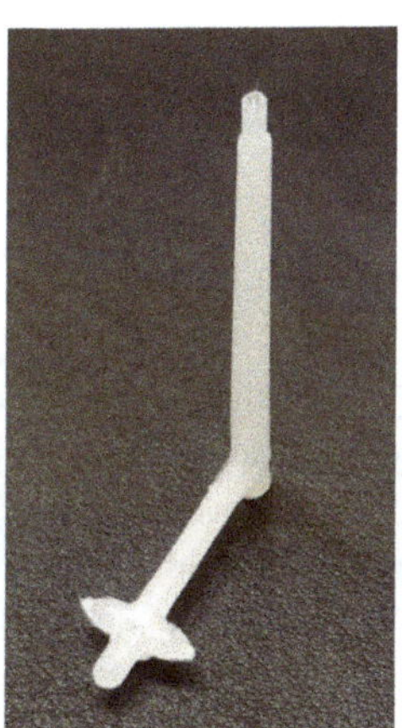

Figure 7.34
Runner

7.11 Decoupled Molding℠

The term Decoupled Molding℠ was coined by Rod Groleau of RJG Inc. During the injection phase, the cavity is filled with molten plastic. The volume of melt injected during the injection phase must be equal to the collective volume of the cavities and the runners. The melt density is lower compared to solid densities. Once this volume of plastic is present in the mold, the injection phase is followed by the packing phase, which is then followed by the holding phase. Before such an understanding was developed by the molding industry, molders would either over-pack or under-pack the mold with molten plastic and would apply the pack and hold phase based on some previous experience. To stress the importance of the separation of the phases and educate the molders, the term Decoupled Molding℠ was introduced (Figure 7.35). Since each phase has a specific reason, they must be decoupled from each other and controlled separately.

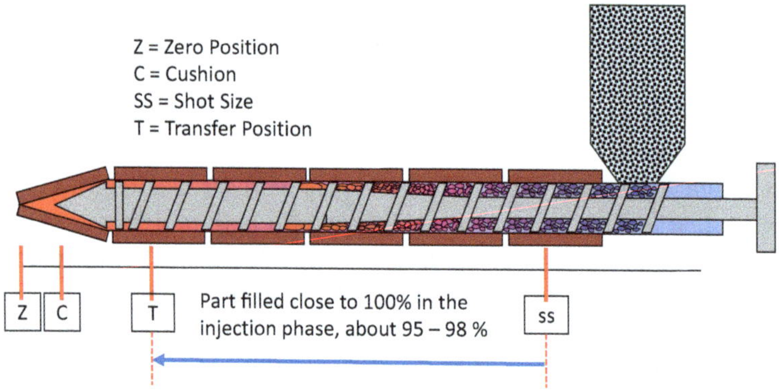

Figure 7.35 Decoupled Molding

Decoupled Molding provides the best control over the molding process and offers the most consistency. It is common practice to fill the cavity to 95–98% capacity during injection before entering the pack and hold phases because, in Decoupled Molding, the goal is to make sure the mold is not filled more than 100% during the injection phase. Targeting 95–98% provides a small margin of safety to make sure that the part is not excessively filled. The second reason is to compensate for the momentum of the screw during injection, which allows the melt to slow down before the start of the next phase. The melt is also compressible and therefore the cavity can get over-packed to more than a 100% fill. The packing phase has a slower velocity because we are only packing at the rate of shrinkage that is taking place in the cavity. A slowdown is therefore essential to insure the mold is not over-packed and or blown open and flashed. Slowing down the injection speeds just before hitting the transfer position also helps in the consistency. This is described in the next chapter.

7.12 References

[1] Kulkarni, S. M., Hart, David, *SPE ANTEC Tech Papers* (2003) p. 736

[2] Mertes, S., Carlson, C., Bozzelli, J., Groleau, M., *SPE ANTEC Tech Papers* (1997)

Suggested Reading

Osswald, T. A., Turng, L., Gramann, P. J., *Injection Molding Handbook* (2007) Hanser, Munich

Beaumont, J. P., *Runner and Gating Design Handbook* (2007) Hanser, Munich

Beaumont, J. P., Nagel, R., Sherman, R., *Successful Injection Molding* (2002) Hanser, Munich

Rosato, D. V., Rosato, D. V., *Injection Molding Handbook* (2000) CBS, New Delhi, India

Kulkarni, S. M., *SPE ANTEC Tech Papers* (2003) p. 736

Cogswell, F., *Polymer Melt Rheology* (1981) John Wiley, USA

Dealy, J., Wissbun, K., *Melt Rheology and its Role in Plastic Processing Theory and Applications* (1990) Van Nostrand Reinhold

8 Introduction to Process Development

8.1 Introduction

Developing a molding process is often not done by following a systematic procedure, which leads to a process that is not robust. With increasing pressure to deliver products to markets, the engineering staff often skip steps that produce parts of substandard quality, increase scrap rates, increase workloads of personnel, and reduce the overall efficiencies of the molding process. A process that needs to be adjusted even once a day is not a robust, repeatable, and reproducible process. The next three chapters will provide an insight to the development of the most efficient process for a given mold and will also provide techniques to improve the process even further with recommendations for mold design, part design, and material selection.

8.2 Process Development Fundamentals

Consider that one needs to make a road trip from Denver, Colorado to Las Vegas, Nevada. An online program estimates this to be a distance of about 1200 km (750 miles). The next step is to check that the car making the trip is capable of covering a distance of 1200 km. The engine, the tires, the windshield wipers, and other items would need to be checked and verified by a mechanic that they would perform during a 1200 km road trip. Once it is determined that the car is capable, the driver can punch in the address in a GPS system and drive in the required direction. There were two parts to this activity: First, the equipment (the car) needed to be verified for its capability, and second, was to steer the equipment in the required direction (a southwest direction) toward Las Vegas. Refer to Figure 8.1.

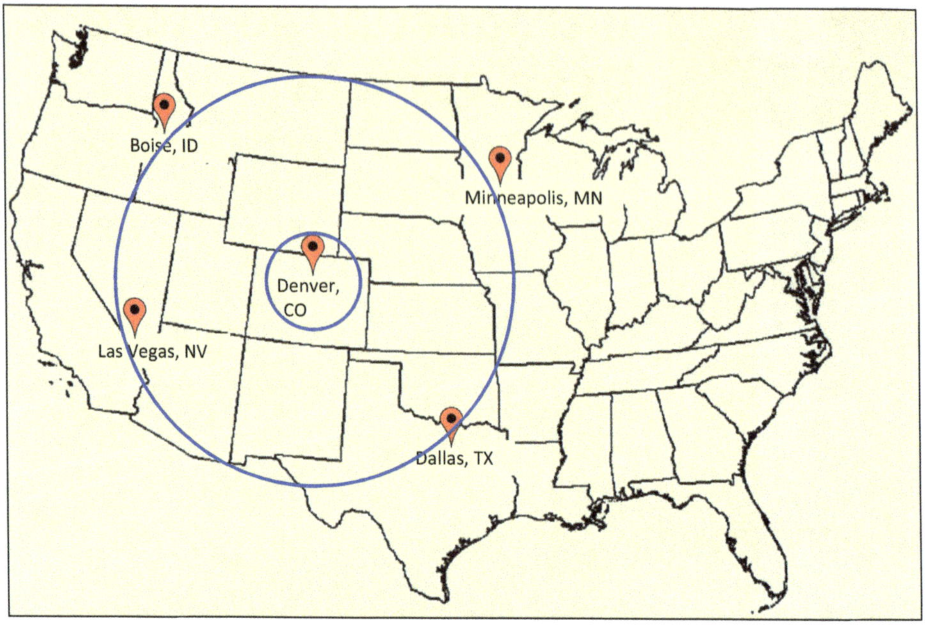

Figure 8.1 Cities equidistant from Denver, Colorado

Once the car is ready, if for whatever reason the driver decided not to drive to Las Vegas, he can confidently make a journey in a direction of his choice up to a distance of 1200 km. Looking at the map, this means he could either drive southeast to Dallas, Texas or northeast to Minneapolis, Minnesota or northwest to Boise, Idaho. All these cities are approximately 1200 km from Denver.

The injection molding process should also be developed in a similar two-part activity. First, the mold should be capable of consistently molding cosmetically acceptable parts with a wide cosmetic process window. Once the preferred cosmetic process is determined, the dimensions should be measured. If the dimensions are not acceptable, then changes to the process within the cosmetic process window should be examined. A process setting on the edge of a cosmetic window should never be selected. In such cases, and in cases where the dimensions do not fall within the cosmetic window, the most robust cosmetic process must be selected and mold cavity dimensional changes and/or part specifications and tolerances must be made. Going back to the driving example above it would be best to drive long distance with a cruise control setting. But this is only possible when the lane widths are wide enough to compensate for any varying road and driving conditions. It is impossible to drive on the edge of a cliff under cruise control. Figure 8.2 shows the recommended flow chart for a two-stage process for developing a robust molding process.

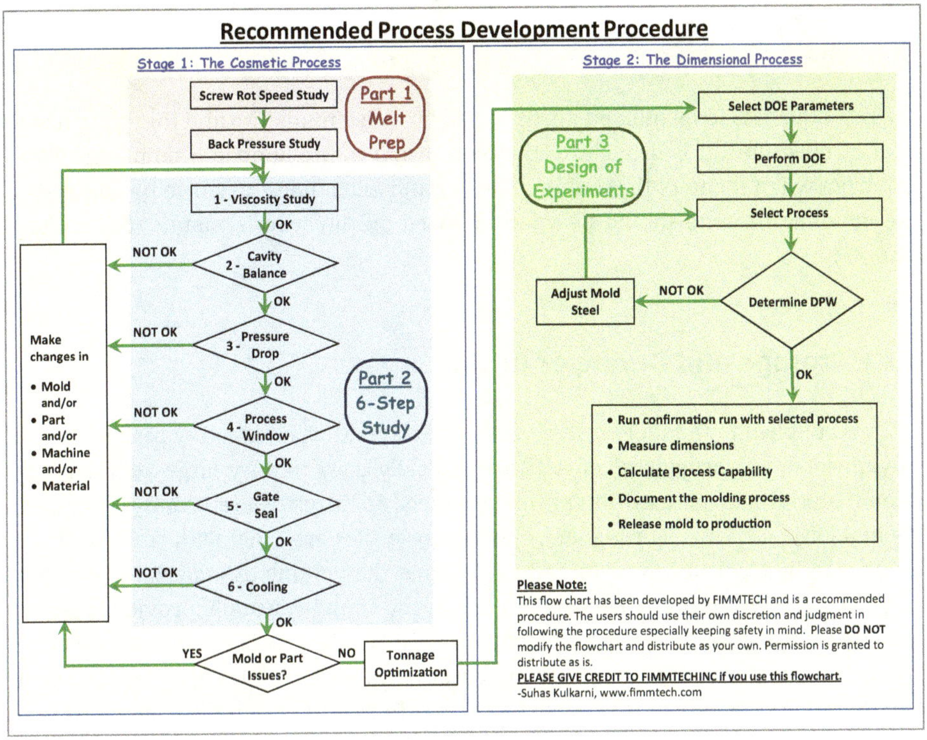

Figure 8.2 Mold qualification and DOE flowchart

In Stage 1, the process parameters are explored, and the most robust areas of the process are selected as the preferred process setting. Dimensions are not necessarily examined, although by doing so, a good idea of how close the dimensions are to the specifications can be obtained. Stage 1 consists of two parts, the melt preparation and the 6-step study, also known as the scientific molding study. In Stage 2, the technique of Design of Experiments (DOE) is used to determine the dimensions at the preferred process and to determine the relationship between the process parameters and the dimensions.

This chapter will discuss the melt preparation and the following two chapters will discuss the 6-step study and design of experiments. Before the discussing of melt preparation there are a few important topics that need discussion. A summary of the preparation to start the process development activity is also presented.

8.3 Premolding Setup

The resin that has to be molded needs to be of consistent quality and therefore has to be controlled until it is brought to the feed throat of the hopper. This means every step from when it enters the molding facility and exits that it needs to be controlled. The molding machine should be selected based on the criteria mentioned in Chapter 6.

8.3.1 Storage and Drying of Resin

Resin is supplied to the molder in a variety of ways. Most commodity resins, such as polyethylene, polypropylene, or ABS are typically used in very large volumes. The same material may be used in various products. Such resins are usually shipped to the molder in gaylords. For higher volumes, large silos are filled with resin, which is brought in by trucks or even by rail cars. When the volume of material is not very large, such as for most engineering materials, the supplier usually provides resin in prepackaged bags of 25 kg each. In some materials, a loss of properties can occur if the resin is not stored correctly. Some polyurethanes are examples of resins with shelf life limitations. Therefore, it is important to use this resin within the prescribed time specified by the manufacturer. In resins with shelf life limitations it is typically not the polymer that degrades, but rather the additives that can be time sensitive.

Other resins, such as acrylic, must not be stored such that they are exposed to sunlight because this may result in a loss of color or transparency. The resin must also be stored in a dry area to avoid unnecessary absorption of moisture. Some resin manufacturers prepackage hygroscopic resins in their dry/processing state and ship them to the customer in sealed bags. Such resins do not need drying and can be used directly at the molding press. However, unused resin that was stored back on the shelf must be redried before being used again. Clear identification of the resin and keeping a log of all the materials in use are considered good manufacturing practices. Many times, a sample resin is brought in, and some of it is used and the rest stored for later reuse. A date, project ID, and material ID must be recorded on the sample bag to ensure the history of this material can be clearly identified.

8.3.2 Plastic Drying

The topic of plastic drying has been discussed in detail in Chapter 4. Hygroscopic resins must be dried in order to avoid loss of properties, surface blemishes, and/or internal defects. The loss of properties in some resins is a result of the hydrolytic degradation of the resin. In other resins, the presence of the moisture in the resin results in surface defects, such as splay or internal defects like voids. Processing difficulties, such as gassing and foaming of the melt, as in the case of nylons, may be encountered. For these reasons, resins must be dried before processing. The specific amount of drying time and the correct drying temperature is dependent on the base polymer resin. The times and temperatures for each resin is dependent on the strength of the chemical bond that the water forms with the polymer. For example, PBT needs to dry at 80 °C (180 °F) for 4 hours, whereas polycarbonates need to dry at 121 °C (250 °F) for about 3 hours.

Additives are often added to plastic to enhance their properties for specific applications and/or to reduce their cost. Fillers such as glass and minerals are the most widely used additives by volume. Other additives including plasticizers, lubricants, flame retardants, heat stabilizers, colorants, blowing agents, and biocides are added to polymers in small percentages. These are usually low molecular weight compounds and/or oligomers. If the drying time is extended beyond the manufacturers' recommended limits, it is possible to degrade these additives and/or to cause them to leave the resin. For example, with PBTs and nylons it was found that if the suggested drying times were exceeded, there was a danger of producing defective or out-of-specification parts. Drying PBT resin for more than 8 to 10 hours produced brittle parts and loss in surface appearance. Parts that normally appeared black with high gloss now appeared dull gray. Figure 8.3 shows that PBT exhibited a drastic drop in the impact strength after it was dried for more than 12 hours [1].

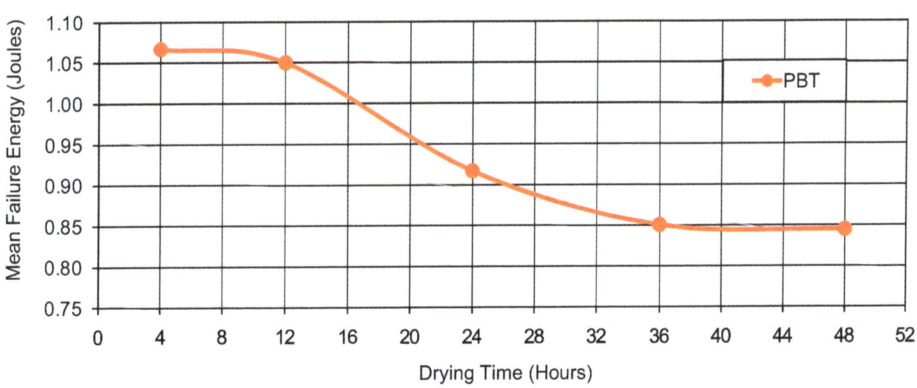

Figure 8.3 Effect of drying time on the impact strength of PBT [1]

Drying nylons for more than about 12 hours steadily increased the viscosity of the plastic, causing constant adjustments to the molding process. The viscosity graph for nylon at different drying temperatures is shown in Figure 8.4 [1]. In this graph, nylon-4 represents nylon that has been dried for 4 hours and nylon-48 nylon that has been dried for 48 hours. Figure 8.5 shows the parts molded in the 'injection only' phase at various drying times. Parts molded within the 'injection only' phase are those that are molded with both pack and hold time and pack and hold pressure set to a zero value. Therefore, the pack and hold phase are absent. With the same set process at the molding machine, as the drying time increases, the amount of part filling is less. This is a result of the increase in viscosity, which reduces the flow of plastic. Overexposure or 'over drying' must be prevented.

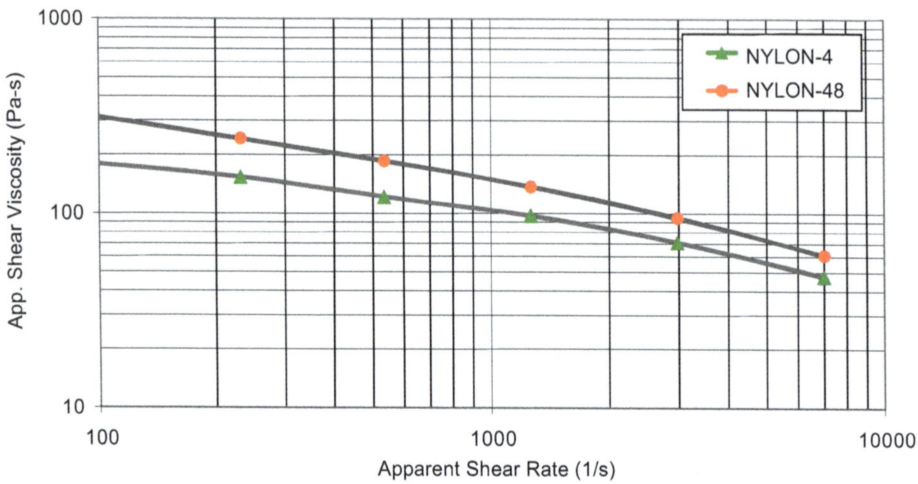

Figure 8.4 Viscosity of nylon dried for 4 and 48 h [1]

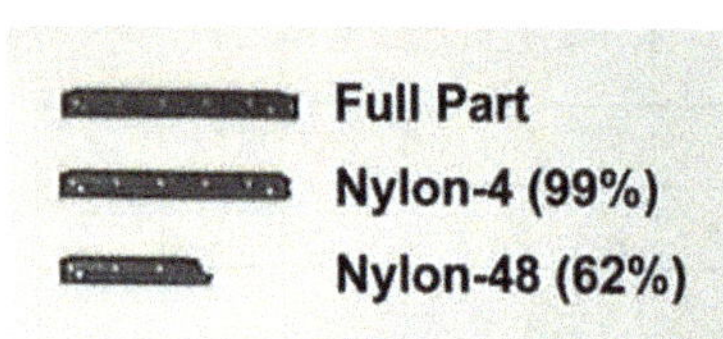

Figure 8.5 Parts molded using the 'injection phase only' with nylon dried for 4 hours and 48 hours

The most efficient ways to prevent over drying are as follows:

- If the molding operation is not ready to be started after the material is dry, turn down the dryer temperature to about 80 °F, but keep the dryer running. Supplying the hopper with room temperature dry air will keep the moisture out and will not have detrimental effects on the resin.
- Size the hopper dryer such that the residence time of the plastic does not exceed the manufacturers' minimum/maximum recommended drying times for the plastic. When the available hopper sizes are larger than required, it is recommended to retrofit the dryers with an adjustable level sensor and adjust the sensor to load only the required amount of material.
- If the machine is going to be down for a long time (several shifts or days) after the drying has taken place, shut the hopper dryer off as soon as possible.

Precautions: Drying may have a cumulative effect in some resins, such as PBT. With PBT, the drying process permanently removes the low molecular weight additives, which are not recovered when the drying is stopped, and the resin is returned to the shelf for the next run. With nylons, moisture can be reabsorbed back into the resin. The effects of over drying for PBT and nylon that are mentioned here must not be considered typical results for other materials. The effects of the over drying of other resins have yet to be studied because each resin and its additives are unique. Customized experimentation and testing will best suggest the type of drying control that is required. However, there is no real need to conduct this type of expensive and time-consuming experimentation as long as the material is not exposed to excessive drying times.

8.3.3 Machine Selection

Machine selection was discussed in detail in Chapter 6. Following are the main requirements for the given mold:

- The mold must physically fit in the machine with the molding area always within the machine platens.
- The required mold tonnage should be less than the machine tonnage.
- The percentage of the shot size should preferentially be between 20 and 80%.
- The residence time of the plastic in the machine should be less than the maximum allowable residence time of the plastic.

8.4 Importance of Adding Charge Delay Time

The three phases of mold fill are injection, pack, and hold. The pack and hold phases are pressure and time controlled. As soon as these timers time out, the screw begins to rotate under back pressure to pick up the next shot. At the end of the injection phase, the gate is not frozen and the mold is about 95–98% full, or in other words, it is 2–5% unfilled. If the timers for the pack and hold phases are set to zero, then as soon as the injection phase is done the screw begins to rotate. The applied back pressure now applies pressure on the melt, and since the cavity is still not full and the gate not frozen, the part begins to fill to 100%. The screw recovers to the shot size only after the cavity is full. If one is trying to find the decoupled point of the 95–98% fill, then it becomes an impossible task to set the transfer position. However, if one adds a screw rotate timer, more commonly known as a charge delay or screw recovery timer, then the screw will wait for the specified time and then begin to rotate. The idle time will give a chance for the gate to freeze off. Once frozen, the plastic cannot enter the cavity, therefore, one can truly determine the percentage fill in the injection phase. Refer to Figure 8.6.

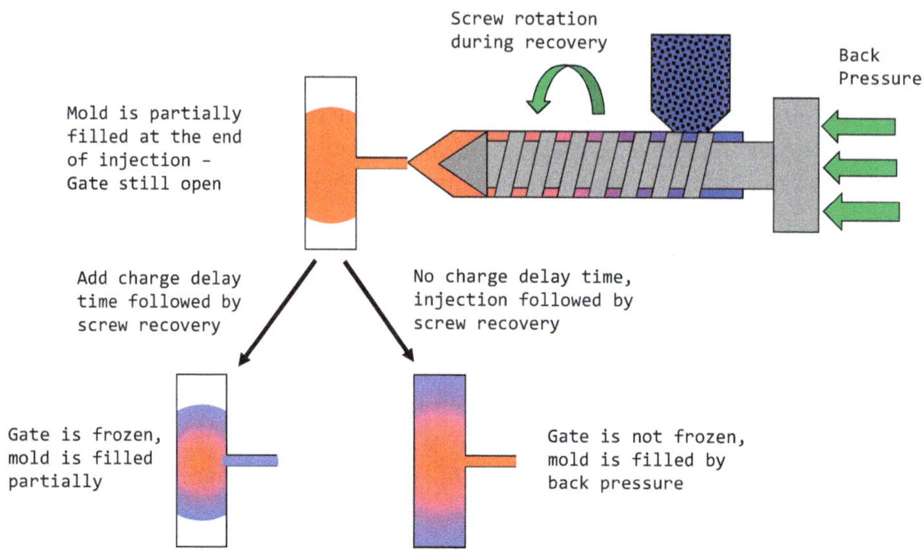

Figure 8.6 Importance of charge delay time to determine percentage fill during the injection phase

To understand this further, a quick experiment can be performed at the molding machine. The results of such an experiment are shown in Figure 8.7. Choose a simple one- or two-cavity mold that has a history of running consistently with no issues. Care must be taken during this experiment not to over-pack the mold and any changes made must be in small steps. When running a low-viscosity material, such as a nylon or LDPE, set the pack and hold times to zero, and set the back pressure to very low values of around 3 to 4 plastic MPa, or about 300 to 400 plastic psi. Please note these are plastic pressures and not hydraulic pressures. Increase the cooling time by an amount equal to the addition of the initial pack and hold times. Set the charge delay time to about 4 to 6 seconds. Mold a part and examine to see if it is not filled to 100%. If the mold is filled, adjust the transfer position to mold a short part that is filled to about 70 to 80%. Next, remove the charge delay time and increase the back pressure to about 14 plastic MPa, or about 2000 plastic psi, and mold a part. Weigh the parts that were molded with and without the charge delay time. The parts without the charge delay time would be heavier because the back pressure fills the parts via the unfrozen gate. This experiment must be conducted with those who have substantial molding shop floor experience. Parts could get over-packed and get stuck in the mold if care is not taken. In the example shown it is clear that the part without the charge delay time fills more than the one with. Also notice that there is a clear distinct horizontal line that shows where the flow stopped in the injection phase. As the screw started to rotate, the back pressure was pushing the material at a lower rate. The flow front stopped maybe for a fraction of a second and then started. The stop and start manifests as the stalled flow front showing up as the flow line.

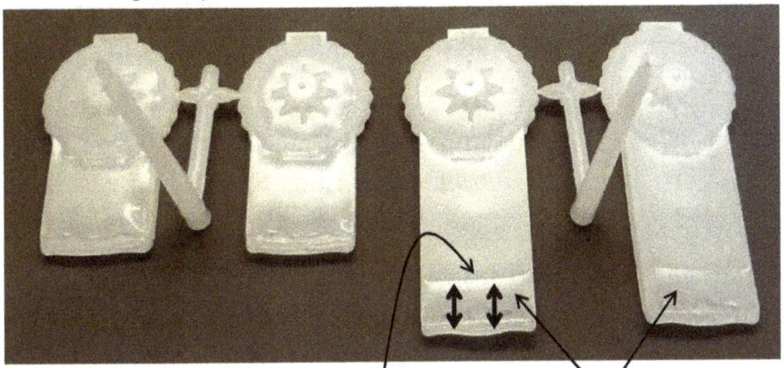

Figure 8.7 Parts molded with and without charge delay time

8.5 Filling in Injection: Weight or Volume?

Filling the part up to 95 to 98% in the injection phase is a technique that was described in the previous sections. There is always a confusion between processors whether the mold should be filled to that limit by volume or by weight.

The formula for calculating weight from volume is

Weight = Volume × Density

Because the melt density is going to be constant, it does not matter if one decides to use weight or volume, so 95% by weight = 95% by volume. The key is to always use the same factor and finally weigh the part using this weight as the 'Injection Only Part Weight' value. This weight must be recorded on all process sheets. Every time a mold is started up, the injection only part weight must be matched. This greatly enhances the consistency of the molding process especially regarding lot-to-lot variations.

On thick parts the plastic fills the walls first to form a skin and then the inside will get packed out. There are no visual signs of the part being filled to a certain number and could get confusing to a processor.

Comments on the % number: The 95 to 98% number is only a recommendation. There is no number or a range that must be followed. Here are some general guidelines:

- The thinner the part, the closer one will need to get to the 100%. Because there is not much shrinkage and/or because the end of fills are thin sections, the mold will need to get filled completely in the injection phase with little amount of pack and hold pressure and some hold time to seal off the gate.
- The thicker the part, then most of the work is done in the pack and hold phase and therefore the mold may need to fill about 90% in the injection phase and the rest in the pack and hold phase.
- For better consistency, it helps to slow down before hitting the transfer position. As an example, on one particular part, the part was filled to 85% and was then transferred to the pack and hold phases.
- For crystalline materials that freeze off fast, the number should be closer to a 100%. Amorphous materials are more forgiving.

Suggested Reading

Osswald, T. A., Turng, L., Gramann, P. J., *Injection Molding Handbook* (2007) Hanser, Munich

Rosato, D. V., Rosato, D. V., *Injection Molding Handbook* (2000) CBS, New Delhi, India

9 Process Development Part 1: The Melt Preparation

Before the plastic melt is injected into the mold the melt should have reached the desired melt temperature, and should be a homogeneous mix free of any unmelted plastic, bubbles, unmixed colors, or nonblended fillers. The melt density should also be consistent from shot to shot and run to run. There are three settings on the machine that contribute to the condition of the melt. These are the settings of the barrel heaters, the screw rotation speed, and the back pressure. Each of these settings have both an individual and an interactive effect on the melt quality. There is one initial step to profile and set the barrel temperatures followed by two studies to optimize and set the screw rotation speed and back pressure.

It is very important for the processor to refer to the processing data sheet before the start of molding. Two materials from the same chemical family can have completely different processing temperatures. For example, the OQ family of polycarbonate supplied by Sabic Innovative Plastics has a recommended melt temperature range of 305 to 332 °C (580 to 630 °F), whereas for the SP family of polycarbonate from the same company, the recommended temperature ranges from 248 to 271 °C (480 to 520 °F). Setting the temperature too high for the low melt temperature resin will degrade the resin and produce unacceptable parts. Setting the melt temperatures too low for the higher melt temperature resin may result in equipment damage, such as a broken screw or screw tip. Unfortunately, such failures are not uncommon on the production floor.

The processor must first select a target melt temperature from the material datasheet. For example, Cycolac X37 from Sabic is recommended to be processed between 245 and 275 °C. The starting preferred melt temperature should be towards the lower end of this range. It may need to get adjusted at a later stage during process development. In the case of this ABS, we would choose the target to be 250 °C. There are two reasons for choosing the lower end of the temperature range. First, the higher the temperatures, the lower is the maximum residence time in the barrel, increasing the chances of material degradation. There is always some amount of molecular weight deterioration going on even at lower melt temperatures but that may not be significant. Second, molding is a heat transfer process. The more heat one provides to the melt, the more heat that will need to be removed to get the molded part down to the required ejection temperature. This will increase the cooling time, resulting in higher cycle times, affecting the overall efficiency of the molding operation.

An important note about measuring melt temperature is required here. The temperature of the melt is measured by taking a purge shot and inserting a melt temperature probe into the purge. During the purging process as the melt leaves the nozzle and is exposed to the outside, the melt stream begins to cool instantaneously and continues to cool as it falls on to the surface, usually a piece of cardboard. During the melt temperature measurement process, different parts of this purge will have varying temperatures at varying times. This makes the whole measuring process unreliable and unrepeatable. It is therefore important to set a procedure that everyone can follow consistently. Training of all the processors in the manufacturing location on this procedure would be essential. There are some systems that involve having thermocouples in the nozzle and/or on the mold that could improve reliability and repeatability but these may not give the temperature of the bulk. The goal should be to repeat these values during each shot and then in each run.

Procedure to check melt temperature: There are three common tools that are used for measuring the melt temperature. The melt immersion probe works the best and is the most reliable. The IT gun method and the IR camera method are both not very reliable because they work more on measuring the surface of the melt. The surface tends to cool down as soon as it is exposed to the outside air.

The probe used to measure the melt temperature must be preheated. Inserting a cold probe into the melt will quickly cool the melt on the surface of the probe creating an insulator and giving false readings. An easy way to preheat the probe is to perform the initial purging on a large sheet of paper and then leave the probe underneath this sheet for preheating. The paper will prevent the melt from sticking to the probe.

Figure 9.1 shows a simple preheating unit that was quickly built in-house using a used sprue bushing and a heater band connected to a temperature controller. The melt probe can be inserted in a sleeve that is fitted into the sprue bushing. The temperature can be set to a desired value, usually about 5 to 10 °C above the expected melt temperature. Once the plastic is purged, the probe can be inserted quickly in the melt to record the temperature. This is a better and reliable method to preheat the probe.

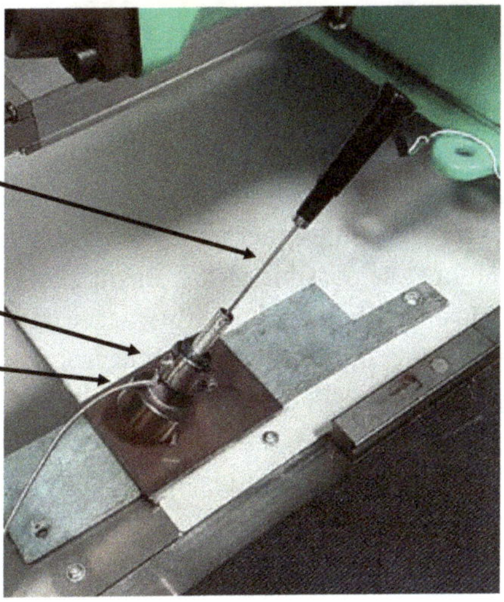

Figure 9.1 Probe preheating unit build built in-house

The following is a recommended procedure to measure melt temperature:

- Get an initial estimate first by purging some material and immersing the melt probe into the melt.
- Start the machine and let it run twice as many shots as are contained in the barrel.
- Preheat the melt probe to close to the expected temperature.
- Place the machine in semi-auto mode, pull the barrel back, and purge the melt.
- Immediately look for the 'thickest' area of the melt, place the probe in there, move it around in the melt, and record the maximum temperature as the melt temperature.

9.1 Setting of the Barrel Temperatures

The heater bands encapsulate the molding barrel and provide external heat to the plastic inside the barrel. At the same time, the screw is turning inside the barrel to auger the material and help melt it with the shear energy. Consider a mold running at a cycle time of 30 seconds. If the percentage usage of the barrel is on the lower side, for example at 20%, there are 5 shots in the barrel and therefore, based on the cycle time, the residence time of the material in the barrel is 30 multiplied by 5 equaling 150 seconds or 2 minutes 30 seconds. If the percentage usage of the barrel is high, for example 75%, then there are 1.3 shots in the barrel, resulting in a residence time of 40 seconds. The above two calculations demonstrate that the lower the percentage usage of the barrel, the higher the residence time, and the higher the percentage usage of the barrel, the lower is the residence time. Since higher residence times can cause the material to degrade, it is best that the barrel temperatures are kept on the lower side as much as possible. Refer to Figure 9.2, in which the target melt temperature is 250 °C. With low percentage barrel usage, the front zones are kept around or just higher than the target melt temperature and with a higher usage of the barrel, in order to get as much heat as possible to the plastic as quickly as possible, most of the zones can be kept closer to or just above the target temperature. With a percentage usage of the barrel of 20%, the settings of the barrel from the metering zone to feed zone would be 250, 250, 245, 240, and 235 °C. In this case, the melt should not be subjected to the higher temperatures until it is ready to be injected. If the percentage usage of the barrel is 80% then the settings would be 255, 255, 255, 250, and 245 °C. In this case the plastic melt needs to reach the desired melt temperature within almost one cycle time and therefore higher values are set for the barrel temperatures.

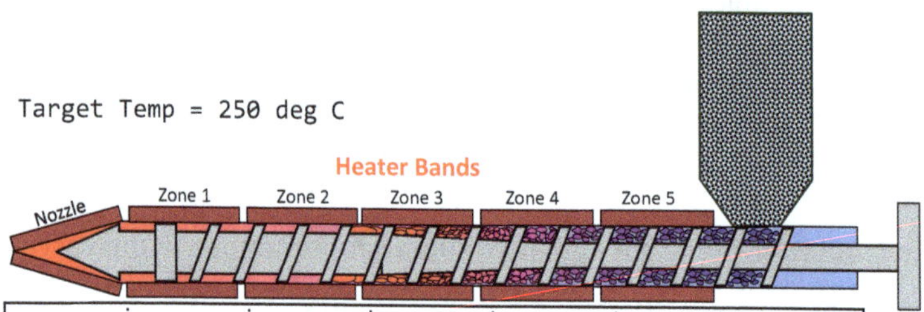

% Barrel Usage	Zone 1	Zone 2	Zone 3	Zone 4	Zone 5	Feed
20	250	250	245	240	235	90
50	250	250	250	245	240	90
80	255	255	255	250	245	90

Figure 9.2 Barrel settings based on percentage shot size used

The profiles must be set based on the percentage usage of the barrel and the maximum residence time. In some cases, a reverse profile, also called a hump profile, may need to be used, especially for crystalline materials. The function of the back zone of the screw is to auger and feed the plastic to build up the shot. The plastic pellets should only be softened and never melted in this zone and so the barrel temperatures in this zone should never be kept high enough to melt the plastic. If the plastic melts here, the pellets will stick to the screw and prevent any further feeding of the plastic. A danger also exists if the zone setting is too low. In such a case, as the screw is mechanically turned the cold plastic pellets will start damaging the screw flights. See Figure 9.3, which depicts an extreme case of screw wear where all the flights were damaged and eventually wiped out.

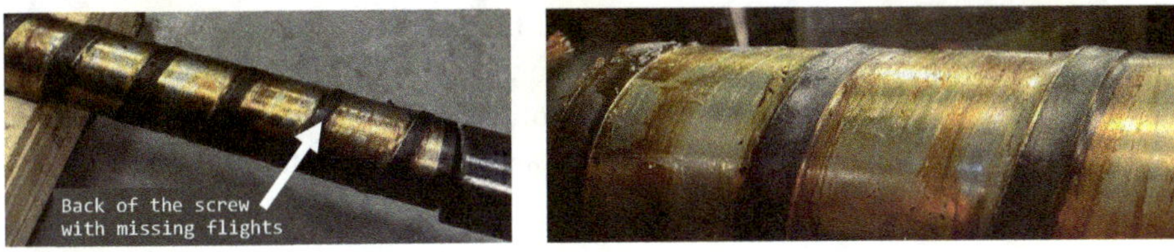

Figure 9.3 Missing flights as result of low temperatures in the feed zone

Residence time values for crystalline materials are lower than those of amorphous materials. Because of the presence of the crystallites, they also need a higher amount of energy to melt. If the melt temperatures are kept on the higher side in the feed zone and if the percentage usage of the barrel is low then the material is likely to degrade. For this reason, the feed zone and/or the zone following the feed zone are kept at higher temperatures but the temperature is then dropped for the subsequent zones since the softening of the pellets is done and the melting can continue with the rest of the settings. Refer to Figure 9.4. This is called a reverse or hump profile. When a molding machine is running, if one hears a grinding noise as the screw is turning then one must increase the temperature of the back zones of the barrel. The noise is an indication of the plastic pellets not being soft and possibly beginning to damage the screw and the barrel.

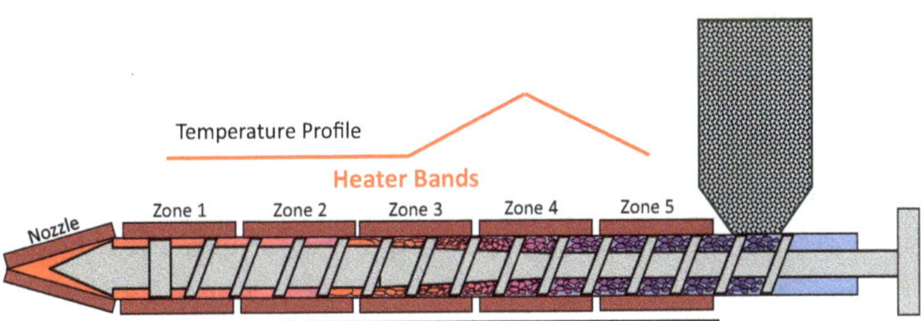

Zone 1	Zone 2	Zone 3	Zone 4	Zone 5
240	240	240	250	240

Figure 9.4 Example of a reverse profile

The setting of the nozzle tip temperature is critical but does not have to align with the above discussions. The mold is colder than the nozzle tip and the two are in contact with each other during the molding cycle. Refer to Figure 9.5. After the required amount of plastic is injected into the mold and the screw now begins to turn, the plastic in the nozzle tip will start to cool down because of the cold mold. If this plastic freezes then it will not be possible to inject the next shot into the mold. If the temperature of the nozzle tip is kept high, then it is also likely to decrease the plastic viscosity leading to drooling of the plastic. This will cause shot-to-shot inconsistencies and visual defects in some molded components. The temperature of the tip should therefore be set such that there is no freeze off or drool of the plastic. Due to the fountain flow effect, the plastic in the nozzle tip will end up on the skin layer of the sprue and possibly runner for a cold runner mold. The type of nozzle tip will also matter. The right type of nozzle with the right orifice diameter also plays a role in the freeze off and drool of the plastic.

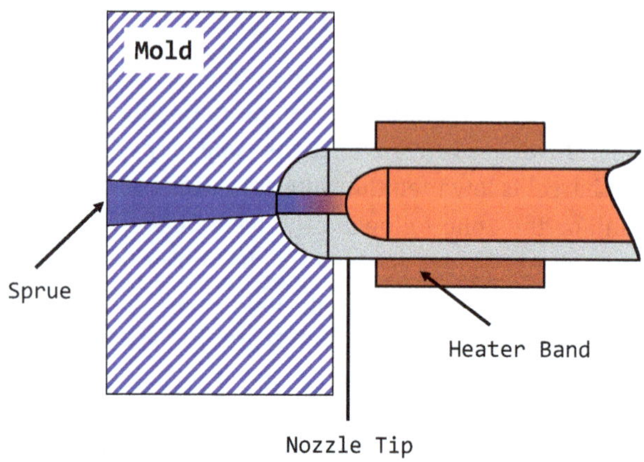

Figure 9.5 The mold is colder than the nozzle tip

In the case of hot runner molds it will mainly be the plastic in the manifold that will comprise the complete shot. The number of shots in the manifold should also be calculated. The residence time of the plastic in the manifold should be added to the residence time in the barrel. There is no mixing or homogenization of the melt happening in the manifold, and therefore a good quality melt coming from the barrel is important. Good quality here refers to being homogeneous, and free of unmelted plastic and/or volatiles. The temperatures of the manifold can be set very close to the desired melt temperature. The logic of profiling the temperatures from the hot tip to the sprue is the same as that for the barrel temperatures discussed above. If there are 3 shots in the manifold then set the zones closer to the cavity closer to the target melt temperature. Here too as is in the case of the nozzle tip for cold runner molds, set the nozzle tip temperatures to avoid freeze off or drool.

The next step is to begin with optimization of the rest of the molding parameters. During the optimization, the molding machine will need to mold parts, which means there has to be a value that should be set for each of the molding parameters. To begin with, these will need to be set based on the experience of the processor but will get optimized during the step-by-step optimization process discussed in the following sections and chapters. For example, a cooling time of 15 seconds may be an acceptable value for a part that is 2 mm thick and is about 80 cm by 80 cm in the other dimensions. At the end of the cooling time study this value will be optimized and the final value may not be 15 seconds.

9.2 Optimization of Screw Rotation Speeds

Once the barrel temperatures and profiles have been established the next step would be to optimize the screw rotation speeds. The barrel temperatures and the shear from the screw rotation are the two sources of heat to the plastic. The barrel heats will give the heat for the plastic closer to the barrel walls but it is the shear from the rotating screw that will give the heat to melt the plastic on the inside. The higher the screw rotation speed, the greater will be the shear and the higher will be the temperature. There may be some exceptions that will be discussed towards the end of this section. The screw also performs the role of homogenizing the melt.

Refer to Figure 9.6. Continuing with the discussion for the ABS material from the previous section, we select the target melt temperature to be 250 °C and set the barrel heater settings accordingly. Considering the shear from the screw rotation, at lower screw rotation speeds there is not enough shear and therefore the resultant melt temperature will be lower than the target temperature of 250 °C. At higher than required screw rotation speeds the shear heat will be excessive resulting in melt temperatures higher than the target of 250 °C.

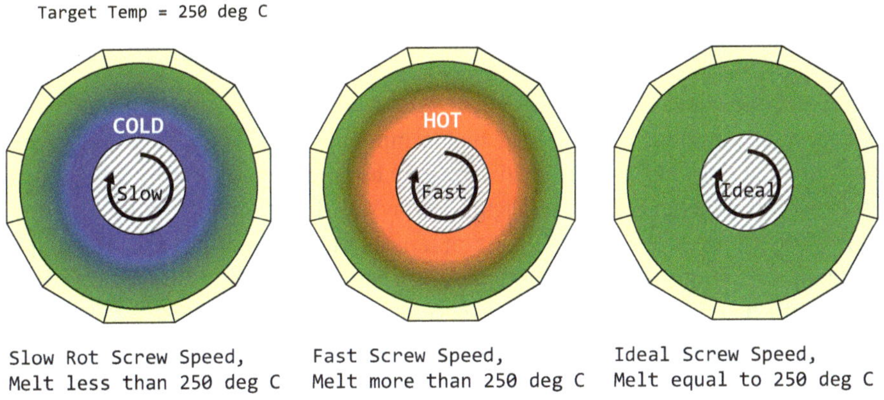

Figure 9.6 Effect of screw rotation speed on melt temperature

Keeping this in mind, a logical study to find the optimum screw rotation speed is therefore to check the melt temperatures at different screw rotation speeds and select the rotation speed that will result in the desired target melt temperature. The results can be plotted as a graph of melt temperature as a function of the screw rotation speed. Refer to Figure 9.7, which shows such as graph where the melt temperature increases with increase in screw speeds for the ABS under discussion. In this case, the screw rotation speed of 170 rpm provides a melt temperature of 250 °C. The additional advantage of performing this test is to also be able to determine the range of screw speeds one could use. If the acceptable melt temperature tolerance is +/– 5 °C, then the graph also shows us that screw rotation speeds from 125 rpm to 230 rpm will be acceptable. This gives the processor flexibility in setting the screw rotation speeds based on experimental data. In some cases, as is the case in thin wall molding the cooling times are very low and low screw rpms will give high screw rotation times that can dominate and lengthen the molding cycle. In such cases, the screw rpms can be increased since the processor now has the data to back the decision.

Screw Rotation Speed (rpm)	Melt Temperature (deg C)
75	236
150	248
225	254
300	259

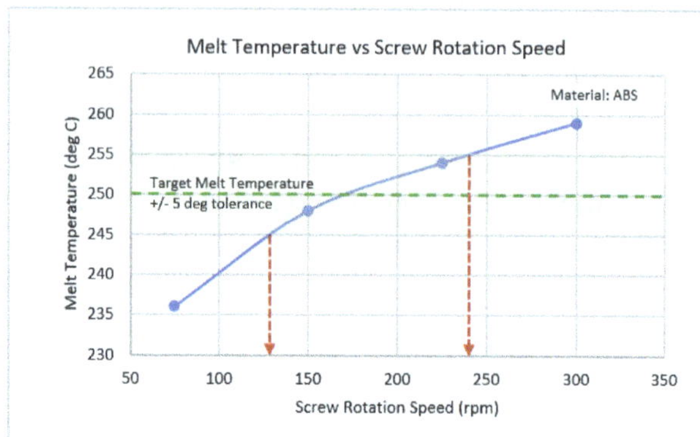

Figure 9.7 Graph of melt temperature versus screw rotation speed for ABS

The profile of the curve in the graph above is typical for shear sensitive materials, meaning that as the shear increases the melt temperature increases. However, some materials may not be very shear sensitive or may be shear sensitive only up to a certain screw rotation speed. Refer to Figure 9.8, which is a graph for the melt temperature study for a polyethylene. The target temperature here was 190 °C. The graph shows that increasing the screw rotation speed beyond 200 rpm did not have any significant effect on the temperature of the melt. The maximum melt temperature that was achieved was 177 °C. In such cases, in order to achieve the desired melt temperature the barrel heater settings must be increased. In this case the actual melt temperature was 13 °C lower than the target and so increasing the temperatures of the barrel heaters by 15 °C increased the melt temperature to the target value. As previously discussed, the percentage shot size usage of the barrel will play a role in deciding which zones would be increased.

Screw rotation Speed	Melt Temperature
100	161
150	168
200	176
250	179

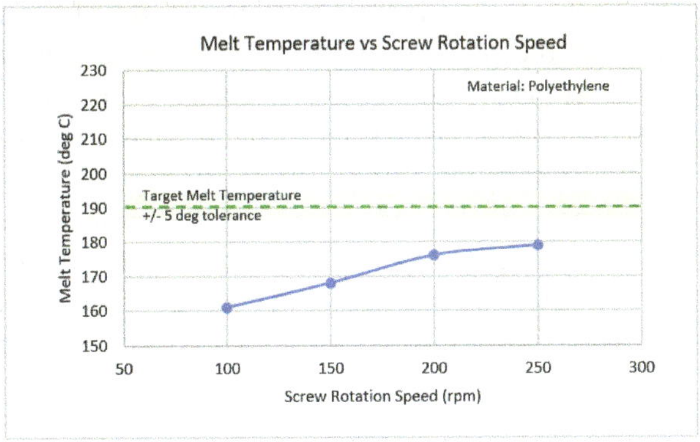

Figure 9.8 Graph of melt temperature versus screw rotation speed for polyethylene

In the following are some guidelines and a procedure to perform the screw rotation study. This procedure should be modified and applied to the mold, material, and machine on which the study is being conducted.

Consider the target melt temperature to be 250 °C.

- Set the barrel profiles based on the percentage usage of the barrel.
- Set the screw rotation speeds to a lower setting than you would normally set as a starting point. For example, if for a given material the normal setting is 150 rpm set the screw speed to 50 rpm. Set it at the lowest practical speed to begin the experiment. The screw should build the shot in a reasonable amount of time.
- Start the molding machine and mold about 20 shots, stop molding, and take a melt temperature reading. Record the screw rotation speed and the measured melt temperature.
- Increase the screw rotation speed to the next value desired, mold 20 shots, and record the melt temperature.
- Repeat until at least four data points are generated and plot the graph. Determine the optimum screw rotation speed and the range based on the data.

The above study was conducted for the same grade of polypropylene material but on two different machines with two different screw diameters. The screws were 30 mm and 46 mm in diameter. The screw rotation speeds were set using the units of circumferential speeds as mm/sec. This unit of measurement is now a common option on most molding machines. Refer to Figure 9.9, which shows two curves for the melt temperature, one for each screw diameter. The curves are almost parallel to each other with the melt temperature for the 46 mm screw diameter higher than that for the 30 mm screw diameter. The average difference in the temperatures is about 5 °C across the range studied. It would have been assumed that if the shear from the screw is matched, then the melt temperatures should be identical regardless of the screw diameter. In other words, the two curves should have overlaid each other and that is clearly not the case. This data brings up an interesting element that has far been ignored. As the screw diameter increases, the barrel diameter also increases. As the barrel diameter increases, the shear between the barrel and the material also increases adding to the heat generated by the screw. This explains the fact that even if the circumferential speeds are matched on different screw diameters, the melt temperatures can differ. The difference may need to be compensated by either changing the screw rotation speed and/or the barrel heater zones.

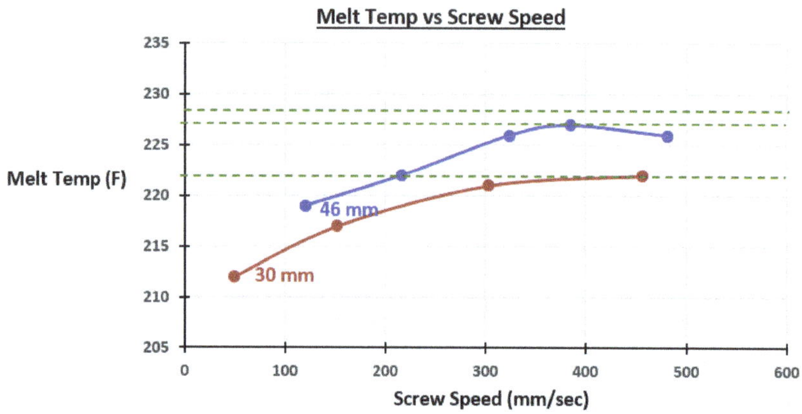

Figure 9.9 Graph of melt temperature versus screw rotation speed for polypropylene at two screw diameters

As the screw rotation speed increases, the screw recovery time will decrease. Any value of the screw rotation speed chosen should be such that the screw recovery time is less than the required cooling time for the part. If it is more than the set cooling time, then the mold close time is now dictated by the screw rotation times and the actual cooling time will be higher than the set value. Molding machines by default do not open until the screw recovery is done. Some machines do have the option to bypass this function. This is common in the packaging industry where cycle times are very fast. The screw may still be turning during mold open or even mold close. The machines have shut-off nozzles or the molds are valve gated and this helps with the drool during the screw recovery process.

9.3 Screw Rotation Speed Rule of Thumb That Should Not Be Used

There is a rule of thumb written in several texts that states that the screw rotation speed should be set such that the screw is turning until 2–3 seconds before the cooling time is done. This rule of thumb is not accurate and should never be used. Screw rotation speed is material specific and has nothing to do with the part design. Refer to Figure 9.10. Assume some fictitious numbers for the sake of understanding the topic. There are two parts with identical weights equal to 10 grams. The wall thickness for one is 1 mm and other is 5 mm. Consider both parts being molded from the same grade of a nylon. The 1 mm part will need a lower cooling time of 10 seconds and the 5 mm part will require a higher cooling time of 30 seconds. Going by the rule of thumb, the screw should rotate for 8 seconds for the 1 mm part, and for 28 seconds for the 5 mm part. This will lead to different shear heats in each of the shot buildups. The shot built up for the 1 mm part will have a lot more shear heat as compared to the 5 mm part.

Part	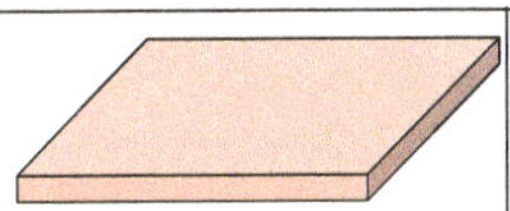	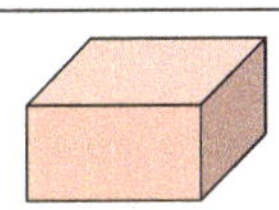
Part Weight :	10 grams	10 grams
Wall Thickness :	1 mm	6 mm
Cooling time :	10 seconds	30 seconds

Figure 9.10 Identical part weights with different cooling time based on part thickness

As discussed in the sections above, screw rotation speeds should be set such that the material gets the right amount of shear and such that the material melt temperature is close to what is desired. Melt homogeneity must be examined. In case of the 5 mm part, if the screw rotation speeds are very slow, there may not be enough shear heat, and the part could suffer in its final properties and may not perform as intended. Setting of screw rotation speeds must be done independent of cooling times.

Assuming that the above-mentioned 10 gram parts are molded on the same machine, since both the parts weigh 10 grams and need identical melt temperatures, the barrel temperatures and screw rotation speeds would be identical and therefore the screw rotation times for both molds would be 6.0 seconds. In the case of the thicker part with the 30 second cooling, the plastic will be picked up in 6.0 seconds and will now wait for 24.0 seconds before it gets injected into the mold for the next shot. The front heater zones are at a higher temperature than the rest, and if the residence time is on the higher side there is now a chance of degrading the plastic that is waiting to be injected.

In such cases addition of a screw recovery delay time will help. For this example, add a 20 second screw recovery delay time. Once the delay timer times out, the screw will begin to rotate for the 6 seconds. This is now a total of 20 + 6 = 26 seconds. The plastic will therefore have to wait for only 30 – 26 = 4 seconds before it gets injected into the mold. This can help in the prevention of degrading the plastic in the screw.

9.4 Optimization of Back Pressure

Back pressure is the pressure applied to the back of the screw during the screw recovery process. It performs three main functions. First, it helps in the consistency of picking up the material from shot to shot. Second, it helps to squeeze out any volatiles that are present in the melt during the melt formation process. This usually takes place in the compression zone of the screw. Third, the back pressure helps compress the melt to the same consistent melt density from shot to shot. These reasons were discussed in detail in the section of the parameter inputs for the molding machine.

As we increase the back pressure the melt density will also increase. To test the consistency of the melt density a simple test would therefore be to mold parts with different back pressures and plot the range of part weight versus the back pressure. The one factor that will complicate this is the pack and hold pressures which help in compensating for some of the variation. For this reason, the pack and hold pressures should be kept to zero and parts with no pack and hold phase should be molded and used for the data. These would be what are called 'injection only' or 'fill only' shots, where the screw moves from the shot size to the transfer position and does not move further. These are essentially parts molded only in the injection phase and they should look visibly short in most cases. Thick parts will show sinks. The weight of the runners should also be included since it is part of the injected shot. We calculate the range as the difference between the highest and the lowest shot weight and then plot a graph of range versus back pressure settings.

Refer to Figure 9.11. The back pressure study was performed at 4 back pressure values and the range in the part weights was plotted as a function of the back pressure. The machine was an electric machine and so the values here are in plastic pressure. The weight range at 100 psi was 0.738 grams, which then dropped to 0.071 and 0.032 grams for 300 and 500 psi, respectively. This was expected as the melt density became more consistent with increasing back pressure. However, at a back pressure of 700 psi, the range went up to 0.12 grams, indicating that the consistency had decreased. At higher back pressures, the pressure prevents or adds resistance to the return movement of the screw. Visualize extremely high back pressures where the screw is pushing forward as it is turning but the high pressure is preventing the plastic to collect in front of the screw to build the required amount of plastic for the next shot. This is what is happening at a lower intensity at 700 psi where the plastic is slipping on the screw causing the shot-to-shot inconsistency. This is similar to when a race car driver gives a sudden acceleration and the tires spin in place causing the rubber to burn, with very little movement of the car in the forward direction. Analogous to this increase in the tire temperatures, high back pressures will also increase the melt temperatures. If the back pressure values are set with the optimum range, an increase in melt temperature should not be seen.

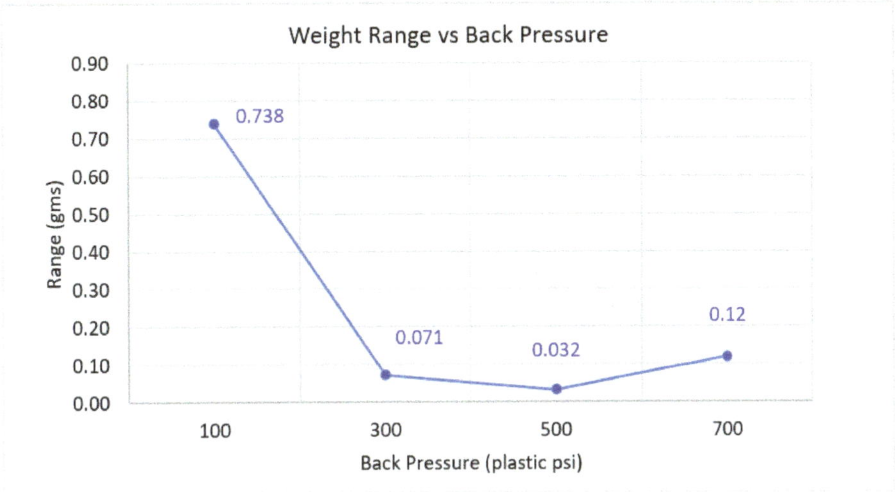

Figure 9.11 Weight range versus back pressure

With the data from the graph, any value from 300 psi to 500 psi will be acceptable for the setting of the back pressure. Again, the advantage with the study is that the processor now has a range of back pressures to work with. A center value of 400 psi will be acceptable. As discussed above, the back pressure also helps in squeezing out the volatiles out of the melt. Once the back pressure is selected and the rest of the molding process has been optimized or is being optimized and if one notices splay on the parts, the back pressure may be increased to eliminate the splay. In such a case, increased back pressure is acceptable since it is important to get rid of the splay.

Refer to Figure 9.12. This shows the increase in the shot weight as the back pressure is increased. The volume of injection is fixed since the screw starts the injection from the shot size and stops at the transfer position. This is the fill only shot weight. Weight equals volume multiplied by melt density. Since the melt density increases with increasing back pressure, the shot weight also increases. The melt density is an important number that is used in several calculations. With the information above, since the shot weight is known and the volume is known, the melt density can be calculated for the given melt temperature and back pressure:

Melt density = Shot weight / Volume of injection

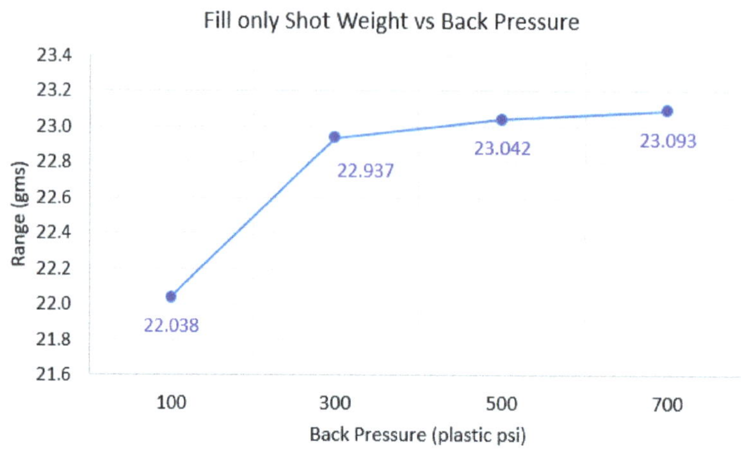

Figure 9.12 Injection only shot weight versus back pressure

Refer to Figure 9.13. The back pressure study was done on two different machines that were molding different components but with the same grade of ABS discussed above. The size of the parts and molding parameters were different on each machine but the profiles of back pressure curves were identical and parallel to each other.

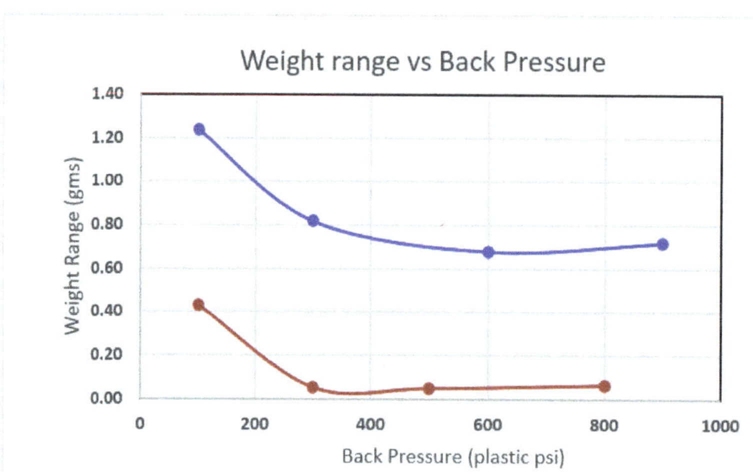

Figure 9.13 Weight range versus back pressure for non-identical machines and molds molding the same material

It can be noticed in the studies above that the point of focus has been the melt. There has been no discussion about the quality of the parts. These studies can be time-consuming but necessary and therefore beg the question if there is a way to reduce the work. This can certainly be done if enough data is available. The goal of a molding company should be to collect data from various machines and various molds for the same given material and analyze it for commonalities. There are several machine learning tools that can also be used for this purpose. It would then be possible to generate home grown process datasheets for back pressure and screw rotation speeds.

Suggested Reading

Osswald, T. A., Turng, L., Gramann, P. J., *Injection Molding Handbook* (2007) Hanser, Munich

Rosato, D. V., Rosato, D. V., *Injection Molding Handbook* (2000) CBS, New Delhi, India

10 Process Development Part 2: The 6-Step Study – Exploring the Cosmetic Process

Chapter 8 introduced the process development procedure. There are three parts to process development. They are the melt preparation, the 6-Step Study, and Design of Experiments. Melt preparation was discussed in Chapter 9. This chapter will discuss the 6-Step Study, also known as the Scientific Molding Studies (SIM).

Of the 13 molding parameters, the melt preparation studies have optimized 3 inputs. They are the barrel temperatures, the screw rotation speed, and the back pressure. We will now move onto the optimization of the other 10 inputs.

10.1 The 6-Step Study

In Part 2 the remaining 10 process inputs of the molding machine will be optimized. Consider a product that is to be molded where dimensions are not important. Examples of such parts are buckets or forks and spoons. At this time our goal is to find a robust process and not consider dimensions. Dimensions will be considered in Part 3, which deals with Design of Experiments (Chapter 11).

10.1.1 Step 1: Optimization of the Injection Phase – Rheology Study

All plastic melts are non-Newtonian. This means that their viscosity does not remain constant over a given range of shear rates. In the strict sense, the rheological behavior of a plastic is a combination of non-Newtonian and Newtonian behavior. At extremely low shear rates, which are rarely encountered in injection molding, the behavior of the plastic is Newtonian; but as the shear rate increases, the plastic tends to exhibit non-Newtonian behavior. Interestingly, as the shear rates increase further, the plastic tends to act in more and more Newtonian manner after an initial steep drop in viscosity. In Figure 10.1, the plastic viscosity is plotted on a log-log scale, and in Figure 10.2, the plastic viscosity is plotted on the linear scale. Both graphs are for PBT and are plotted from identical data, the only difference being in the scale type.

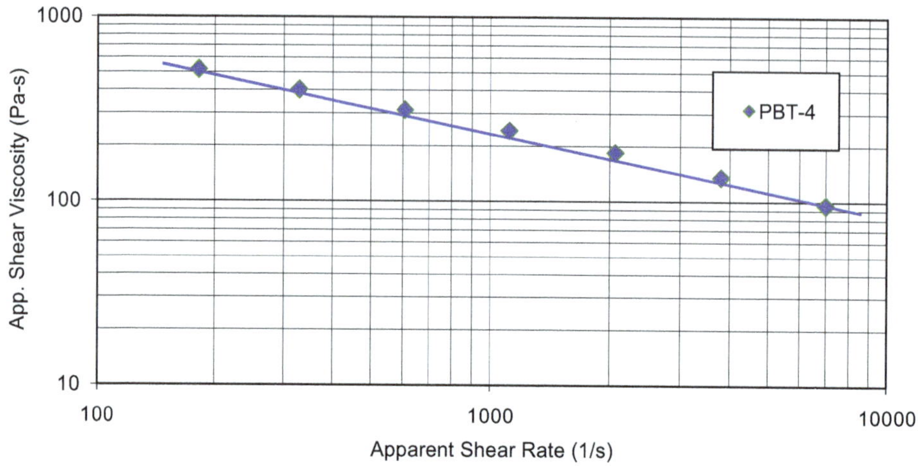

Figure 10.1 Viscosity of PBT represented on a logarithmic scale

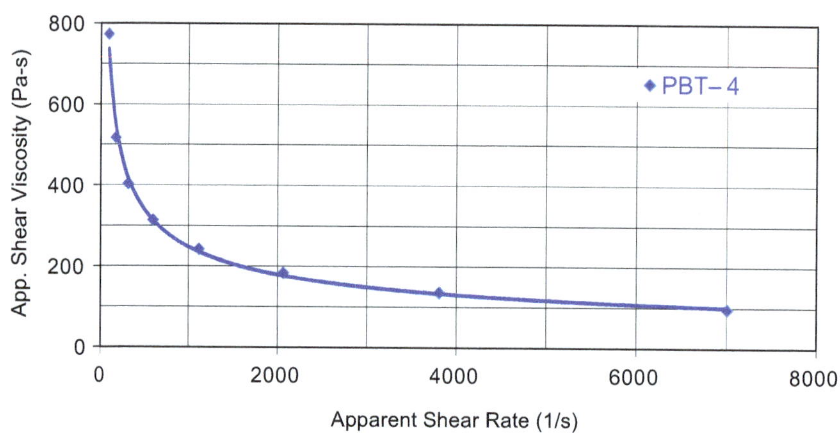

Figure 10.2 Viscosity of PBT represented on a linear scale

On the linear scale graph, it can be seen that the change or drop in viscosity is much greater at lower shear rates as compared to higher shear rates. This happens because with increasing shear rate, the polymer molecules start to untangle from each other and start to align themselves in the direction of flow. This reduces the resistance to flow (the viscosity). The plastic tends to get more Newtonian at higher shear rates. Although there is still a continuing drop in the viscosity, the change is not as significant as at the lower shear rates. Figure 10.3 illustrates this phenomenon. For the sake of discussion, we shall refer to these regions as non-Newtonian and Newtonian regions. These are no truly Newtonian regions and the slope of the line depends on the nature and properties of the plastic. Typically, crystalline materials tend to have a flatter region compared to amorphous materials because the viscosities for crystalline materials are lower, facilitating the ease of orientation.

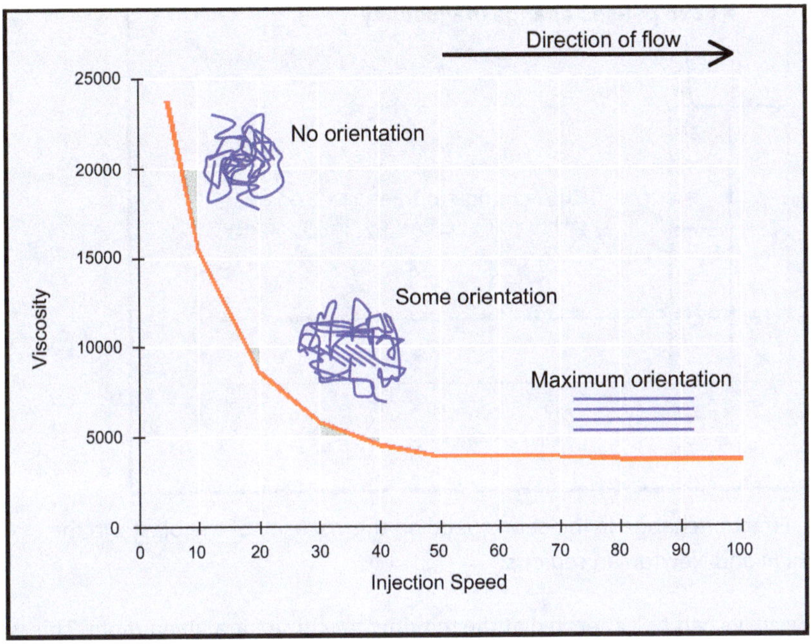

Figure 10.3 Orientation of the molecules in the direction of flow at different injection speeds (shear rates)

During injection molding, the material is subjected to high shear forces during the injection phase. The shear rate is proportional to the injection speed. If the shear rates are low and are set in the initial non-Newtonian region of the curve, then small variations in the shear rate will cause a large shift in the viscosity. Because there is always some natural variation, the mold filling will be inconsistent and will therefore result in shot-to-shot inconsistencies. However, if the injection speeds are set to higher values, the viscosity tends to be consistent. At high injection speeds, the shear rates are high and the effect of shear rate on the viscosity is not as significant as it was at low injection speeds. Small changes in injection speed result in small or almost no change in the viscosity of the melt. Figure 10.4 illustrates this concept using data that was generated on a molding machine. Any natural variation in the speeds will not have a significant effect on the cavity fill in the Newtonian region of the curve and it is therefore important to find this region of the curve and set the injection speed, and with that the shear rate, here.

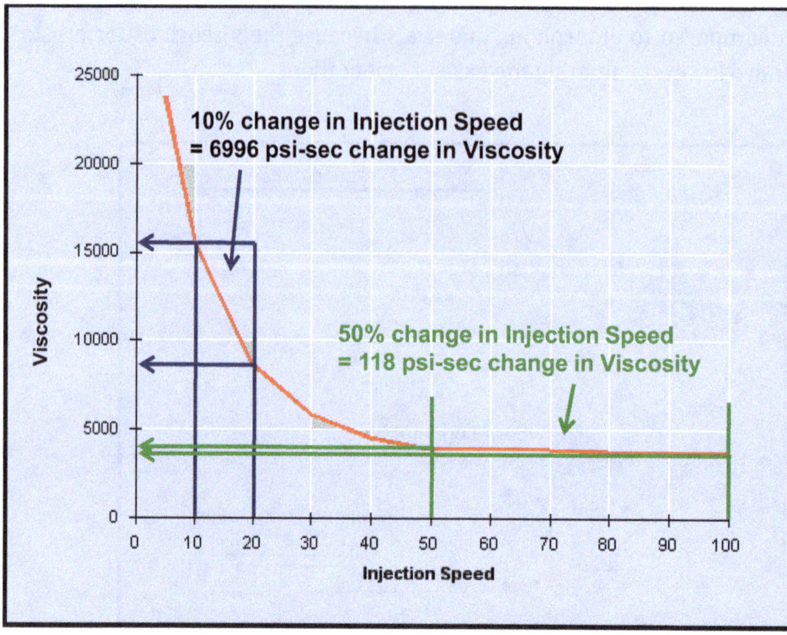

Figure 10.4 Effect of change in injection speed on the viscosity of the plastic in the non-Newtonian and Newtonian regions

The viscosity curve can be generated at the molding machine for a given mold. This is known as an 'in-mold rheology study' or simply 'developing the viscosity curve.' A study by Mertes et al. [2] demonstrated that at high speeds the variations in the end of fill pressure were smaller compared to the variations at low speeds. Perturbations were introduced on purpose and two polypropylenes with high and low viscosities were used in the study. The results are shown in Figure 10.5.

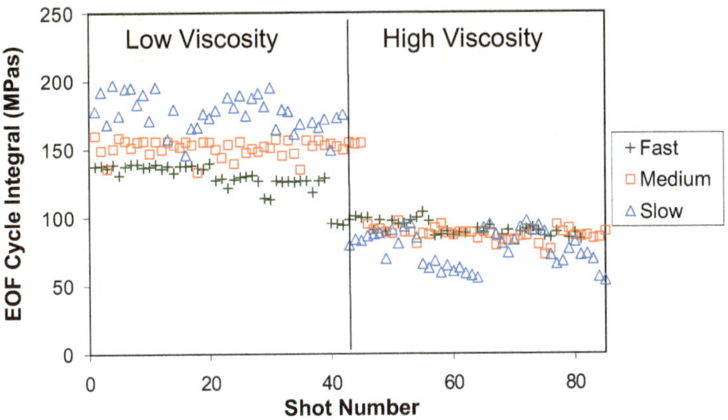

Figure 10.5 End of fill cycle integral data at various injection speeds for two polypropylenes of different viscosities [2]

The other advantage of using high injection speeds is the reduction of the effects of lot-to-lot variations. Figure 10.6 is a compilation of viscosity tests performed on various lots of a certain grade of polycarbonate supplied to a molder on a specific project. Different lots of the same material can have different viscosities at the same injection speed. However, the difference is greater at the lower injection speeds.

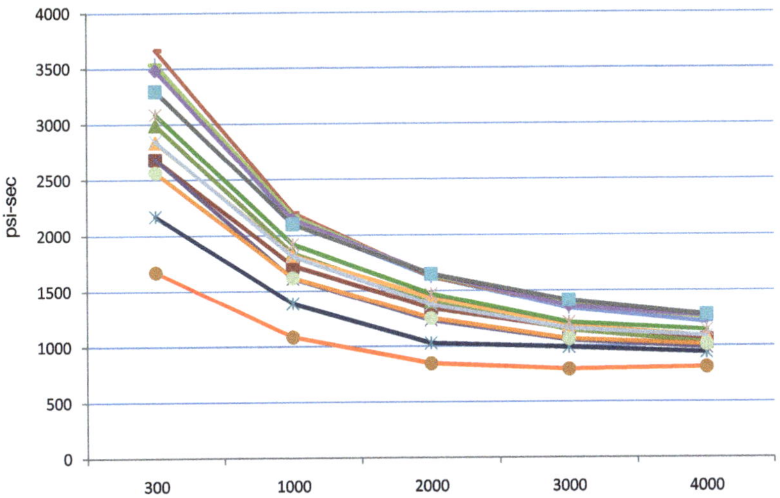

Figure 10.6 Viscosity of polycarbonate of the same grade but from different lots

The procedure to determine the viscosity curve was first developed by John Bozzelli. The basic principle is based on the melt rheometer that is used to study the viscosity of plastics (see Chapter 3). The injection molding machine can also be treated as a rheometer, where the nozzle orifice is the die of the rheometer and the screw is the piston. The hydraulic pressure is applied to the molten plastic with the help of the screw. The pressure required to move the screw at a set speed is recorded.

10.1.1.1 Procedure to Determine the Viscosity Curve at the Molding Machine

Safety should always be given first priority when conducting any experiments:

- Set the melt temperature to the one recommended by the manufacturer. If there is a range, set the temperature to the center of the range. However, as previous discussed, because we are looking for the profile of the curve and not the actual values, the selection of the temperature does not matter. See Section 10.1.1.6 for an explanation.
- Set all the pack and holding phase parameters to zero. This means that there will not be any holding phase but rather only the injection phase.
- Provide a screw recovery delay time of about 4 seconds.
- Set the injection pressure to the maximum available value. There is an assumption here that the mold and the machine are compatible in terms of shot size and plastic pressure usage. If the shot used is small or if the required pressure is considerably less than the maximum available, then set the pressure to lower values to avoid any mold damage or over packing of the cavity.
- Set the cooling time to a safe value such that the part will be cool and has reached the ejection temperature before mold opening.
- The machine must be set to transfer on screw position. Set the injection speed to 'slow' and make a part. The part should be short. If not, adjust the transfer position to make the part such that it is filled only to about 50%.
- Incrementally increase the speed as close to the maximum as possible in small steps and make sure that the parts are still short. If the part is full, adjust the transfer position, such that the part is about 95–98% full by volume. This means that at close to the maximum possible injection speed, the parts are 95–98% full with no holding time or pressure. If this is a multicavity mold, the part that fills the most must be the one that is 95–98% full. If at higher speeds, cosmetic issues or burning issues are seen, do not increase the speeds any further.
- Make another shot and record the fill time and the peak hydraulic pressure at transfer. For example, the machine is set to 2200 psi but may require only 1850 psi to move the screw at the maximum speed of 5 in/s. If the hydraulic pressure was set to a value other than the maximum available, and during the experiment the peak pressure matched the maximum value, the set pressure must be increased. The process must not be pressure limited.

- Next, lower the speed by a small amount, for example from 5 in/s to 4.5 in/s or from 90% to 80%. Note the fill time and the peak injection pressure.
- Repeat the above steps until you achieve the lowest injection speed possible. Divide the range of available injection speed into about 10–12 speeds so that you get the most data points possible.
- Obtain the intensification ratio of the screw from the machine manufacturer. If this number is not available, use a value of 10. It is important to understand that the intensification ratio is necessary to calculate the plastic pressure. However, this should not be a show stopper for the experiment or the optimization process. Since this is a constant used in the equation, the graph will shift up or down respectively, but the profile of the graph will remain the same, which is of primary interest to us.
- To get the viscosity, use the following formula:

  Viscosity = Peak injection pressure × Fill time × Screw intensification ratio

Plot the graph of viscosity versus injection speed.

Figure 10.4 shows a typical viscosity curve generated at the molding machine. Table 10.1 shows a sample worksheet that can be used for the study.

Table 10.1 Viscosity Curve Worksheet (Screw Intensification Ratio = 10.5)

No.	Inj. speed (in/s)	Fill time (s)	Peak hyd. Press (psi)	Viscosity (psi-s)	% Drop in viscosity
1	0.20	5.75	735	42263	-
2	0.50	2.37	805	19079	54.9
3	1.00	1.28	963	12326	35.4
4	1.50	0.87	1112	9674	21.5
5	2.00	0.68	1240	8432	12.8
6	2.50	0.57	1339	7632	9.5
7	3.00	0.50	1427	7135	6.5
8	3.50	0.44	1504	6618	7.2
9	4.00	0.41	1588	6511	1.6
10	4.50	0.38	1663	6319	2.9
11	5.00	0.36	1669	6008	4.9
12	5.50	0.35	1631	5994	0.2

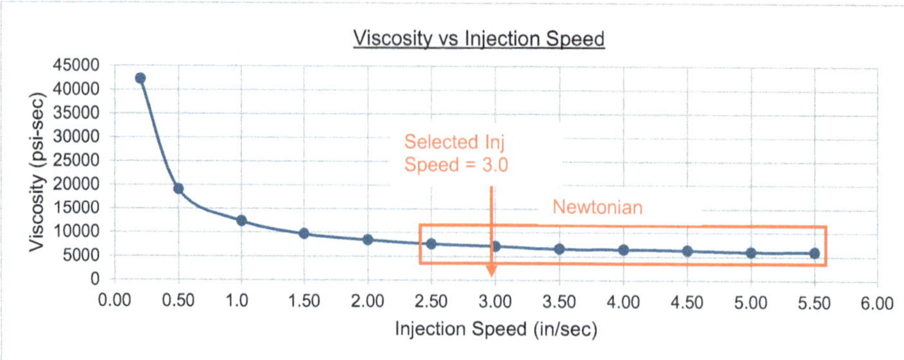

Figure 10.7 Viscosity graph generated at the molding machine

10.1.1.2 Interpreting the Data

Notice in the data table that the percentage drop in the viscosity after 2.5 in/s is less than 10% between every step. This drop of less than 10% also reflects in the generated curve and it can further be noticed that the viscosity stays fairly constant after an injection speed of about 2.5 in/s. Therefore, setting the injection speed to 3.0 in/s would ensure that the filling stage of the process will stay consistent. Any small natural variations will not cause large changes in viscosities that result in shot-to-shot variations as they would at lower injection speeds. The injection speed must be selected closer to the 'knee' part of the curve. This is where a shift in the viscosity and greater consistency can be observed. It is not advisable to inject the plastic as fast as possible for a couple of reasons: First, plastic materials can be shear sensitive and will degrade at higher shear rates or injection speeds. Second, it is always difficult to get the best venting possible in a mold because of the very valid concern of flashing plastic into the mold vents. For this reason, mold makers are very conservative with vent depths and reliefs. A mold maker will typically machine in the smallest vent clearances possible, as vents are not a steel safe condition. Venting is also sometimes difficult in areas such as deep pockets or corners that are not on the parting line of the mold. A slower injection speed helps in such cases. Optimizing the injection speed through in-mold rheology is the first step to achieving a robust process.

One can also plot a graph of viscosity versus shear rate. Because shear rate is proportional to injection speed, the curve profile stays the same. It is easier to read an injection speed graph so that one can directly select the injection speed from the graph rather than having to cross-reference in a table of shear rates.

When observing the short shots from a viscosity study, it is evident that the parts get shorter and shorter as the injection speed is lowered. The question that usually comes to mind here is why are the parts getting smaller if the shot size and transfer position is not changed?

There are two phenomena explaining this condition. First, the viscosity of the plastic is being reduced with the increased injection speed. One must remember that when the viscosity of the melt is lowered, its flow rate also becomes lower. Second, the momentum of the screw and the melt also reduces with decreasing injection speed. When the hydraulic pressure is abruptly cut off at the end of the injection phase, the screw and/or the melt can still travel further and fill the cavity due to the inertia of the screw. At higher injection speeds the momentum is higher and the viscosity is lower.

The position of the screw can also be tracked by recording the furthest forward position it reaches during the injection phase. The cushion value displayed on the machine can be misleading because this number represents the position of the screw at the end of the holding phase and may not necessarily represent the furthest position of screw travel. In some cases, a screw bounce-back can cause the cushion value to be higher. In cases of electric machines, there is no hydraulic pressure but instead a servo motor that controls the position of the screw. The servo system will stop the screw instantly with almost no momentum, but the melt will still have some momentum. Pictures of a fill progression during a viscosity study series are shown in Figure 10.8.

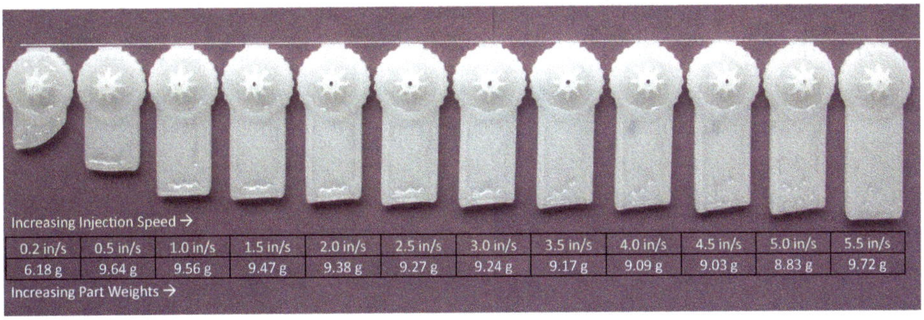

Figure 10.8 Progression of fill during an in-mold rheology study from a slow to fast injection speed

10.1.1.3 Cautions and Exceptions

Generating the viscosity curve is a great tool to find out the effect of injection speed on the viscosity of the plastic. However, it must not be used in all cases; there are exceptions:

- Insert molded components typically have inserts that need to be partially or wholly encapsulated with the plastic. An example of this is a power cord where the pins are placed in the mold and the plastic is molded over it. Injection of plastic at high speeds can easily dislocate and/or deform the insert, thus molding an unacceptable part.

- High speeds also tend to shear the material and in some cases cause gate blush or splay. Optical components can sometimes exhibit this problem, which necessitates that lower speeds be used. PVC is a prime example where faster speeds will cause burning.
- At higher injection speeds, gas must be vented from the mold very efficiently. Venting some sections of the mold can become tricky and difficult. In such cases, only the fastest possible speed without burning should be used. In these molds, the vents must be cleaned very frequently because any buildup will result in burning of the plastic.
- If the required pressure to fill the mold equals the maximum available pressure, then the process is said to be pressure limited. In such cases, the injection screw will never reach the set injection speed for lack of pressure. Therefore, the values of the fill time will not be accurate and the obtained viscosity graph will be of no value. Viscosity studies must not be done for pressure limited processes.
- To get accurate fill time and peak pressure readings, the machine must have enough time to respond. Using lower that 20% of the barrel does not allow for the screw to have enough time to consistently build the plastic pressure and/or output the fill time accurately or consistently. The plastic melt is compressible, and this also adds on to an inconsistency of pressure readings.
- When the parts are small (less than about 5 to 8 mm) a viscosity curve may not be accurate. This follows the same reason of the previous bullet point.

In all such cases remember, "Inject as fast as required and not as fast as possible or what the viscosity curve provides." Fast injection speeds provide better consistency. For example, if at a speed of 2.5 in/s there is gate blush, reduce the speed to 2.0 in/s and look at the parts. If this is not acceptable, try lower speeds until the defect is eliminated. Giving a window for any variation will always help. If 1.5 in/s was good, set the speed to 1.25 in/s.

10.1.1.4 Profiling of Injection Speeds

Refer to Figure 10.8. When the screw of the molding machine moves at predefined injection speeds in predefined segments, the injection speed is said to be profiled. Profiling of injection speeds is common in injection molding. For example, if the plastic is injected at fast speeds at the gate, a cosmetic defect such as gate blush could develop. For this reason, processors will slow down the injection speed in the corresponding segment and then speed it back up. Because plastic flow is laminar, as soon as the skin around the gate area is formed, changing the injection speed will not change the cosmetics in the gate area.

In the last section of the injection phase, the screw must always be slowed down. See velocity V4 in Figure 10.9. To achieve part consistency, the point where the process transfers from the injection phase to the pack and hold phase must also be consistent.

For example, in Figure 10.9, if the transfer position is 20 mm, then the screw must start to slow down at around 30 mm in order to consistently hit the 20 mm mark. Overshooting or undershooting will inject more or less material into the cavity causing a variation in part quality. This slowdown is analogous to coming to a red traffic light where the driver starts to slow down a few meters before the signal line in order to avoid over passing the signal line.

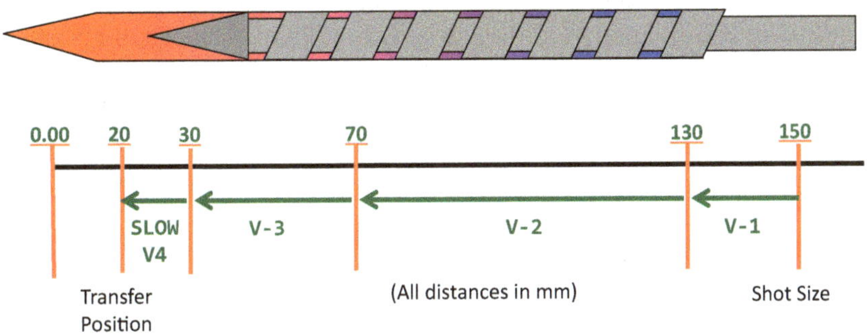

Figure 10.9 Injection speed profiling

Profiling must be done within reason. One must think of the distance traveled and the number of profiles added. The melt is a viscous mass that will keep moving if it has the momentum, even if the screw has stopped moving. The valves and hydraulics also will need to react to the changes. If the total movement of the screw from the shot size to the transfer position is, for example, 25 mm, adding 5 velocity profiles with a segment length of 5 mm each will not be effective. The ineffectiveness is worse with fast injection speeds. One must keep in mind the melt momentum and response times of the valves and hydraulics when setting the profiles and the velocities.

10.1.1.5 When a Short Shot Sticks

In some molds it is not possible to continually make short shots because the parts can stay in the fixed side of the mold. In this case, an attempt must be made to find the 95–98% full cut-off position at close to the maximum injection speed. There are a couple of ways this can be done. The first is the method described earlier and would involve removing the part every time the short shot occurs. This will dictate that the machine be run in a semiautomatic mode. If this is quick, without the use of excessive times of more than about 10 seconds, it is acceptable. If it takes time to get the parts out, then the procedure below should be used:

- Start molding with the least amount of pack and hold pressures and times such that the part moves with the movable side of the mold.
- Change the transfer position such that the part looks under-packed and/or slightly short.

- Increase the injection speeds to the maximum possible with care so as to not over-pack or burn the plastic. At each speed, adjust the transfer position such that the part looks under-packed and/or slightly short.
- Start the viscosity study and collect the fill times and peak pressures.

The reason this technique will work is the following: In this procedure, it is acceptable to add any holding pressure to fill the part so that it pulls away from the fixed side of the mold and stays on the ejection side during mold opening. The two values that are required in the calculation of the viscosity are the peak injection pressure and the fill time. Therefore, what happens in the holding phase, which follows the injection phase, does not affect these values. Whether or not the parts get filled and packed out is not of concern. Not having any holding pressure or time and producing a short shot has its value; however, in cases where this is not possible, it is acceptable to fill the parts.

As the injection speed increases, the fill time will decrease and the shear rate will increase. However, the pressure required to maintain the injection speed is an aggregate of the individual pressures of the various phenomena taking place as the plastic melt travels to the end of fill. Therefore, with increasing injection speeds, the required pressure can either increase or decrease, depending on the dominant phenomenon. These phenomena and the resulting required pressures are described in the following:

- As the melt hits the mold, it starts to cool, resulting in an increase in the viscosity of the plastic. This will result in an increase in the required pressure.
- The flow channels in the mold typically call for progressively smaller cross sectional areas towards the end of the flow. Smaller cross sections result in higher required pressures. However, if the cross sections are generous, the resulting pressure may not change.
- When the plastic first enters the mold, the layer of plastic next to the wall forms a frozen layer and the hot melt that is entering the mold now flows in between these frozen layers. This is also called fountain flow. This frozen layer gets thicker and thicker during the filling phase, again increasing the required pressure to move the screw at the set injection speed. With thick parts or generous flow channels, this effect is not significant because the filling may be completed before the frozen layer is sufficiently built up.
- Because plastics exhibit non-Newtonian behavior, with increasing injection speeds, the viscosity drops and the pressure required to move the screw will reduce.

10.1.1.6 Selecting Melt Temperature for Viscosity Graphs

Referring to Figure 3.9, the lines of the viscosity graphs at various temperatures as a function of shear rate are all parallel to each other. A reason to perform the viscosity study on a molding machine is to find the injection speeds at where the viscosity remains fairly consistent. The actual viscosity number is not of any value at this point. If the temperature increases, the viscosity drops, and if the temperature decreases, the viscosity increases, but the profile of the viscosity curve stays the same. Refer to Figure 10.10, which shows the viscosity study done on a HDPE material at 175 °C and 220 °C. It is clear that the two profiles of the curves are identical, the only difference is that the viscosity is higher at 175 °C and is lower at 220 °C, therefore, the curves run parallel to each other.

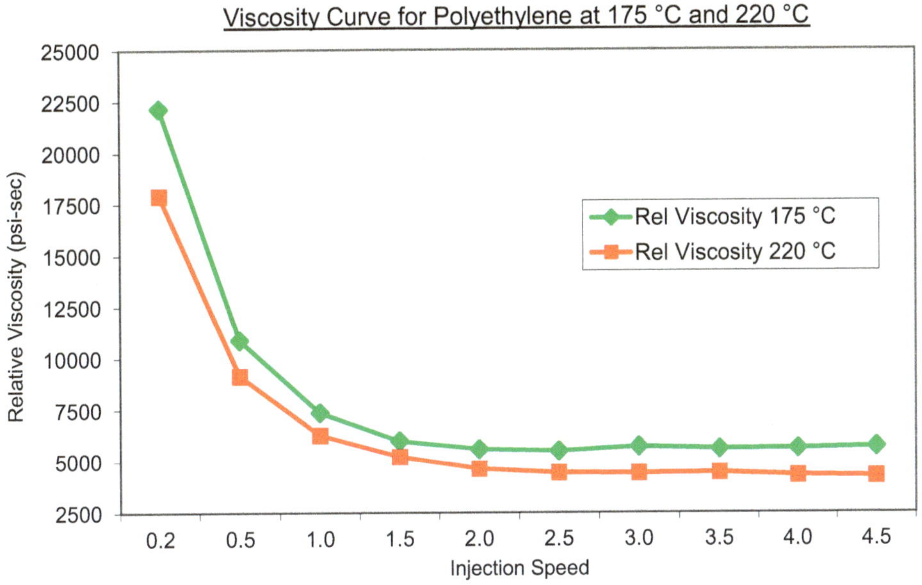

Figure 10.10 Injection speed

Because the reason for developing the viscosity curve is to find the region of the curve that is most consistent, the profile of the curve is all that matters. Because the profiles are identical, the selection of the temperature for the study is not critical.

10.1.2 Setting the Shot Size and the Transfer Position

At the end of the viscosity study the injection speed was selected. In the above example, the injection speed selected was 2.5 in/s. The part weight at the 2.5 in/s is 9.27 grams. In the injection phase, the part is filled close to being a 100% full, typically 95–98%

full. During the viscosity study we start the study at the highest injection speed with the shot size and the transfer position adjusted such that the part is 95–98 full. In the current example, that part weight is 9.73 grams. These are shown in Figure 10.11.

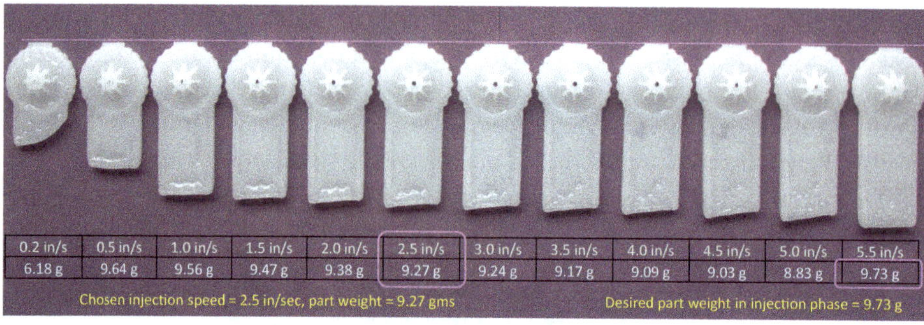

Figure 10.11 Part weights at chosen injection speed and at close to end of fill

The desired part weight in the injection phase is 9.73 grams, but the part weight at the set injection speed is less than desired and is at 9.27 grams. Therefore, to achieve the desired part weight, either the shot size must be increased or the transfer position should be decreased. At the same time, as discussed in Section 10.1.1.4, in the screw speed should be profiled to slow down before hitting the transfer position to achieve better shot-to-shot consistency. In this particular example, during the viscosity study the shot size was 2.10 inches and the transfer position was 0.535 inches. To achieve the increased part weight of 9.73 grams and adding a position to profile the injection speeds, the new transfer position was moved to 0.515 inches. The slow down position was 0.815 inches with a speed of 1.0 in/s. Refer to Figure 10.12 for the before and after positions of shot size, slow down, and transfer.

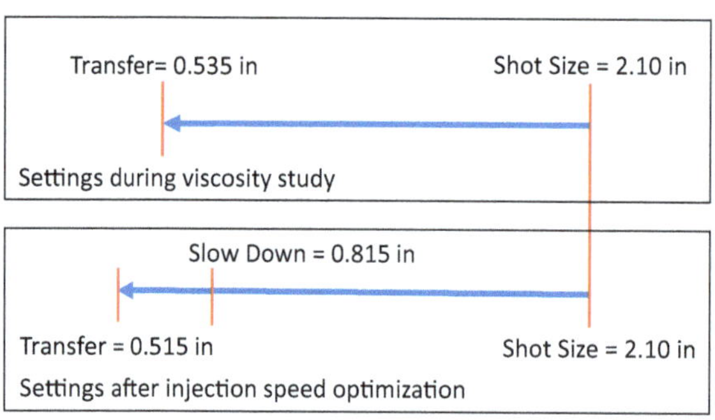

Figure 10.12 Setting of transfer and slow down speed positions after injection speed optimization

10.1.3 Step 2: Determining the Cavity Balance – Cavity Balance Study

A specific volume versus temperature graph was discussed in Section 2.8. The graph is shown in Figure 10.13 with the areas that correspond to the different phases of the injection molding cycle. The injection phase corresponds to the plastic being in the melt phase and as the melt starts to cool, the pack and hold phase comes into effect. When the pack and hold is done, the plastic is cooled down below the ejection temperature, at which point it can be safely ejected out of the mold.

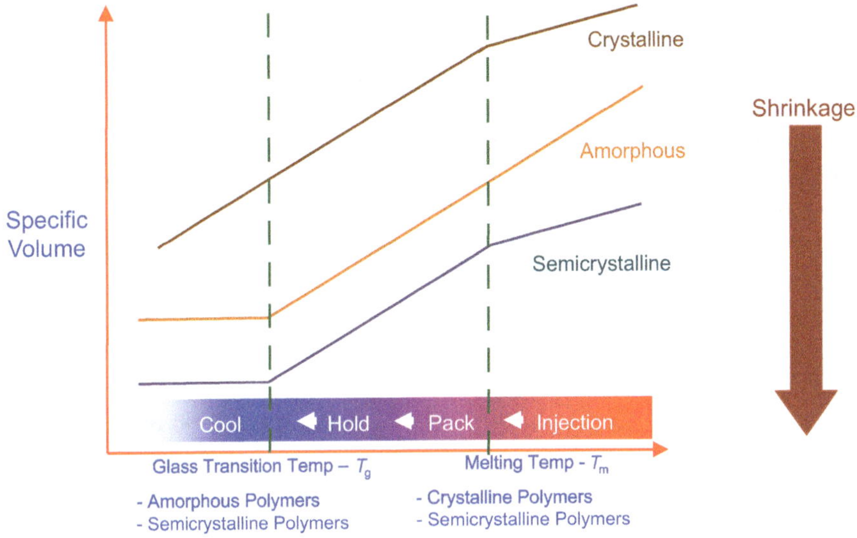

Figure 10.13 Effect of temperature on specific volume and the application to injection molding

Shrinkage is directly related to specific volume. As the pressure increases, the specific volume decreases. The packing pressure therefore affects the shrinkage of the part. The higher the packing pressure, the higher is the pressure of the plastic inside the cavity and the lower is the shrinkage. Therefore, to control the dimensions of a part, the packing pressure or cavity pressure must be controlled. Incidentally, this is the principle behind the use of cavity pressure sensing technologies that will be described in Chapter 12.

Figure 10.14 shows the effect of cavity pressure on the length of a tensile bar. The tolerances are also shown on the graph. As the cavity pressure increases, the length of the tensile bar increases.

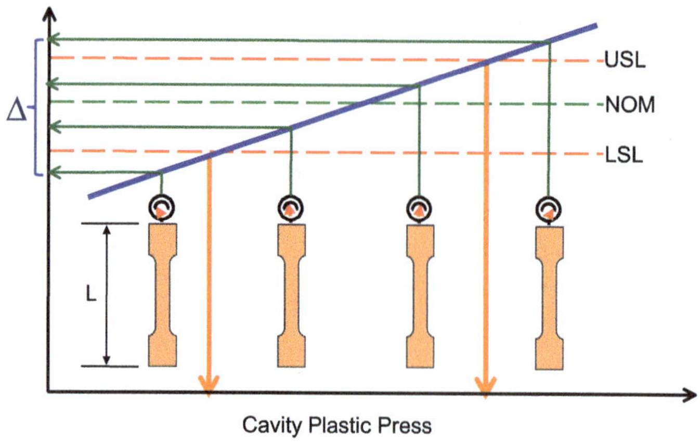

Figure 10.14 Effect of cavity pressure on the length of a tensile bar

Now let us consider a two-cavity tensile bar mold. To mold tensile bars of equal length, the plastic pressure in each of these cavities must be identical. When the cavity fill is identical, the pack and the hold phases that follow will have the same effect and produce identical cavity pressures. If one cavity happens to fill less than the other, then at process switch-over from pack to hold, the cavity that fills first will get pressurized more, leading to higher cavity pressure and therefore a larger dimension. This will result in cavity-to-cavity dimensional variation. One cavity may replicate the first tensile bar dimension in the figure and the other may replicate the last one. When one is within specifications, the other may be out of specifications.

In addition to the dimensions, when the process windows are small, the cavity that fills first may end up with flash, even when the other cavity is still unfilled (short). In such cases it will be impossible to mold cosmetically acceptable parts. An example is shown in Figure 10.15.

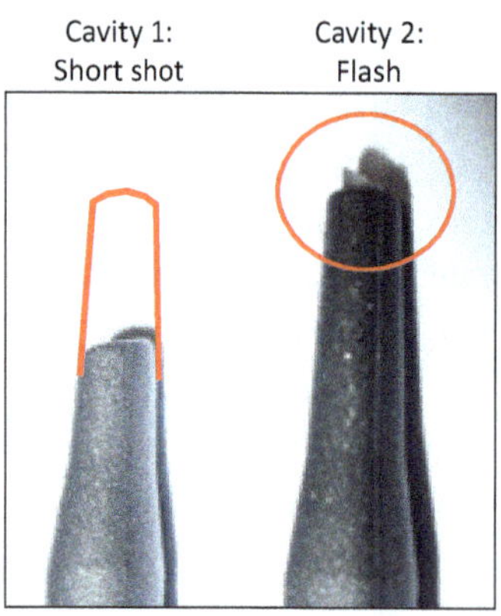

Figure 10.15
Two-cavity mold with one part short and the other with flash

10.1.3.1 Procedure to Determine the Cavity Balance

In order to achieve cavity balance, it is important to determine the reason(s) for any existing imbalance. First, the actual steel temperature in each cavity must be checked via a reliable method. Contact type probes work best. A thermal imaging system can also be very useful because it provides a quick picture even while the mold is running. Any difference in the cavity temperatures should be close (within a temperature difference of less than 2 °C or 5 °F).

Next, the incoming inspection of the mold should include a record of the runner and gate sizes, which must be identical. If no such record exists, the runner and gates sizes must be inspected and recorded. Runners, and especially gates, wear over time. Tracking these data over time is very valuable. These two factors will determine if cooling and flow channel variations could pose a problem. The procedure outlined in the following will determine if any flow imbalance exists because of rheological or venting factors.

- Set the holding pressure to zero.
- Set the holding time to zero.
- Set the screw recovery delay time to a value close to an estimated holding time, similar to the step in generating the viscosity curve in the previous section.
- Set the cooling time to a value that will ensure that the part will be cool enough to eject.
- Set the injection speed to the value obtained from the viscosity curve study.
- With the rest of the settings identical to the viscosity study, start molding. Because this test is done immediately after the viscosity study, the settings must yield a short part. If there is a visible cavity imbalance, the cavity that fills the most of all cavities must still be unfilled. For example, in a four-cavity mold, if cavity number 4 fills first, then in this study start molding parts so that cavity 4 is about 98% filled. At this setting, the other cavities will now have filled less than the 98%. All cavities will now be underfilled. Save the short shot and record the weights of each cavity. Take an average of at least two shots.
- Next, generate a short shot series and record the cavity weights for each shot as in the step above. Depending on the size of the part, at least 3 to 4 progressing shots must be recorded, starting with the smallest short shot parts that are possible to make. Next, the data must be plotted on a chart similar to the one shown in Figure 10.16.

Table 10.2 shows an example of the worksheet that can be used to document the cavity balance study.

Table 10.2 Cavity Balance Worksheet

Cavity ID	Part weight (g)				
	10% Part	25% Part	50% Part	75% Part	End of fill
1	1.94	3.15	6.12	8.34	12.94
2	1.92	3.14	6.20	8.21	12.85
3	1.95	3.25	6.92	8.86	12.12
4	1.95	3.36	6.55	8.52	12.82

Figure 10.16 shows that at lower percentage fills all the cavities are filled evenly.

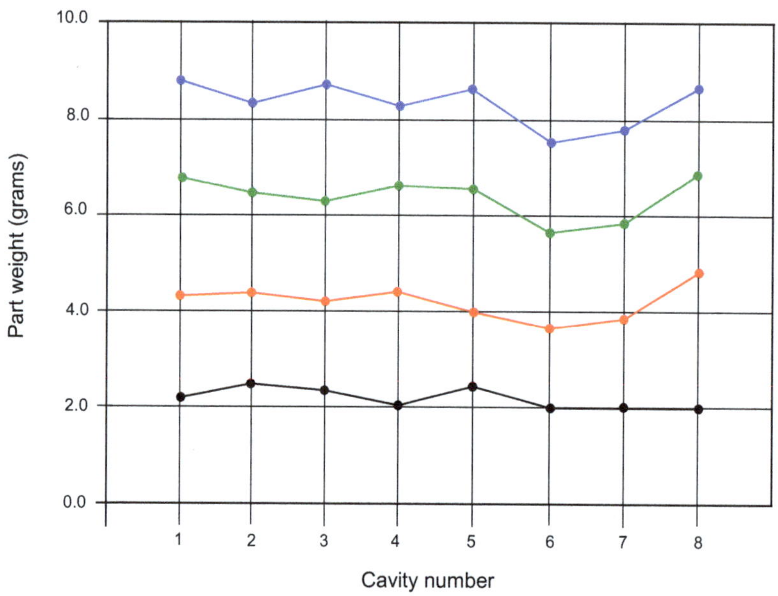

Figure 10.16 Cavity balance graph at various fill percentages

10.1.3.2 Reasons for Cavity Imbalance

Refer to Figure 10.17(a)–(e), which show a two-cavity mold that has an imbalance in the fill, where Cavity 1 fills less than Cavity 2. The possible reasons for the imbalance between the two cavities are as follows:

1. Nonuniform flow channel dimensions: The runners and the gates are the flow channels that feed the cavity. If they are not of identical dimensions, then the melt flow rate into the cavities will not be identical resulting in the imbalance. In Figure 10.17(a), either the runner or the gate of Cavity 2 could be larger than that of Cavity 1.

2. Nonuniform venting: The air has to escape out of the mold in order for the plastic to get in. There can never be too much venting and the more vents, the better. In Figure 10.17(b), if Cavity 1 has fewer vents than Cavity 2, then the air does not escape as easily as in Cavity 2, and therefore the plastic flow rate into Cavity 1 is less than that of Cavity 2.

3. Nonuniform cavity wall dimensions: Refer to Figure 10.17(c). This issue can show up mainly in thin wall parts and is not an issue with thicker wall parts. If the wall thickness decreases, a couple of phenomena can happen. First, a decrease in wall thickness will reduce the cross section and will cause that cavity to fill less than the other cavity. However, it could also happen that the drop in the wall thickness can also increase the shear in the plastic, which will in turn reduce the plastic viscosity filling that cavity more than the other. Assuming there is a wall thickness issue, if the thinner part fills more than the other, it is a rheology difference, whereas if the thinner wall fills less than the other, it is an issue of flow rates.

4. Nonuniform cooling: Refer to Figure 10.17(d). If the water flowing in Cavity 2 is warmer than the water in Cavity 1, the plastic is going to flow more easily into Cavity 2 and cause the imbalance. In a production environment, it is common that a line can get plugged up or the water line was not opened during startup because of an error. Cavity imbalances due to cooling usually do not show up until a few shots have been molded, and the cavity has got warmer. Water flow also takes the path of least resistance. If the pressure drop is higher in one cavity than the other, there can also be an imbalance, especially when the flow is closed to being a laminar flow. Flow types should therefore be checked using the Reynolds number formula.

5. Rheological imbalances: Refer to Figure 10.17(f) and Figure 10.17(e). As discussed in Chapter 3 on *Polymer Rheology*, there can be a rheological imbalance in the mold filling process. The imbalance in a two-cavity mold will be different than what you see in an eight-cavity mold. In a two-cavity mold, there will not be part weight difference in the fill. A short shot in a two-cavity mold will have parts that will represent mirror images of each other. Rheological imbalances can be fixed by altering the melt flow based on the principles and procedures developed by Beaumont Inc. Before the introduction of this technology, it was common to open the flow channels or the gates in order to balance out the flow. When cavity balance is discussed, it is not just the fill pattern that we are interested in. Ultimately, the quality of the melt that is delivered to the different cavities must be identical. The quality is defined not only by the flow rate, but also by the pressure and temperature history of the melt. Although modifying the runners and gates will balance out the flow rate, it will not balance out the pressures or the temperatures.

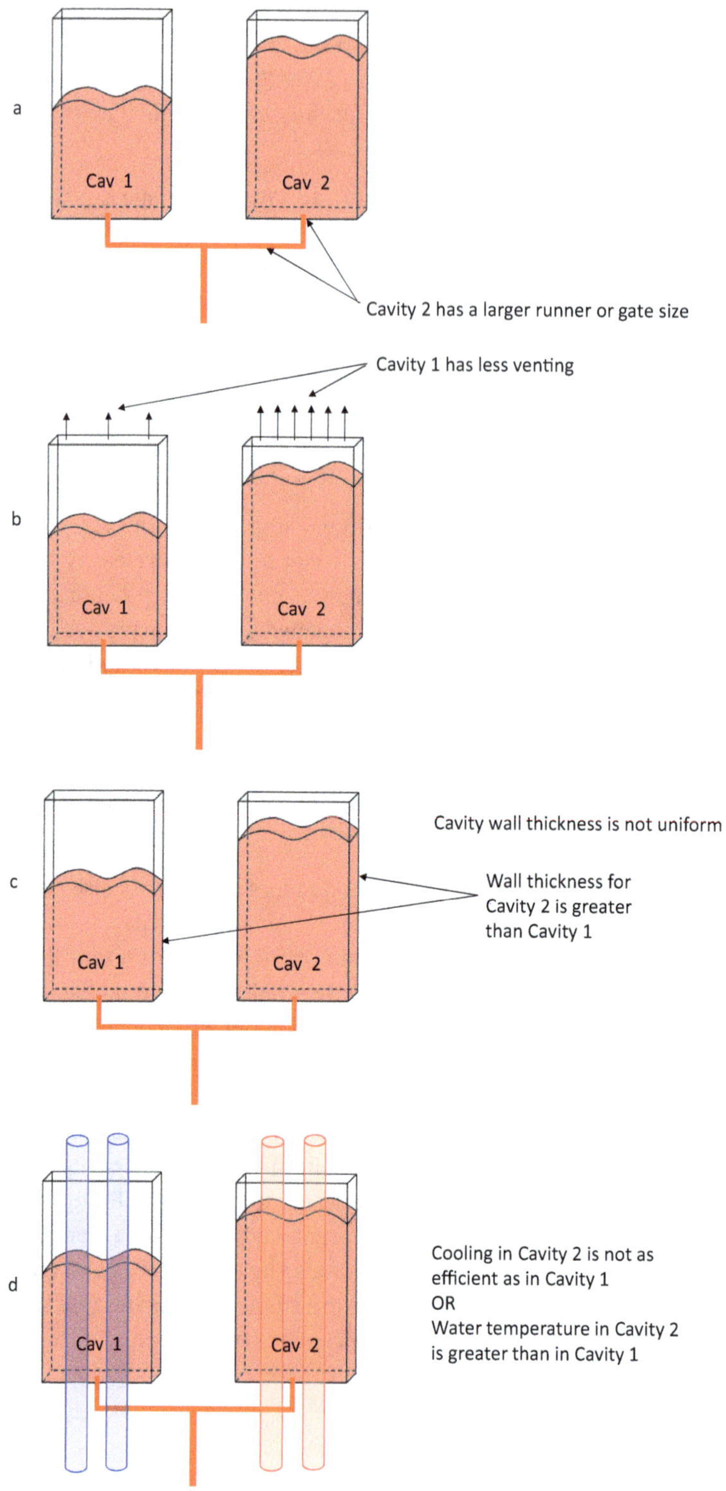
a
Cav 1
Cav 2
Cavity 2 has a larger runner or gate size
Cavity 1 has less venting
b
Cav 1
Cav 2
c
Cav 1
Cav 2
Cavity wall thickness is not uniform
Wall thickness for
Cavity 2 is greater
than Cavity 1
d
Cav 1
Cav 2
Cooling in Cavity 2 is not as
efficient as in Cavity 1
OR
Water temperature in Cavity 2
is greater than in Cavity 1

e

f

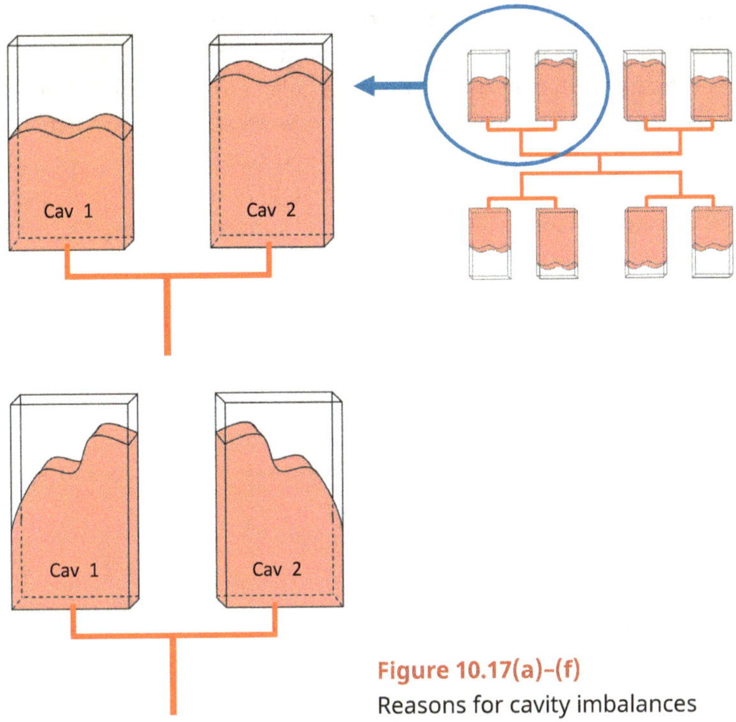

Figure 10.17(a)–(f)
Reasons for cavity imbalances

10.1.3.3 Determining the Cause of Cavity Imbalances

If a mold has an imbalance, the reason could be any one of the ones described in the previous section. To determine the reason, a short shot series fill study is performed and the cavities are filled progressively and collected. A short fill of about 10% of the part, a medium fill of about 50% of the part, and an almost full part of about 95% filled shots are collected. If the 10% fill shows an imbalance, it is most likely a flow channel variation. Venting imbalances, wall thickness imbalances, and rheological imbalances usually do not show until more of the cavities are filled. Filling the cavities further to 50% and then to the end of the fill will provide the rest of the possible reasons. If the initial 10% is balanced, but the 50% and 95% are not, then it could be a venting issue. If it is a thin walled part, the part thicknesses must be checked. If there is a rheological imbalance, then depending on the number of cavities, there will be two or four or eight groups of cavity weights. For example, if it is an eight-cavity mold, then there will be two groups of parts with similar weights in each group. After running the mold for a few shots, if one cavity starts to fill more than the other, the cooling must be checked. Performing a short shot series will provide the answers to the source of the imbalance. An example of a short shot fill experiment is shown in Figure 10.18. In this case, the parts form series 1 had similar part weights. This indicates that the gate and runner sizes were identical. As the parts filled, it was cavity 2 that filled less than the others. If this was a rheological imbalance then there would have been two parts that had similar weights and the other two would have had similar weights. Rheological imbalance is therefore ruled out. The parts were not thin wall parts and therefore wall thickness variation was ruled out. The cooling lines and the actual mold cavity temperature were checked and were all within an acceptable temperature range. By the process of elimination, it was determined that this could be a venting problem. When the mold was checked in the tool room, it became apparent that it was in fact a venting issue. Adding more vents to the short cavity helped the issue and improved the balance.

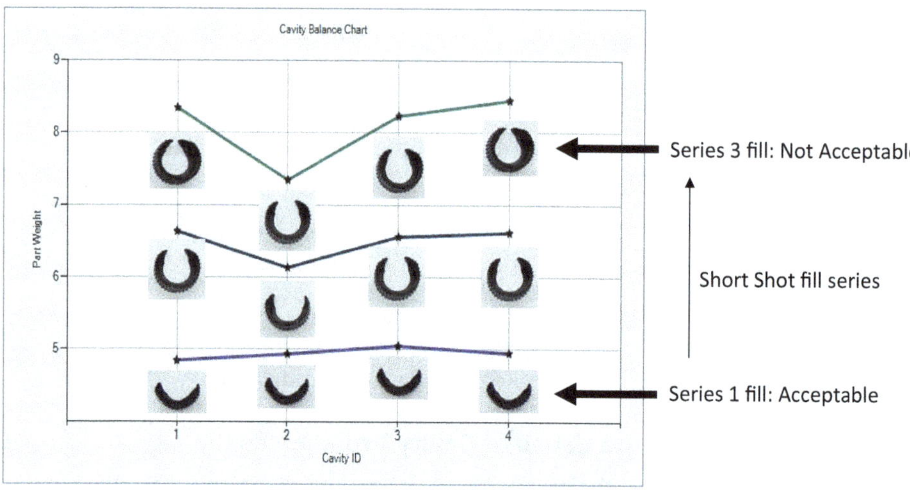

Figure 10.18 Shot shot series fill to determine the cause of imbalance

10.1.3.4 Calculating Cavity Imbalance

The common formula for calculating cavity imbalance is

((Max. cavity weight – Min. cavity weight)/Min. cavity weight) × 100

However, this is incorrect.

One must realize that in the mold filling process, the plastic is entering one cavity in preference to the other when it should have entered both cavities at the same rate. If one cavity was partially filled and the derived part weighed 12 grams and the other was also partially filled and gave a part weighing 8 grams, in an ideal situation they both should have filled to 10 grams each. The total fill was 20 grams and the average is 10 grams. One cavity filled 2 grams more whereas the other filled 2 grams less. Or in other words, the variation from the average was +2 grams and –2 grams, respectively. If a change was made to a gate size, the 20 grams will now get distributed again between the two cavities and the average will still be 10 grams. For example, if the gate size for the 8 gram cavity was opened up, then the part weight could now be 9 grams, but then the other cavity's part would weigh 11 grams. The part weight in the

12 gram cavity changed without any change in its gate size. Variation from the average should always be calculated.

The steps to the calculations are as follows:

- Record the weights of each short filled cavity.
- Add the weights as the total fill.
- Calculate the average fill by dividing the total part weights by the number of cavities.
- Calculate the variation from the average using the formula:

  % Variation from average = ((Average − Cavity weight)/Average weight) × 100

See the example of a cavity balance study in Table 10.3.

Table 10.3 Example of a Cavity Balance Study

Cavity No.	95% Fill Parts	% Variation from Average
1	5.154	5.00
2	5.634	−3.84
3	5.468	−0.78
4	5.446	−0.38
Total weight (g)	21.702	
Average part weight (g)	5.426	

Cavity imbalance numbers in small parts can get very misleading because a small change in the weight can lead to large changes in the percentage numbers. One must be careful in such cases and sometimes a coefficient of variation should be considered.

10.1.3.5 Acceptable Level of Cavity Imbalance

It is impossible to get a near perfect balance in fill. The question then comes to mind as to what is the acceptable level of cavity imbalance. Most companies have a fixed number and most training courses teach that there should be a fixed number. The answer is there cannot be such an acceptable number. For example, consider a molder who is molding forks and spoons, and the percentage imbalance between the least filled and the most filled is 15%. However, when the molder applies the pack and hold pressures and packs the parts out, not only does she get acceptable looking parts, but there is also an acceptable process window. Dimensions on forks and spoons do not matter, so in this case, a 15% or, for that matter, any cavity imbalance would not matter as long as the parts are cosmetically acceptable and the process window is acceptable. In some multicavity molds, if one cavity is not acceptable, it is a common practice to block off that cavity and then mold the parts. If the parts pass the quality inspection with one blocked cavity, that equals 100% imbalance (1 cavity is 0% whereas the other is 100% full). This means that there was enough cosmetic and dimensional tolerance available to the molder to block of a cavity and mold parts.

Keeping in mind the PVT diagram for a plastic one can easily conclude that the higher the pressure on the plastic, the less shrinkage will take place; or in other words, the parts will get larger with increasing pressures. Refer to Figure 10.14, which represents this concept.

An acceptable number for cavity balance depends on the dimensional and cosmetic tolerances for the given part, so the number should be decided on in a case-to-case basis and never as a generic rule. Refer to Figure 10.19 and consider the two scenarios shown. The imbalance between the two cavities is 10%. In scenario 1, the dimensional tolerances are +/−0.010″ on both sides of the nominal and the two cavities are within tolerance. However, in scenario 2, the tolerances are dropped down to +/−0.005″ and now 10% becomes unacceptable.

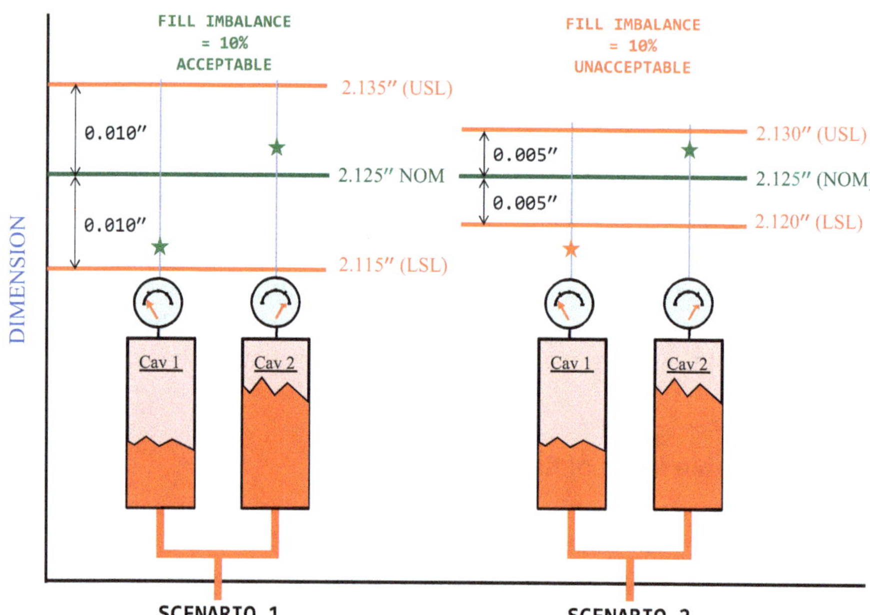

Figure 10.19 Acceptable % imbalances and tolerances

In some molded products, such as the wall clips shown in Figure 10.20, there are very large percentages of imbalances that do not matter. Most common users do not realize that these are actually sold with the runner system that acts as a carrier or a holder. The clip is broken off and used and the runner is discarded. The same is true with several toy kits where children assemble a figure or a vehicle, breaking parts off an array of a grid, most of which is the runner system.

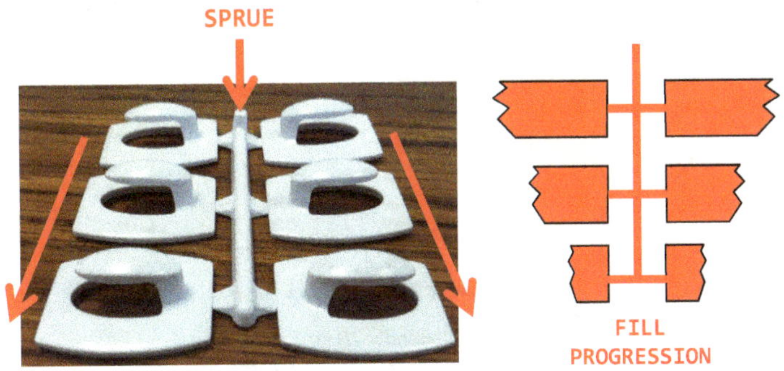

Figure 10.20 Fill progression in a wall clip mold

The polymer morphology also plays a role in cavity balance. Amorphous materials can tolerate more imbalance than crystalline materials. Amorphous materials do not shrink as much as crystalline materials, and therefore cavity-to-cavity dimensional variations are smaller. At the same time, because crystalline plastics flow more easily, they tend to flash easily, reducing the process window. Balancing out the flows between cavities helps improve the process window regardless of the morphology of the material and whether the tolerances are wide or not.

Family Molds

Family molds are molds that do not mold identical parts in all cavities. It is common to build family molds for small production runs of assemblies. The balance of fill is important in family molds too. Again, the concept here is the same as before, where the cavity pressure is the deciding factor for the dimensions. Typically, in these cases, the gate and runner sizes are adjusted to balance out the fill. Depending on the configuration, the number of cavities, and the cavity layout, a decision must be made on the type of cavity fill balance technique.

10.1.4 Step 3: Determining the Pressure Drop – Pressure Drop Studies

Consider a driver driving a car on a flat road at a desired speed of 70 km/h. To achieve this speed, the gasoline pedal is depressed 50% of the maximum possible. As the driver approaches a hill, to maintain the desired speed of 70 km/h the driver will now need to depress the pedal further to provide for more gasoline needed to overcome the excess work to drive uphill. To maintain the desired speed, the position of the pedal is adjusted. If the position of the pedal was not adjusted, and was kept at the 50%, then as the car starts to climb the hill, the speed would drop from 70 km/h to a lower value. In molding, enough pressure must be available to compensate for all the pressure losses and material viscosity variations while keeping the injection speed constant. Refer to Figure 10.21.

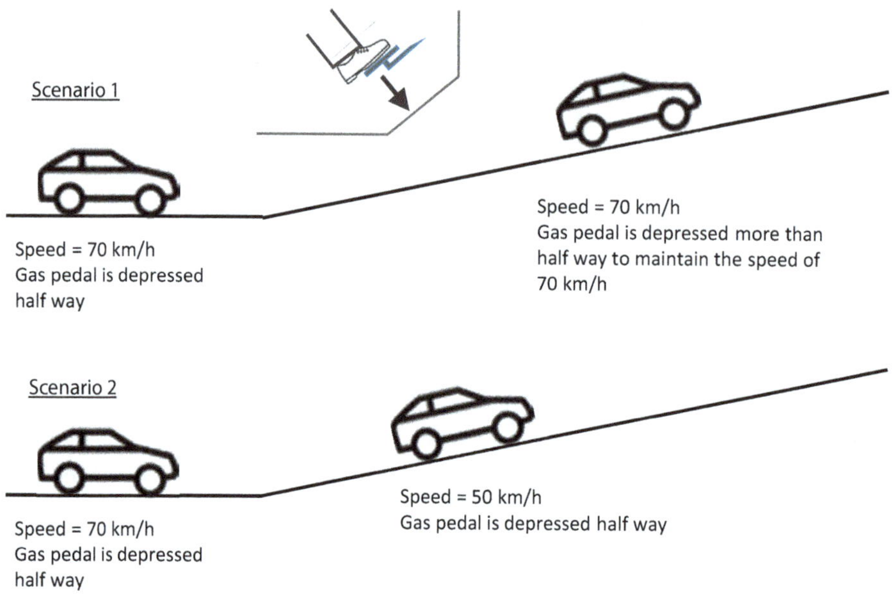

Figure 10.21 Maintaining the speed of car

As plastic flows through the different sections of the nozzle and the mold, the flow front of the plastic experiences a loss of applied pressure because of drag and frictional effects. Additionally, as the plastic hits the walls of the mold, it begins to cool, increasing the viscosity of the plastic, which in turn requires additional pressure to push the plastic. Due to the fountain flow effect, the outer frozen layer also gets thicker and thicker, thereby reducing the available cross sectional area for the flow, and therefore increasing the required force to push the plastic through the flow channel. Depending on the pump capacity of the molding machine, there is a limited maximum amount of pressure available to push the screw at the set injection speed. The required pressure to push the screw at the set injection speed should never be more than the maximum available pressure. If the pressure required is higher, the screw will never be able to maintain the set injection speed throughout the injection phase and the process is considered pressure limited. Initially, the set speed may be reached, but as soon as the process becomes pressure limited, the screw slows down as shown in Figure 10.22.

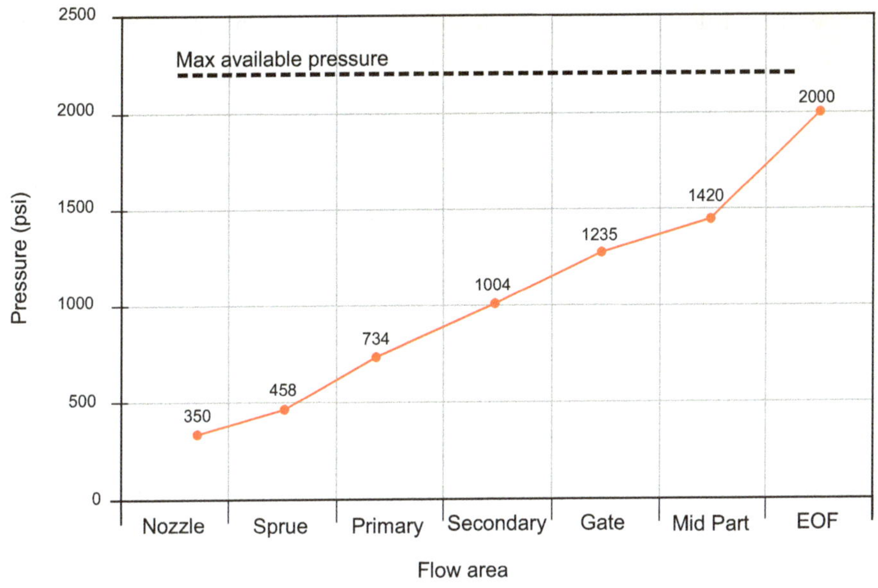

Figure 10.22 Graph to determine the pressure drop through the mold

The term 'pressure drop study' can be confusing. The machine pressure required to push the plastic at a constant speed increases as the flow progresses, but the actual pressure of the melt front drops as the flow progresses. It can easily be understood by looking at a short shot where the pressure at the end of the flow is zero. The words pressure and drop are related to the plastic.

During process development, knowing the pressure loss in every section of the flow path helps to determine the overall pressure loss and the sections where the pressure drops are high. The mold must be modified to reduce this pressure drop, if the process is pressure limited. The runner system of the mold can then be modified to reduce this pressure drop and achieve a better consistent flow. Sudden changes in pressure drop from one channel into another are not desirable. The sections where there are sudden changes in the pressure must be modified to achieve a better consistent flow.

10.1.4.1 Procedure to Determine Pressure Drop

The pressure drop study is performed as the third step and therefore, by now, the injection speed must be fixed based on the viscosity study; the shot size and transfer positions must be fixed based on the 95 to 98% fill. It could be possible that a viscosity study was not done because the process was pressure limited. In such a situation, it becomes even more imperative to conduct a pressure drop study and determine the restrictive sections of the flow. The guidelines and procedure are as follows:

- Consider the plastic flows through the following sections: the nozzle of the machine, the sprue, the primary runner, the secondary runner, the gate, and the end of fill.
- Set the melt temperature, shot size, transfer position, and injection speed as follows:
  - For a non-pressure limited process, set these based on the viscosity study and the 95 to 98% rule as described in the previous sections.
  - For a pressure limited process, set the injection speeds to the center of the range on the machine and find the values of the shot size and transfer where the parts are about 95 to 98% full.
- The injection pressure setting should be set to about 20% higher than the maximum recorded value during the viscosity study.
- Start molding and let the process stabilize. The parts should be short and almost filled. Record the displayed peak injection pressure on the screen of the machine. This value would represent the amount of pressure required to fill the mold almost up to the end of fill.
- Next, adjust the transfer position to fill about 50% of the part and record the peak pressure value again.
- Continue the procedure to mold 5 to 10% of the part, the secondary runner, the primary runner, and then just the sprue.
- After this data is collected, stop molding and pull the barrel back and away from the mold. Build a shot and inject the plastic as one would purge the barrel. This is called an air shot. Record the maximum pressure during this injection. This would be the pressure required to inject the plastic through the machine nozzle.

Plot a graph of the required pressure versus the flow section starting from the machine nozzle. The graph is called the pressure drop study and will show the required pressure progressively increasing as the plastic fills the mold and up to the end of fill. Figure 10.23 shows such a graph.

Note: Data should be collected on stabilized cycles. Therefore, if at any time a section of the part or runner starts to stick on the mold and it takes more than 20 to 30 seconds to remove the stuck part, the study should be stopped because the data will not be reliable. This is very common as one comes to molding just the runner or the sprue. It is not required to collect the data especially if the mold is not pressure limited. If the last data point collected was a number that was high, such as about 50% of the maximum available machine pressure, the runner sizes must be investigated. This is a common problem with some hot runner molds where one can notice that most of the available pressure is taken up to get the plastic to exit the hot runner and into the mold. In the case of the hot runners, the same steps above can be performed and the pressure drop through the manifold can be determined.

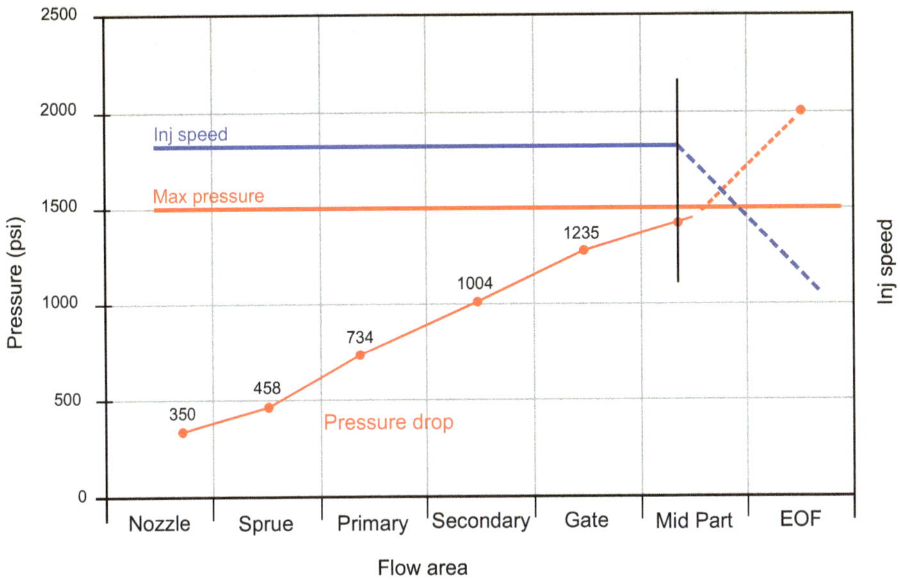

Figure 10.23 Effect of a pressure limited process on injection speed

In all instances, the pressure drop through the nozzle should always be measured and recorded. The pressure drop through the nozzle is inversely related to the size of the orifice. As the orifice size decreases, the pressure drop increases. During mold startup, the actual pressure drop through the nozzle must be compared to the recorded pressure drop. This will give an indication of whether the right size and type of nozzle is installed on the machine. For example, a nylon type nozzle will have a slightly higher pressure drop as compared to a conventional nozzle for the same diameter orifice.

If any problems arise with the part filling during production, one of the first actions must be to pull the barrel back and check the pressure drop through the nozzle. Any debris, such as metal shavings stuck in the nozzle, can cause an increase in pressure drop. This loss of the pressure drop at the nozzle gets added to the sprue, the runner, the gate, and the part. If the process is close to being pressure limited, this can easily lead to short shots and defective parts. Moreover, the loss in pressure in the runner system leads to a loss in cavity pressure and therefore a change in dimensions, as shown in Figure 10.24. The type and orifice size of the nozzle are therefore very important and must be recorded as part of the setup sheet.

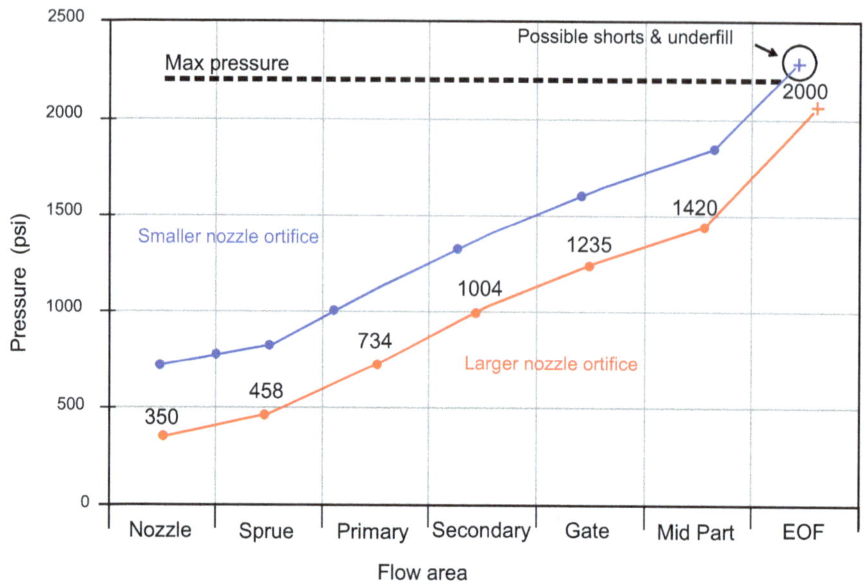

Figure 10.24 Effect of nozzle type and size on near pressure limited processes

In hot runner molds the procedure should start with the nozzle and then continue with the section where the melt is first injected into the mold. The hot runner section of the mold must be considered as one flow section, and then each subsequent flow section must be analyzed. If it is possible to manually inject the plastic at the set speed. An injection of the plastic into an open mold should be performed, and the pressure drop should be recorded. For safety reasons, some machines do not allow the processor to inject plastic at the full set injection pressure. In such cases, the data should still be recorded, and the machine should be clearly identified on the setup sheet. Safety should be of primary concern, and no safety features of the machine should be compromised. In the case of a hot runner mold, injecting through the manifold with the mold open can provide insight into the pressure drop through the hot runner system. Preferably, a piece of thick cardboard should cover the movable side of the mold to prevent any splatter of plastic onto these surfaces. A cardboard shield should also be placed underneath the bottom of the mold for the purge, and another over the top of the mold, to prevent any unexpected burst of high pressure plastic from splattering above the mold and falling onto personnel near the machine.

The pressure drop through the different sections of the part must also be evaluated especially in thin walled parts. Thin sections in a mold require more pressure to fill and can make a process pressure limited. Although the runners and the gates are generously designed, the part can have restrictions that may cause a pressure drop.

Table 10.4 is an example of a worksheet that can be used to document a pressure drop study.

Table 10.4 Pressure Drop Study Worksheet

Section	Peak pressure (psi)
Nozzle	236
Sprue	454
Primary runner	734
Secondary runner	1004
Tertiary runner	1248
10% part	1630
50% part	1837
End of fill	2030

10.1.4.2 How to Use the Information from the Study

The maximum pressure used in the process should never reach the maximum available pressure on the machine used for molding the parts. For example, if the maximum available hydraulic pressure is 2200 psi, the end of fill pressure should not reach 2200 psi. Typically, only up to 90% of the maximum available pressure should be used. In the generated graph, if the process is pressure limited or reaches more than 90% of the maximum, steep increases in the required pressure between two sections must be identified. For example, if the required pressure to fill through the nozzle, sprue, primary, and secondary runner is 900 psi, and if through the tertiary runner, the pressure increases to 1600 psi, this increase is significant as compared to the increases in the previous sections. Modifications to the tertiary runner must be made. Increasing the diameter of the runner will help in reducing the pressure.

Another reason for a pressure drop study is to prevent the mold from being overpacked. Refer to Figure 10.25. Consider that the shot size and the transfer positions were set in accordance to the basis of the 95% fill and that the mold is not pressure limited. The screw will start from the shot size position and will use the pressure to start filling the mold with constant set injection speed. If for some reason during production one of the gates gets plugged, then the other cavity will get filled to 100% first. Because the screw has not reached the transfer position, it will now use all the available machine pressure to try to reach the transfer position and will end up over packing the cavity with the open gate. The cavity can flash, and the mold components can get damaged. To avoid this problem, the injection pressure should be set to about 10 to 15% of the peak required pressure. Some molders set the maximum injection pressures to about 2000 psi plastic pressure (135 bar) above the peak injection pressure. This is also acceptable. The other safety limit that should be set is a maximum limit on the injection time. A tolerance of injection time is +/− 0.02 to +/− 0.05 seconds on machines with screw diameters of less than about 70 mm. A larger screw can use a slightly higher tolerance.

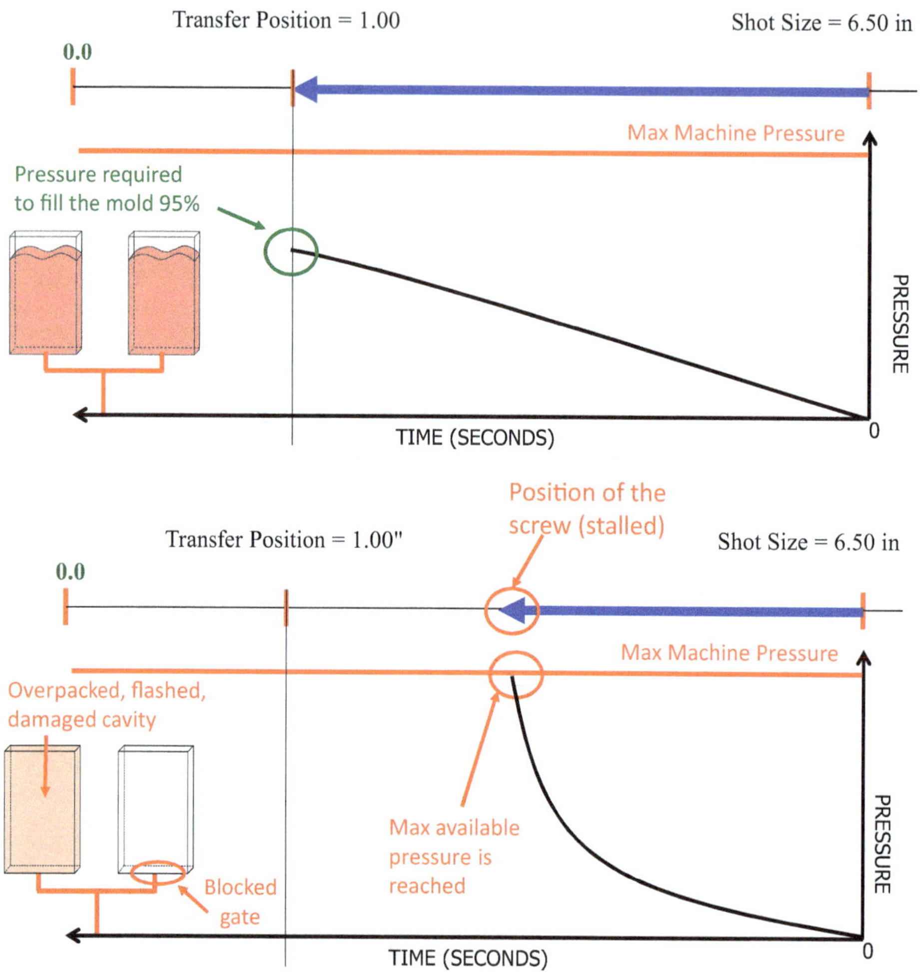

Figure 10.25 Pressure used to fill two cavities; (a) to 95% fill, and (b) when one cavity gets blocked

10.1.4.3 Effect of Pressure Drop on the Pack and Hold Phase

The pack and hold phase are collectively the compensation phase. Pressure must be available so that the plastic in the cavities can be pressurized and the shrinkage can be compensated for. If the process is pressure limited, it is very common to produce parts with sink and/or dimensional issues (typically undersized) and with large variations. If process capability is measured, then the values are typically low. This is because of the unavailability of the plastic pressure to pack the parts to the optimum levels since the pressure is used up in filling the parts and not much remains to compensate for the shrinkage. The plastic also cools down, requiring higher pressures for the pack and hold. Refer to Figure 10.26.

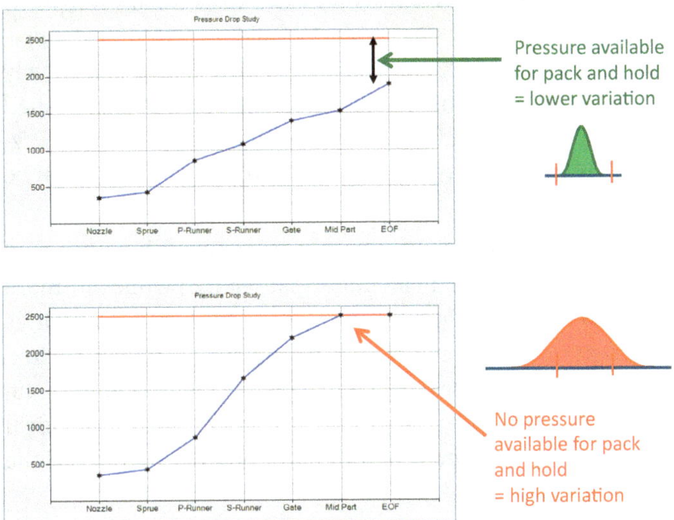

Figure 10.26 Effect of a pressure limited process on quality variation

10.1.5 Step 4: Determining the Cosmetic Process Window – Process Window Study

Of all the six steps used for Part 2 of process development, the step of developing the cosmetic window is the most important step. The aim of a molder should be to set the molding process, start the molding run, and walk away from it, and not have to come back to constantly to adjust the process in order to make acceptable parts. It is similar to wanting to drive a car in cruise control. But one can only drive a car in cruise control when the road is wide enough to allow the minor variations by the driver during a long drive. It will be impossible to apply cruise control if one is driving on a cliff of a mountain! The wider the road, the better the chances of a smooth drive. Similarly, the wider the process window, the better the chances are for a robust process. The process to achieve the required dimensions of the part is described in Chapter 10. The wider the cosmetic process window, the higher are the chances of achieving the dimensions as the dimensional window is a subset of the cosmetic window.

As described earlier, the injection of plastic into the cavity can be divided into two main phases. The first phase is the injection phase, in which the mold cavity is completely filled with the molten plastic. The volume of the melt is equal to the volume of the cavity. The second phase is the compensation phase, which consists of the pack and hold phase. The pack pressure must pack additional plastic material, equivalent to the volumetric shrinkage caused by cooling as the plastic makes contact with the cold mold walls and the hold pressure must hold the plastic inside the cavity until the gate freezes off. The various parameters that need to be controlled during this phase are packing pressure, holding pressure, packing time, and holding time. In most production processes, packing and holding are not differentiated and are collectively called the holding phase. For this reason, for the discussion of this section, we shall call the pack and hold phase as just the hold phase. Section 10.1.7 discusses optimizing the gate freeze times. The topic of differentiating between pack and hold will be discussed then.

The packing phase consists of packing the cavity with plastic equal to the theoretical weight of the part. Any less plastic will result in an under-packed part while any more will result in a part that is over-packed. Under-packed parts show defects such as sinks and internal voids. Such parts usually exhibit considerable post-molding shrinkage. Over-packed parts can have built-in stresses that usually get relieved after molding, resulting in defects such as warpage or premature mechanical failure. The ideal pack and hold pressures are determined by evaluating the process window of the mold followed by the gate seal study.

An example of a process window is shown in Figure 10.27. The recommended melt temperature for the ABS material being molded is 473 °F (245 °C) to 527 °F (275 °C). At the lower end of the melt temperature range, cosmetically acceptable parts were molded at a holding pressure of 4000 psi. Below 4000 psi the parts were short and had sink marks. The pressure was increased in increments and it was observed that above 10,000 psi of pressure the parts had flash. The window of acceptable parts was therefore 4000 to 10,000 psi or a window of 6000 psi. The same experiment was repeated at the higher end of the melt temperature and the low and high limits were 3000 psi and 9000 psi, respectively, also a window of 6000 psi. When these four decision points are joined, they form the cosmetic process window. Parts molded inside this window will always be cosmetically acceptable. The most robust setting to mold parts will be the center of the window, where all the natural variations in the material, the machine, and other factors will all get cushioned and the parts molded will be cosmetically acceptable. An example of a process at the edge of the cliff will be setting the process at a melt temperature of 500 °F and a pressure of 9500 psi. Such a process will produce parts that will be a mix of acceptable parts and parts that have flash. This would therefore not be an acceptable process to set on the molding machine.

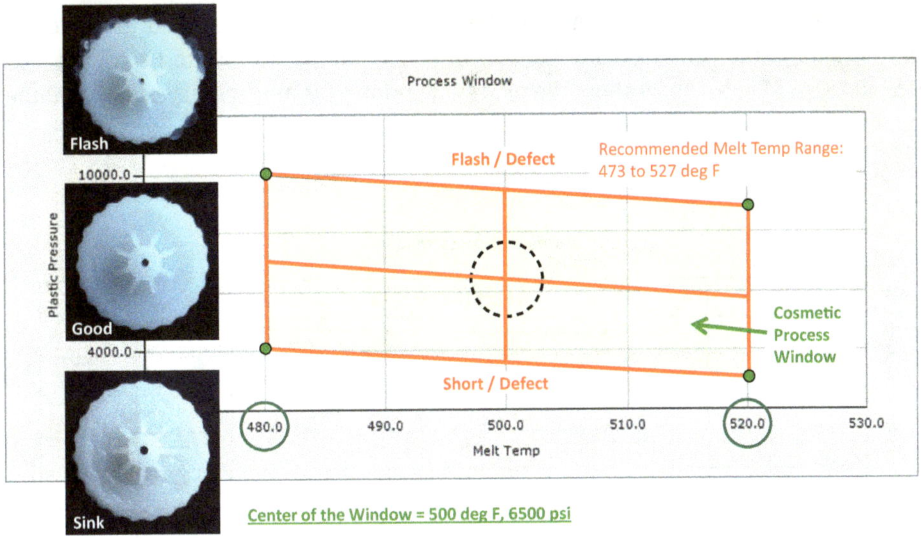

Figure 10.27 The cosmetic process window

Two process variables are varied to establish the process window. For amorphous materials, the holding pressure and the melt temperatures are the common variables that are used. The mold temperatures are set either in the center of the recommended values or at a desired value within the recommended range. This could be based on experience on a similar part or for other reasons such as for better cosmetics.

For crystalline materials, the two variables that are selected are holding pressure and mold temperature. In the case of crystalline materials, the mold temperatures are much more critical for the formation of the crystallites that finally dictate the properties of the part. The melt temperature range for crystallites is also narrow and therefore varying the temperature is not going to give any more information than that obtained at the center of the range. Holding pressure usually has the most impact on the quality of the part because it directly influences the specific volume and therefore part dimensions. Therefore, the significance of holding pressure applies to both types of materials. For amorphous materials, melt temperature usually has the higher impact compared to mold temperature, and in the case of crystalline materials, mold temperatures have a higher impact compared to melt temperatures.

The process window is also called the molding area diagram. This is the area in which cosmetically acceptable parts are molded. Dimensions are not considered. The bigger the window, the more robust is your process. It must be noted that the process window provides only the range of processing parameters within which cosmetically acceptable parts are produced, also called the cosmetic process window (CPW). Chapter 10 is dedicated to the different types of process windows.

Dimensional data must be taken and the mold steel must be adjusted to move the dimensions to the center of the window in order for the process to be robust. Figure 10.28 and Figure 10.29 show the process windows for amorphous and crystalline materials, respectively.

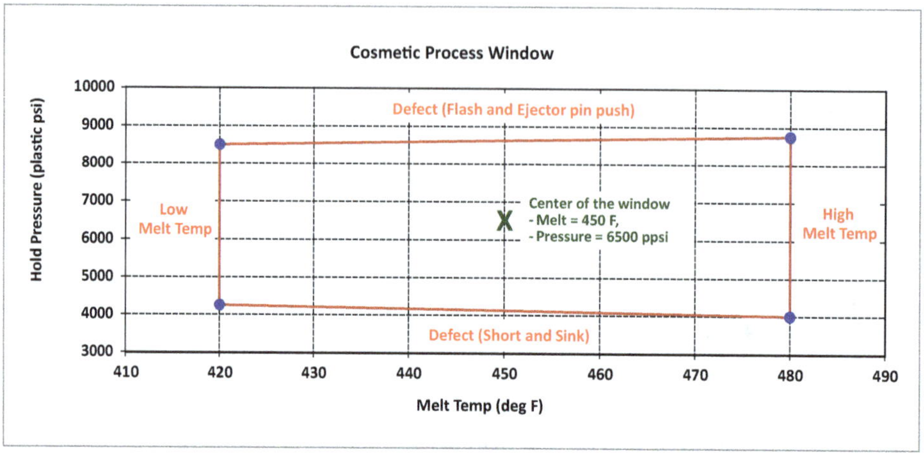

Figure 10.28 Process window study for amorphous materials – hold pressure vs. melt temperature

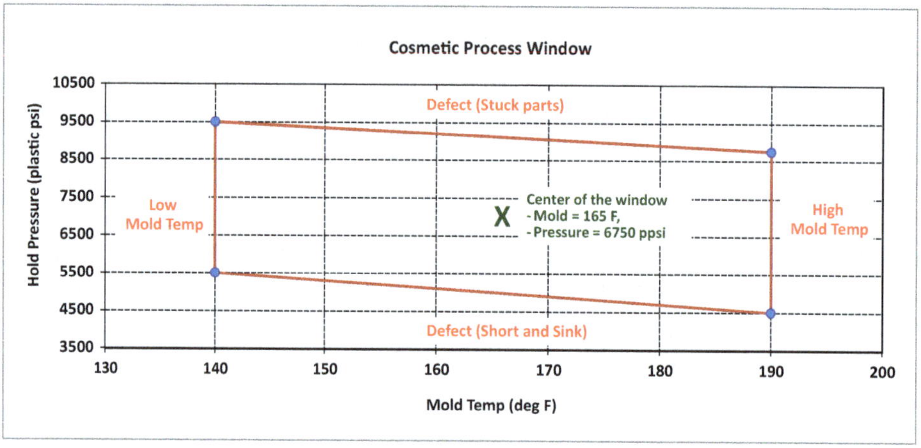

Figure 10.29 Process window study for crystalline materials – hold pressure vs. mold temperature

Outside the illustrated process windows, parts produced will not be acceptable because of defects such as sink, flash, or built-in stresses. Below the melt temperature limits, unmelted plastic may cause problems such as short fill or lack of required properties because of lack of melt homogeneity, whereas above the temperature limit, degraded plastic may lead to parts that are unacceptable in terms of properties, such as mechanical properties. With low mold temperatures, the plastic may not have enough energy for crystallite formation. On the other hand, with mold temperatures higher than necessary for crystallization, cycle time may increase. The process is set to the center of this window so that any variations within the window would still yield cosmetically acceptable parts. The larger the window, the more robust is the process.

Chapter 10 will define and further discuss the concepts of dimensional process window (DPW) and control process window (CoPW).

10.1.5.1 Procedure for Determining the Process Window for Amorphous Materials

The following details the procedure for developing a process window for amorphous materials using holding pressure and melt temperature as the two variables:

- Set the mold temperatures to a medium value or at a desired temperature within the range of recommended mold temperatures.
- Set the barrel temperatures to attain the lower value of the recommended melt temperatures.
- Set the injection speed to the value obtained from the viscosity curve experiment.
- Set all holding times and pressures to zero.
- Set the screw delay time to a value approximately equal to the estimated holding time for the part. Refer to Section 8.4 for explanations.
- Set the cooling time to a value higher than what would be typically necessary (e.g., if the estimated cooling time is 10 seconds, set the cooling time to 20 seconds).
- Start molding and adjust the transfer position to make a part that is 95–98% full. Record the part weight as the 'injection only part weight value.'
- Let the process and the melt stabilize by molding approximately 10 shots of parts.
- Set the hold time to a value that makes sure the gate is frozen. (The next section deals with optimization of this time.) The determination of hold time is typically based on previous experience. For example, for a part being molded with an ABS material having a gate size of 0.040″, the hold time typically ranges between 4 and 6 seconds. For this part of the experiment, we would set the hold time between 8 and 10 seconds.
- Increase the holding pressure in small increments and record the pressure at which the first cosmetically acceptable part is made. There should be no defects such as shorts, sinks, or voids.
- Record this pressure as the 'low temperature – low pressure' corner.

- Increase the pressure further in similar increments and record the pressure at which there is evidence of an unacceptable result, such as part sticking in the mold, flash on the part, or warpage. Record this pressure as the 'low temperature – high pressure' corner.
- Increase the melt temperature to the higher end of the recommended melt temperatures. Drop the hold pressure and hold time to zero and add the charge delay time. Readjust the transfer position to obtain the 'injection only part weight.'
- Increase the holding pressure in small increments, and record the pressure at which the first cosmetically acceptable part is made. There should be no defects such as shorts, sinks, or voids.
- Record this pressure as the 'high temperature – low pressure' corner.
- Increase the pressure further in similar increments and record the pressure at which there is evidence of an unacceptable result, such as part sticking in the mold, flash on the part, or warpage. Record this pressure as the 'high temperature – high pressure' corner.
- Joining these four corners will generate the process window or the molding area diagram.
- Set the process to the center of this window.

10.1.5.2 Procedure for Determining the Process Window for Crystalline Materials

The following is the procedure for developing a process window for crystalline materials using holding pressure and mold temperature as the two variables. The procedure is the same as that used for amorphous materials except melt temperature is replaced by mold temperature in this study:

- Set the melt temperature to a center value or a desired value within the range of recommended melt temperatures.
- Set the mold temperatures to attain the lower value of the recommended mold temperature range.
- Set the injection speed to the value obtained from the viscosity curve experiment.
- Set all holding times and pressures to zero.
- Set the screw delay time to a value approximately equal to the estimated holding time for the part.
- Set the cooling time to a value higher than what would be typically necessary (e.g., if the estimated cooling time is 10 seconds, set the cooling time to 20 seconds).
- Start molding and adjust the transfer position to make a part that is 95–98% full.

- Let the process and the melt stabilize by molding approximately 10 shots of parts. Record this as the injection only part weight value.'
- Now set the hold time to a value that makes sure the gate is frozen based on previous experience.
- Increase the holding pressure in small increments, and record the pressure at which the first cosmetically acceptable part is made. There should be no shorts, sinks, or voids.
- Record this pressure as the 'low temperature – low pressure' corner.
- Increase the pressure further in similar increments and record the pressure at which there is evidence of an unacceptable result, such as the part sticking in the mold, flash on the part, or warpage. Record this pressure as the 'low temperature – high pressure' corner.
- Repeat the last two steps, but use the high end of the recommended mold temperature. This time the two extreme parameter combinations would be the 'high temperature – low pressure' and 'high temperature – high pressure' corners.
- Joining these four corners will now generate the process window or the molding area diagram.
- Set the process to the center of this window.

Table 10.5 is an example of a worksheet that can be used to document the process window study.

Table 10.5 Process Window Worksheet

Melt temperature (°F)	Low hold pressure (psi)	High hold pressure (psi)
430	5500	10500
510	6000	11000

10.1.5.3 How to Use This Information

The size of the process window is an indicator of how much variation the process will tolerate while still producing cosmetically acceptable parts. The aim is to have a wide process window. If the process window is very narrow, there is always a danger of molding parts with defects. For example, the graphs generated via the process window studies show that with a very small process window even natural process variations could cause occasional short shots or flash. A robust process is one that has a large process window and accommodates the natural variation inherent in the system.

A mold that produces short parts and parts with flash at the same time does not have any process window. The mold must be fixed because it will be impossible to make acceptable parts. Molds with extremely small process windows usually have dimensional issues. The process in which the parts are cosmetically acceptable produces parts that are out of specifications. When a certain specification is achieved, the parts are usually not cosmetically acceptable. Achieving consistent production from such molds is a difficult task and the processor often has to continually adjust the process. A few molds with small process windows on the production floor can easily consume available resources and make a plant run inefficiently. Determination of the process window is the most important process engineering study of all and if nothing else, this is the one study that must be done.

Part dimensions are discussed in Chapter 10. The concept of the dimensional process window (DPW) is presented. The DPW is a subset of the CPW. The bigger the cosmetic window, the higher the chances that the dimensions will be met. The DPW is a subset of the CPW. Every effort must be made to have a large CPW.

In the case of amorphous materials, set the mold temperatures on the lower end of the recommended temperatures, remembering that molding is a heat transfer process and so lower temperatures will give faster cycle times. Mold temperatures also dictate the cosmetics of the part. The higher the temperature, the better are the cosmetics. Therefore, one will need to experiment with the cosmetics to select the final mold temperature. Mold temperatures play a role in the dimensions of the parts and therefore can be used as one of the factors when performing the designed experiments.

In the case of crystalline materials, set the melt temperatures to the center of the recommended range. Typically, melt temperatures in the recommended ranges do not have a big impact on part quality since the ranges are narrow.

10.1.5.4 Shape of the Process Window

The shape of the process windows is not of any fixed pattern. It is not necessary that the difference between the pressures on the low temperatures is the same as the difference on the high temperatures. The windows can be of any shape, as shown in Figure 10.30.

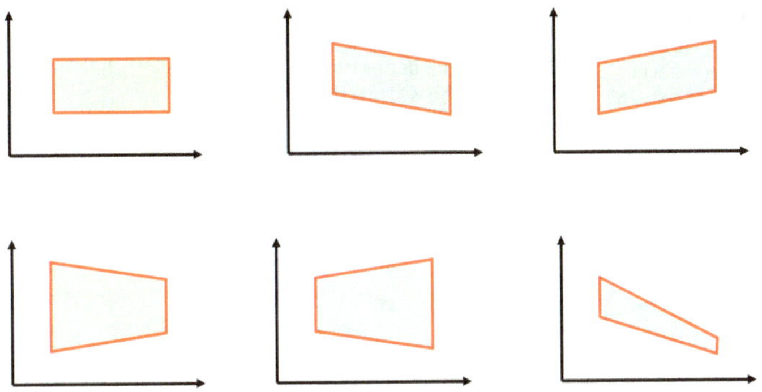

Figure 10.30 Shapes of cosmetic process windows

10.1.5.5 Cautions and Exceptions

The process window described in this chapter is the cosmetic process window. Within this processing window, parts can be molded that are cosmetically acceptable. The procedure calls out for increasing a certain process parameter, such as the holding pressure, to a limit where the parts would have a cosmetic defect such as flash or the parts stick in the mold due to over packing. However, consider a case where the parts do not flash even at high holding pressures or do not show any signs of over packing. In these cases, although there is no evidence of failure, the parts could very well be over packed. In such cases, the higher pressure at which cosmetically acceptable parts can be molded must not be taken as the upper limits of the process. An evaluation of the parts and the high pressure limits is required and the process must then be set accordingly. In such cases, cosmetic process windows can be misleading because part failures may not be seen until they are used in an assembly. Parts with large gates, hot runner mold, and valve gated molds fall into this category. The concept of pack and hold that will be described in Section 10.1.7 should be used in such cases.

10.1.5.6 Relationship between Cavity Balance and Process Windows

Refer to Figure 10.31, which is a cavity balance study graph for a four-cavity mold. Cavity 2 is the last to fill and Cavity 4 is the first to fill. This indicates that Cavity 2 will get packed out last and therefore will determine the low pressure value of the process window. Because Cavity 4 gets packed out first, it will be the first one to show the defects, therefore, it will determine the high value of the process window. The higher the imbalance between cavities, the smaller is the process window. The ideal situation would be that all the cavities fill at the same rate and therefore will pack the same rate. This will lead to a wider process window and better the chances of a robust process. Cavity balance plays an important role not just in cosmetics but also in process robustness and process capability.

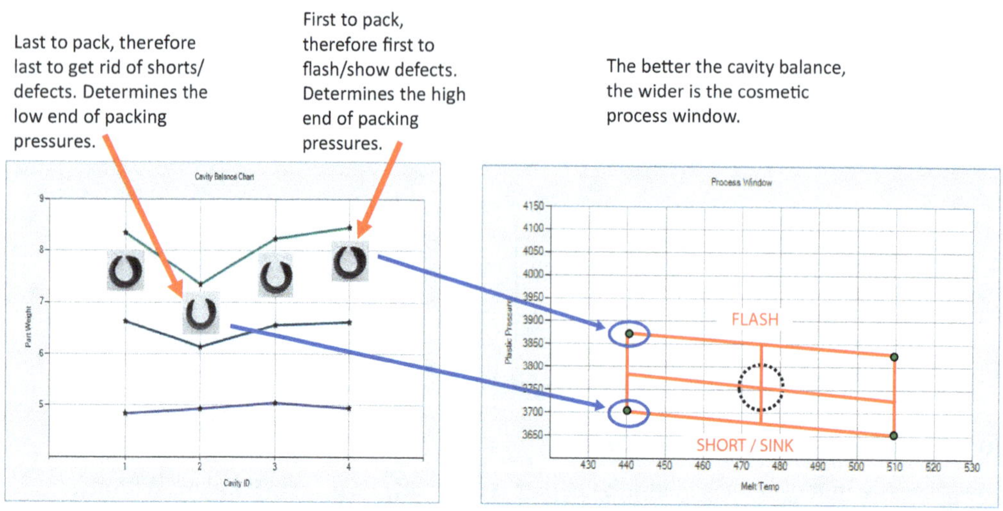

Figure 10.31 Effect of cavity balance on the cosmetic process windows

10.1.5.7 The Pack and Hold Pressure Rule Not to Be Used

There are some molders who suggest that the pack and hold pressures should be set to 50% of the injection pressure. In reviewing the previous sections, it can be deduced that such a rule is impossible to follow, not just for cosmetic reasons, but as will be discussed further, for dimensional reasons also. Further, consider the molding process of a thin part such as a laptop cover and a thick part such as a lens for a large optical equipment. In the case of the laptop cover, the required injection pressures will be very high to get the plastic flowing through the thin walls, but the required pack and hold pressures will be very low because the shrinkage will be minimal. In the case of the lens, the injection pressures will be very low because the flow into the mold cavity will be very easy due to the thick sections. However, high pressures will be required to pack and hold the plastic in there because of the large amount of shrinkage taking place in the thick section. The 50% rule or any such percentage rule should never be followed.

10.1.6 Step 5: Determining the Gate Seal Time – Gate Seal Study

The molten plastic enters the cavity through the gate. The mold filling phase is dynamic, during which melt temperature, pressure, and flow velocity are all changing with time. For the mold fill phase, time begins with injection, which is the start of the forward movement of the screw. As the cavity begins to fill and is nearly full, the pack and hold phases start. The melt flow velocity is reduced and the melt temperature simultaneously drops. This causes an increase in viscosity of the melt. The gate area has the smallest cross-sectional area is the mold. When the viscosity of the plastic in this area drops to a value at which the plastic cannot flow anymore, the gate is considered frozen. The plastic molecules in the gate area are now immobile and cannot flow into the cavity anymore. The time it takes to reach this stage is called the gate freeze time.

For an injection molding process, pressure must be applied to the melt until the gate is frozen. If pressure is not applied for a sufficiently long time, either the part will be under-packed resulting in internal voids or sinks, or the plastic pressure inside the cavity is high enough for the plastic to flow back out of the cavity, which will also result in under-packing. The second phenomenon usually takes place when the holding time is just a little shorter than the gate freeze time, while the cavity is full of pressurized plastic. Gate freeze time is a function of the type of plastic, gate size, gate design, and the processing parameters of the machine. A gate freeze study must be completed for every mold. A gate freeze study is a graph of part weight versus holding time. Once the gate is frozen, the part weight stays constant because plastic can no longer get into or out of the cavity. A constant part weight is an indication of gate freeze. Figure 10.32 shows a typical gate seal graph.

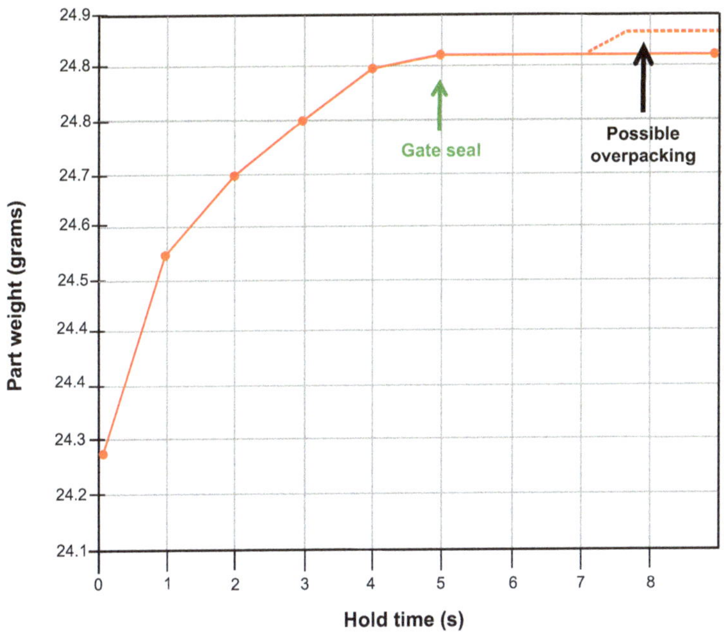

Figure 10.32 Gate seal graph

Procedure to Determine Hold Time

- Set the injection speed to the value obtained from the viscosity curve experiment.
- Set the process at the center of the process window or towards the upper corner of the right hand side upper quadrant of the process window generated in the previ-

ous section. Holding pressure, melt temperature, and mold temperature values may influence the gate seal time. The higher the pressures and temperatures, the easier is the movement of the plastic in and out of the cavity. Higher temperatures will also increase the time it will take to freeze the gate. The aim should be to make sure that anywhere in this process window, the gate must always be sealed and, therefore, conditions towards the higher extremes of the pressures and temperatures must be considered.

- Set the cooling time to a value to ensure that the part is cooled before ejection.
- Drop the holding time to zero and again set the screw delay time as described in S.
- Start molding and mold approximately 10 shots.
- Collect one shot.
- Increase the holding time to 1 sec, decrease the cooling time by 1 sec, and collect a shot. Decreasing the cooling time by 1 sec will ensure that the cycle time will remain unchanged.
- Increase the holding time in increments of 1 sec and collect the shots of up to a holding time value at which the gate should be frozen. As the holding time is increased by 1 sec, drop the cooling time by 1 sec.
- Weigh the individual cavities and plot a graph of part weight versus time.
- Determine the gate seal time. The gate seal time is the time after which the part weight stays constant.

Table 10.6 is an example of a worksheet that can be used to document a gate seal study.

Table 10.6 Gate Seal Study Worksheet

Hold time (s)	Part weight (g)
1	12.98
2	13.26
3	13.84
4	14.06
5	14.12
6	14.23
7	14.23
8	14.23

10.1.6.1 How to Use This Information

If the cavity balance is such that all the parts fill at the same rate, then the graphs for gate seal should be similar. In such cases, weighing the parts from one cavity and plotting the graph is acceptable. The hold time must be set to approximately 1 second longer than the gate seal time. This will ensure that the gate will always be sealed before the end of the holding phase. The set cooling time starts at the end of the holding time. The plastic begins to cool as soon as it hits the walls of the mold. Therefore, actual cooling time is the sum of the injection fill time, pack and hold time, and cooling time. This total cooling time is one of the factors that define the quality of the part. If the holding time is reduced by 0.5 seconds, this time must be added to the cooling time and vice versa. It is therefore best to add 1 or 1.5 seconds to the hold time to ensure gate freeze and then reduce the cooling time by the same amount, keeping the total cooling time the same. In an attempt to reduce cycle time, some processors add only about 0.5 seconds to the gate seal time. This may not be the most optimum setting.

10.1.6.2 Cautions and Exceptions

With a gate seal study, the time at which the gate seals off or is frozen is determined. However, if the part needs a large gate, the seal-off time can be very long. In such cases, if the pressure is applied for longer than the required time, there is the risk of over packing the part. Sprue gated components are good examples of this problem.

If the part weight stays constant during an initial section of the holding time and if with a further increase in holding time there is a sudden increase in the part weight, this can be evidence that the part is being over packed, especially in the gate area. Because the plastic behind the gate and in front of the gate is probably still soft, the extra time causes the frozen gate section to dislocate, forcing more plastic into the cavity. Because the rest of the plastic in the cavity is cooling down and viscosity is very high, the excess pressure from the fresh plastic does not get distributed or compensated for and builds up stress in the gate area. This weight increase is shown with the dotted line in Figure 10.32.

10.1.7 Differentiating between the Pack and the Hold Phase

In most cases, molders do not differentiate between the pack phase and the hold phase and will typically have only one pressure setting and one time setting for both phases. They call this collective phase as the hold phase. The optimization of this phase is done by the gate seal study described in the previous section. However, in some cases, a part weight versus time study will never show a flat region such as the one shown in Figure 10.33. In such cases it is easy to over pack the parts causing failures and, therefore, in these parts, there must be a differentiation between the pack and the hold phases.

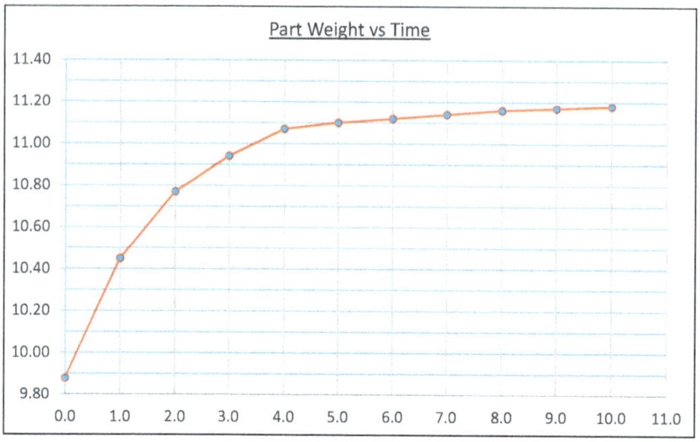

Part Weight vs Time

Time	Weight 1	Weight Increase [gms]	% Weight Increase
0	9.88	-	-
1	10.45	0.57	5.769
2	10.77	0.32	3.062
3	10.94	0.17	1.578
4	11.07	0.13	1.188
5	11.10	0.03	0.271
6	11.12	0.02	0.18
7	11.14	0.02	0.18
8	11.16	0.02	0.18
9	11.17	0.01	0.09
10	11.18	0.01	0.09

Figure 10.33 Gate seal graph for a part molded with low-density polyethylene

Here are some typical scenarios where pack and hold will need to be differentiated:

- In parts with large gate sizes the gates will take a long time to seal off since the plastic will keep flowing inside the part until the inside of the gate freezes off.
- In softer materials such as polyolefins, TPEs, and TPUs, the part weight continuously increases and does not stabilize in the gate seal test.
- In sprue gated parts the gate size is big and waiting for this to freeze off could be impractical. Part weight will also increase with increasing time and will lead to stress buildup in the gate area.
- In some parts there is stress failure at the gate, such as in trash cans where the part is center gated, and the most likely failure is a crack or break through the gate.
- In hot runner molds, because the gate is always molten, the gate will never seal as long as there is movement of plastic.
- In valve gated molds there would be only pack and no hold because the gate is mechanically closed.
- Sometimes, cosmetic defects such as a gate vestige can be solved by adding two stages. (Gate vestige is a mold issue and must be fixed in the long run.)

In the above cases, the phases of pack and hold must be differentiated from each other. In the pack phase the required amount of plastic must be injected and in the hold phase this plastic must be held in there until the gate freezes off. If the hold phase is terminated before the gate is frozen, then the pressurized plastic in the cavity will flow back out of the cavity, often causing sink and/or dimensional variations and issues. This is the reason why a molder will notice sink on parts with high pack and hold pressures. When the molder lowers the pressure, the sink disappears, often baffling the molder because it is opposite of what he expects.

The technique described next can be used to differentiate between the pack and the hold phases. It is best to illustrate this with an example. Consider the graph shown in Figure 10.34. It can be observed that the part weight increase after 5 seconds is about 0.03 grams and tapers down even more as the time increases. The 0.03 grams is about 0.25% of the final part weight (11.23 g) and 0.02 grams is about 0.17% of the final part weight. We therefore can consider that the part has reached very close to the required part weight, or in other words, the pack phase has been completed at 5 seconds. It is similar to visualizing when someone is packing a travel bag. Initially clothes can be placed inside until the bag seems physically full, but the remaining clothes can only be put inside after compressing the clothes that were first put inside. As the bag is filled more and more, lesser and lesser amounts of clothes can be packed in there. So after the initial quick fill further additions slow down. The pressure used during this initial phase is the pack pressure, and the time that this pressure is applied is the pack time. The pack pressure is 8000 psi plastic pressure, and the pack time is 5 seconds.

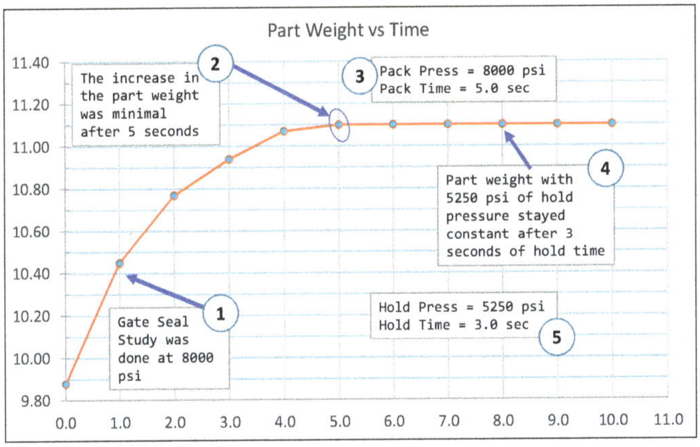

Part Weight vs Time

Time	Weight 1	Weight Increase [g]	% Weight Increase
0	9.88	-	-
1	10.45	0.57	5.769
2	10.77	0.32	3.062
3	10.94	0.17	1.578
4	11.07	0.13	1.188
5	11.10	0.03	0.271
6	11.10	0	0
7	11.10	0	0
8	11.10	0	0
9	11.10	0	0
10	11.10	0	0

Final Settings: Pack Pressure = 8000 psi, Pack Time = 5 seconds, Hold Pressure = 5250 psi, Hold Time = 3 seconds

Figure 10.34 Final settings for the pack and hold phases

Going back to the travel bag analogy, once we have packed the required amount of clothes, we must now zip the bag up in order to hold the clothes in there. If not, the bag top or cover will not be able to keep the clothes inside. Similarly, once the required amount of plastic is now present inside the cavity, it must be held in there. This is done by applying another pressure setting that will be lower in value than the pack pressure for a time until the part weight stabilizes, or in other words, until the gate freezes off. The target part weight here will be the same part weight that was obtained at the end of the pack time. In the experiment above, this hold pressure was 5250 plastic psi and the hold time was determined to be 3 seconds. The total of pack and hold time was 5 + 3 = 8 seconds. The final part weight was 11.10 grams.

The procedure is described in the following steps:

- Set only one pressure collectively for the pack pressure and hold pressure and call this the compensation pressure. Therefore,

  Compensation Pressure = Pack Pressure + Hold Pressure. Consider this pressure to be 8000 psi plastic pressure.
- Set only one time collectively for the pack time and hold time and call this the compensation time. Therefore,
- Compensation Time = Pack Time + Hold Time. Consider this time to be 15 sec.
- Start with the compensation time of 0 sec, generate a graph of 'Part Weight versus Compensation Time' of up to 15 sec.
- Observe the graph and the part weight table to estimate where the change in the part weight begins to slow down. A change in the slope of the graph can be seen. In Figure 10.33, this time can be considered as 5 sec. Based on this, Pack Pressure = 8000 psi and Pack Time = 5 sec.
- Record the part weight and this will be the Pack Only Part Weight = 11.10 g.
- Initially there was only one value for pressure and time, called compensation pressure and time. Split the compensation pressure and time into two phases and identify them as pack pressure, pack time, hold pressure, and hold time. On the molding machine add another pressure and time profile. The first profile will be for the pack phase and the second will be the hold phase. Set the first pressure to 8000 psi and the first time to 5 sec.
- The second set of pressures are the hold pressures and hold time. Because the compensation time was set to 15 sec and the pack time was set to 5 sec, set the hold time to 10 sec (15 – 5 = 10).
- Set the hold pressures equal to the value of pack pressures = 8000 psi, and then mold parts.
- Record the part weight. This should be the same as the 15 sec value above and should therefore be equal to 11.23 g. It will be higher than the pack only part weight of 11.10 g.
- Drop the hold pressures in steps of about 250 plastic psi and keep checking the part weight at every reduction. The pressure at which the part weight equals the pack only part weight of 11.10 grams will be the holding pressure. In this example, the value was 5250 psi. This indicates that at 5250 psi, the plastic is not entering the cavity nor is it existing the cavity. Or in other words, the plastic is held in place.

- Next reduce the hold times in steps of 1 sec at a time, and note the time where the part weight drops below 11.10 grams. In this example, this time setting was 2 sec, because at 2 sec of hold time, the part weight was 11.08 grams. This indicates that at 2 sec, the plastic is coming back out from the cavity. Add 1 sec to 2 sec making this time 3 sec, which will bring the part weight back up to 11.10 grams. This time will be the set holding time.

The results are summarized in Figure 10.34. The final settings can therefore be summarized as follows.

- Pack Pressure = 8000 psi
- Pack Time = 5 sec
- Hold Pressure = 5250 psi
- Hold Time = 3 sec

10.1.7.1 Hot Runner and Valve Gated Molds

In the case of hot runner molds, the weight increase with the time increment is usually steeper compared to a cold runner mold and therefore some past molding experience is helpful when determining the change in the slope in the graph. The same technique can be successfully used.

In the case of valve gated parts, since the valve is mechanically closing the gate, once the pack time is determined with the above procedure the valve must be shut. In the case of valve gated molds, the holding phase does not exist.

The technique above is only still used within the cosmetic windows. Dimensions are still not considered. Optimization of the pack and hold phases for dimensions is described in Chapter 10, which deals with *the concept of Design of Experiments.*

Gate seal times depend on the following:

- The melt temperature: the higher the melt temperature, the higher is the gate seal time.
- The mold temperature: the higher the mold temperature, the higher is the gate seal time.
- The pack pressures: the higher the pack pressures, the higher is the gate seal time.

10.1.8 Step 6: Determining the Cooling Time – Cooling Time Study

The plastic starts to cool down as soon as it hits the walls of the mold. Once the pack and hold times have elapsed, the set cooling time counter starts. The mold remains closed until the end of the cooling time. The mold then opens and the part is ejected. Before the mold opens, the part must reach the ejection temperature of the plastic. If the part is ejected before it reaches ejection temperature, the part is too soft and will get deformed during ejection. Warpage can also be an issue. Excessive cooling time is a waste of machine time and therefore profits. Cooling time should also be set so that the part dimensions remain consistent and the process is capable of molding acceptable parts over time. As discussed in Chapter 7, the actual cooling time is the addition of the injection time, the pack time, the hold time, and the set cooling time. The dimensions of the parts will also depend the cooling time. This will be discussed in the next chapter.

The procedure to determine the optimum cooling time is relatively simple and involves molding parts at various cooling times and inspecting them for any cosmetic defects. As can be seen in Figure 10.35 the part has four ejector pins on the rectangular portion. At a cooling time of 9 seconds there was evidence of pin push on the part since the plastic had not cooled down sufficiently. Increasing the cooling time to 11 seconds eliminated the defect. The lowest cooling time would therefore be 11 seconds. However, to have a robust process and not to have a process at the edge of the cliff a cooling time of 12 seconds would be preferred.

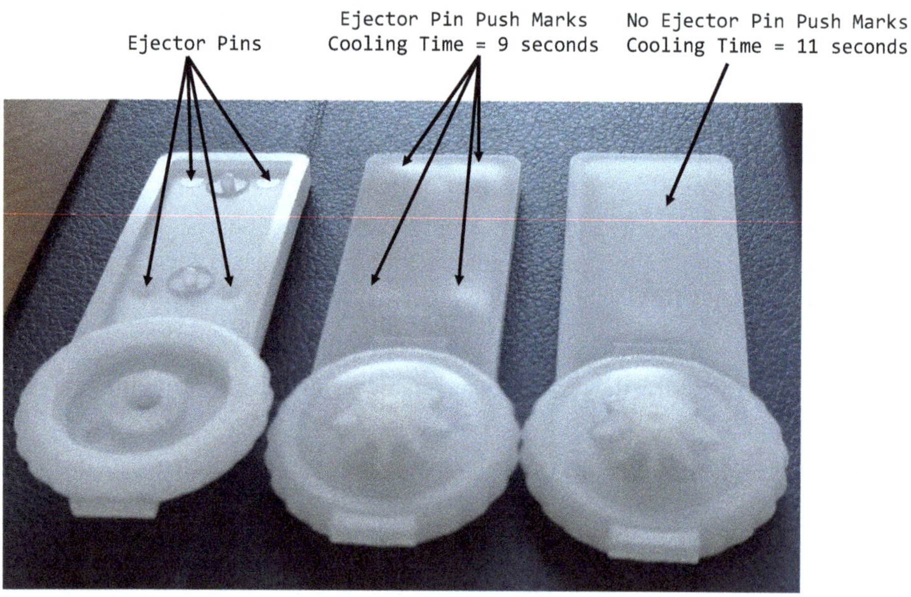

Figure 10.35 Effect of cooling time on the part cosmetics

Although Part 2 of the process development process is strictly for exploring the robust areas of the molding process and not for measuring any dimensions. It is acceptable to inspect the dimensions only to get an idea of their values and ranges. Design of Experiments must be carried out for dimensional process optimizations and to find a robust process. The procedure is outlined in the following:

10.1.8.1 Procedure to Determine Cooling Time

1. Set the process with the process conditions determined by the viscosity study, the process window study, and the gate seal study.
2. Start molding, and once the process has stabilized, collect three shots.
3. Drop the cooling time by one or two seconds, let the process stabilize, and collect another three shots.
4. Continue to drop the cooling time by one or two seconds and collect the shots. Stop when you see a cosmetic defect such as pin push or warpage. Sometimes the parts may not get released from the A-side of the mold at cooling times. This is not an acceptable cooling time. It may be best to perform the study while running the machine in a semiautomatic mode for mold safety reasons in case a part sticks or does not fall off.

The lowest cooling time would be defined as the lowest time at which the parts are cosmetically acceptable. With the parts collected above, there are two different routes that one could take based on whether a Design of Experiments (DOE) would be carried out or not. If a DOE is going to be carried out, then cooling time should be considered as one of the factors, and the least cooling time value will be used as the low value setting of the cooling time. DOEs in injection molding are very simple and must be carried out. Chapter 10 discusses this technique in detail.

If a DOE is not going to be carried out, then the dimensions can be measured, and a process can be selected. The procedure is given below:

1. Measure the critical dimensions on the shots collected. If there are a number of cooling time numbers, it is acceptable to take the low, center, and high value of the cooling, and measure the dimensions at these three settings.
2. Plot a graph of dimension versus the cooling time. Analyze the data to see how the critical dimensions are changing with the cooling time.
3. Decide on a cooling time that best fits the data.
4. Run 30 shots at this cooling time, and perform a statistical analysis to determine the process capability.

Determination of the right cooling time can become complicated. For parts with thick sections, it is difficult to measure the internal temperature in the center of the thickest section. In some parts of the mold, it is difficult to get enough cooling and, therefore, cooling times have to be increased to allow time for additional heat transfer. In some cases, with large molds or thick parts, the mold temperature could take a couple of hours to stabilize.

Figure 10.36 shows that some dimensions may be more sensitive than others. Dimension B is not influenced by the changes in cooling time. However, Dimension A changes with the cooling time. The target value for dimension A is 2.8656″. Therefore, the cooling time can either be set to 16.5 seconds, or steel changes have to be made to run the mold faster to achieve the same dimensions. Identifying the lower and upper limits in Figure 10.36 will also present a graphical representation of where the cooling time can be set.

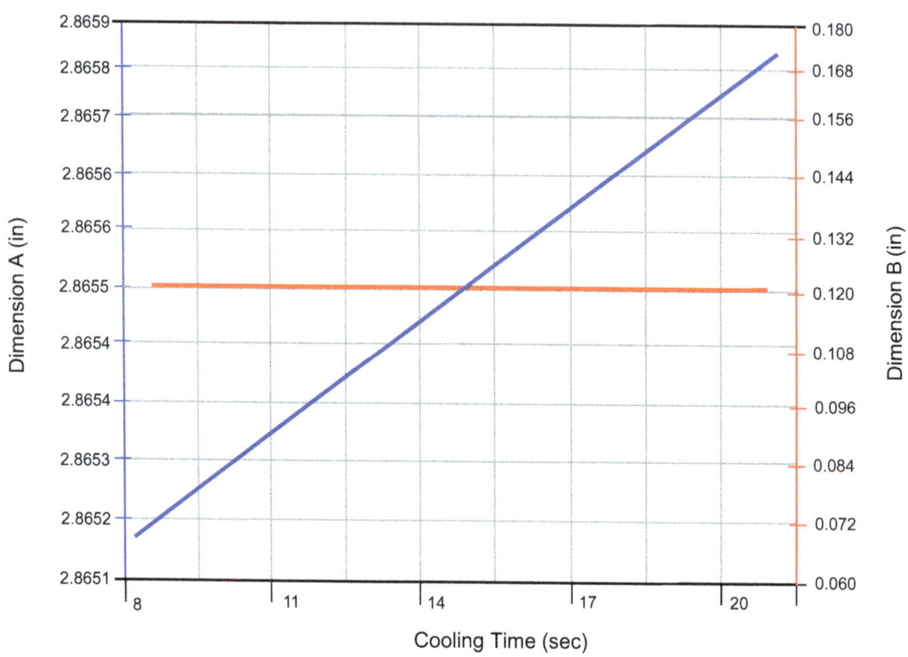

Figure 10.36 Cooling time study graph

Cycle time is the most important factor because it directly impacts the bottom line profit of a molding operation. In most cases, if the process is capable at shorter cooling times, changes in the mold steel may achieve the same dimensions at shorter cycle times. The cooling time is usually the major part of the molding cycle. Therefore, optimization of cooling time is critical for profitability. Figure 10.37 shows the typical cycle time breakdown for all phases of the molding process. Part design and mold design can have a significant impact on the actual cycle time required for each phase shown in the pie chart of Figure 10.37. A long, thin part may only need a short cooling time; however, the mold may need to stay closed much longer to complete the screw recovery phase. Ideally, the recovery time would not exceed the time required to cool the part for ejection from the mold. However, part design, mold design, and machine selection can all have significant influence on the actual cycle times required for each phase of the injection process.

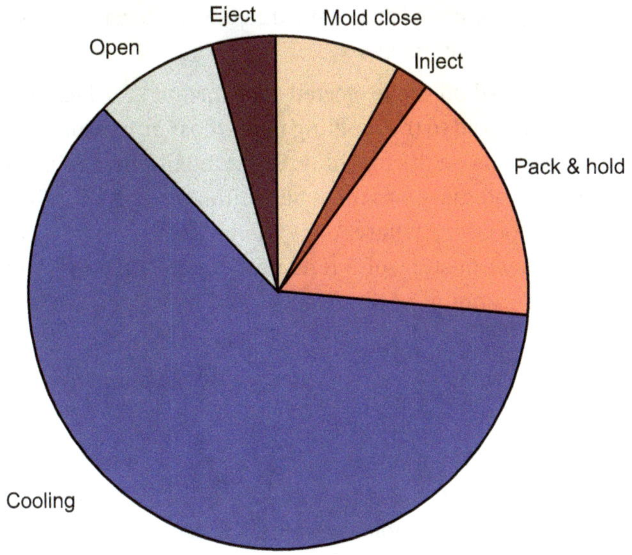

Figure 10.37 Typical cycle time breakdown

Table 10.7 is an example of a worksheet that can be used to document a cooling time study.

Table 10.7 Cooling Study Worksheet

Cooling time (s)	Dimension 1 (in)	Dimension 2 (in)
8	1.552	0.253
10	1.553	0.253
12	1.554	0.254
14	1.554	0.254
16	1.555	0.254

10.2 Optimization of Tonnage

As was discussed in Chapter 6, the tonnage calculation should not be based on projected area alone. There are several other factors that must be considered. These are part thickness, flow lengths, position of gates, number of gates, gates type, gate size, gating mechanisms, orientation of the part in the mold, presence of slides, the type of slide mechanism used for the mold, and the extent of venting. The formula for required tonnage based on the multiplication of projected area and material tonnage specification is therefore not accurate by any means.

Most molders will select a machine based on the projected area formula for lack of a better resource. When the mold is delivered to the molding floor, most processors will set the tonnage to the maximum available on the machine. Using maximum available tonnage will result in several defects and therefore must be optimized. The optimization procedure is a simple, logical procedure where, once the process is developed, the processor should lower the tonnage in steps of 5 tons until a defect is observed and then increase it by a comfortable amount of 5 to 10 tons and use that as a final setting.

Below are the guidelines that one can follow and tailor to a specific mold and quality requirements.

- For the mold under consideration, set the tonnage based on about 4.0 to 4.5 tons per square inch of projected area. For the example that will be discussed, the machine tonnage was set to 60 tons and the molding process was developed using the scientific molding and Design of Experiments principles.
- Optimize the cushion value. For example, for a 35 to 60 mm screw diameter, cushion values should be between 5 to 8 mm (around 0.15 to 0.25 inches). Increase or decrease the shot size, the transfer position, and any other speed transition positions proportionately to achieve the desired cushion value.
- Start molding and once the process is stable, collect and inspect two shots for any defects such as flash at the parting line. These parts should be defect free since these were molded with the optimized process. Reduce the machine set tonnage by 5 tons and inspect the parts for any defects. If there is no defect, reduce the tonnage by another 5 tons and inspect the parts. Repeat this procedure until a defect is seen. The most common defect during this study is flash. Collect all the shots at the different set tonnages.

The data from these parts can be analyzed in a couple ways. Refer to Figure 10.38, in which the shot weight was plotted against reducing tonnage. The graph shows that as the tonnage is reduced, the shot weight increases very slightly until a tonnage value where the mold halves separate and then there is an easily observable increase in the shot weight. In this case, that value was 40 tons and at 35 tons the parts had noticeable flash. Since the noticeable weight increase occurred at 40 tons, the tonnage should be set between 45 and 50 tons.

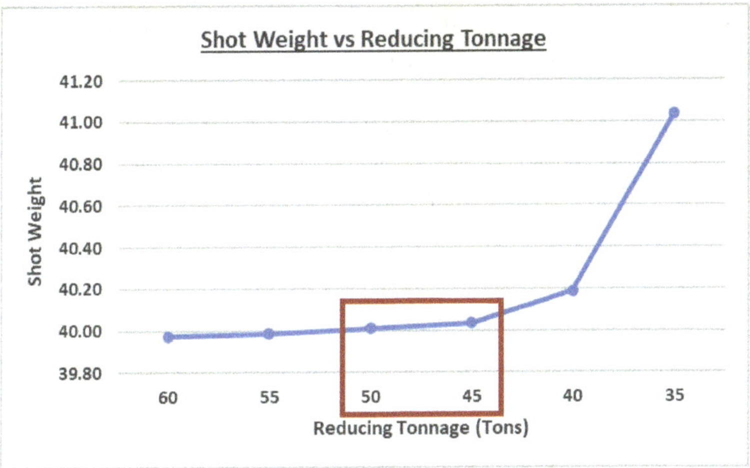

Figure 10.38 Shot weight versus reducing tonnage

One can also plot a part dimension versus tonnage. This is also shown in Figure 10.39. Choose a part dimension in the direction of mold open and a similar trend will be seen where the part dimension will increase when the mold halves separate. Again, the graph shows an acceptable tonnage setting would be between 45 and 50 tons.

This graph also demonstrates how dimensions could change with tonnage. A part may be in spec on a 50 ton press but not on a 75 ton press (or vice versa) even though the plastic processes on the two machines are identical. It is therefore important to match the tonnages when the mold is transferred from one machine to another.

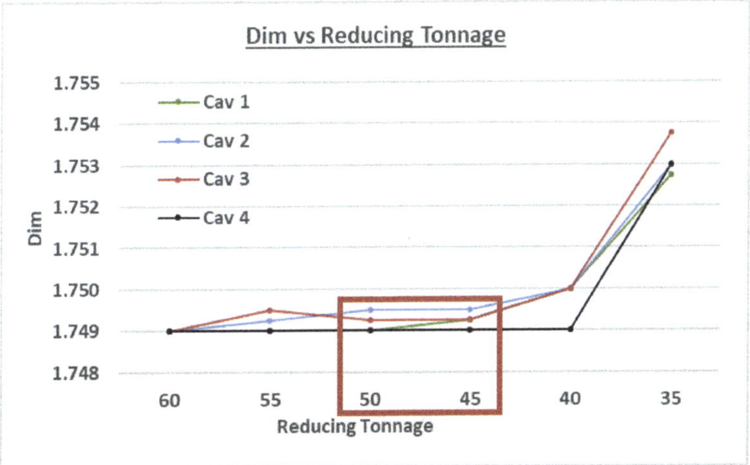

Figure 10.39 Part dimension versus reducing tonnage

The reason for having a low cushion value during these studies is to protect the mold from over flashing with the excess plastic in the cushion. Having a large cushion can create a lot of flash that can end up in slides, vents, and other sensitive areas of the mold. This could damage the mold components and/or will need the mold to taken to the tool room for cleanup and repair.

10.3 The Cosmetic Scientific Process

With the techniques just described, the process that was developed is called as the Cosmetic Scientific Process. Dimensions can be taken but the preferred route to optimize dimensions should be the use of the technique of Design of Experiments. Table 10.8 summarizes the optimization technique of the 11 + 2 molding parameters.

There are no settings that were determined based on any assumptions or any past experiences. All the settings were data driven.

Table 10.8 Summarization of 11 + 2 Molding Parameters

No.	Factor	Method of Optimization
1	Injection speed	Viscosity curve
2	Injection pressure	Pressure drop study
3	Pack pressure	Cosmetic process window study
4	Hold pressure	Cosmetic process window study
5	Pack time	Gate seal study
6	Hold time	Gate seal study
7	Cooling time	Cooling time study
8	Melt temperature	Cosmetic process window
9	Mold temperature	Cosmetic process window
10	Screw rotation speed	Screw rotation speed study
11	Back pressure	Back pressure study
12	Shot size	Part fill of 95–98%
13	Transfer	Part fill of 95–98%

10.4 Post-Mold Shrinkage Studies

When the parts are ejected from the mold, they are below the ejection temperature but are still warm. If the temperature of the part is above the glass transition temperature, the molecules have enough energy to move and settle down into their equilibrium positions. This movement results in additional shrinkage and therefore a dimensional change of the part. If the glass transition temperature (T_g) of the plastic is below room temperature, the phenomenon continues until equilibrium between ambient and glass transition temperature is reached, causing even more excessive shrinkage. Because this shrinkage takes place outside the mold and after molding, it is called post-mold shrinkage. Thermoplastic elastomers are common materials with low glass transition temperatures exhibiting high post-mold shrinkage. The rate of shrinkage is directly proportional to the temperature of the plastic and follows an exponential decline with time after ejection. The highest shrinkage is therefore seen as soon as the part is ejected from the mold, because at this point the temperature is the highest. As the part starts to cool down, the rate of shrinkage also decreases. In some cases, part dimensions can take up to a few days to stabilize.

Because almost all molded parts are used in assemblies after ejection, the fit between the parts is very important. Therefore, part dimensions must have stabilized before assembly. If they continue to shrink after assembly, stresses can easily be induced into the assembly and cause premature failure. Figure 10.40 shows a part that continued to shrink after assembly and therefore induced stress in the assembly.

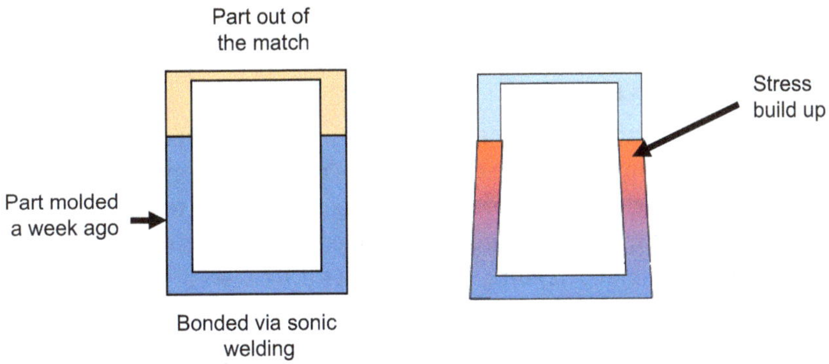

Figure 10.40 Post-mold shrinkage after assembly

There are a number of factors that may affect post-mold shrinkage:

Glass transition temperature: The lower the T_g, the higher is the post-mold shrinkage. In plastics with their T_g above room temperature, post-mold shrinkage is low because there is not sufficient energy for molecular movement. High glass transition temperature plastics, such as PEEK, show almost no post-mold shrinkage.

Fillers and additives: In the case of filled plastics, the presence of fillers prevents the shrinkage. Therefore, resins with a higher filler content exhibit lower post-mold shrinkage.

Part thickness: Part design also plays an important role with respect to the amount of post-mold shrinkage. Thicker part sections retain more heat and therefore can cause additional post-mold shrinkage. This can also lead to warpage, which is the result of uneven cooling in the part. If one area of the part has cooled down while a thicker section of the same part is still cooling and undergoing some shrinkage, the thicker section will influence the thinner section and pull the part in the direction of the shrinkage, thus deforming the part.

Mold temperature: The temperature of the mold provides the energy required to help bring the molecules to their equilibrium state. In the case of crystalline materials, the mold temperature has to be maintained high enough to crystallize the plastic. If the mold temperature is not maintained at a sufficiently high level, the molecules freeze in place without reaching their equilibrium position. When these parts are ejected and are then subjected to higher temperatures during their service life, the molecules acquire the required energy and begin to find their equilibrium positions, thus causing shrinkage. This phenomenon can also occur when molded parts are stored prior to use. If mold temperatures were set lower than room temperature, for example 5 °C, when the parts were produced and then stored at ambient warehouse temperatures of 30 to 40 °C, post-molding shrinkage can occur. When these parts are eventually removed from storage, they could exhibit some amount of warpage that did not exist at the time they were packaged.

Processing conditions: Nonoptimized process conditions can also induce stress in the part and cause post-mold shrinkage and warp. Under-packed parts commonly exhibit post-mold shrinkage while over-packed parts exhibit built-in stresses. These stresses are released after the part is ejected.

Annealing: Parts that have not reached their final state of equilibrium exhibit built-in stresses. The process of annealing can be considered as a forced post-mold shrinkage operation. In annealing, the parts are intentionally taken to higher temperatures to force the movement of the molecules out of their non-equilibrium positions and into their final equilibrium positions. In doing so, all stresses are relieved. Annealing will result in some amount of shrinkage, depending on the factors described earlier.

A typical post-mold shrinkage graph is shown in Figure 10.41. The shrinkage exhibits an exponential drop: it is higher when the part is first ejected from the mold but then stabilizes over time.

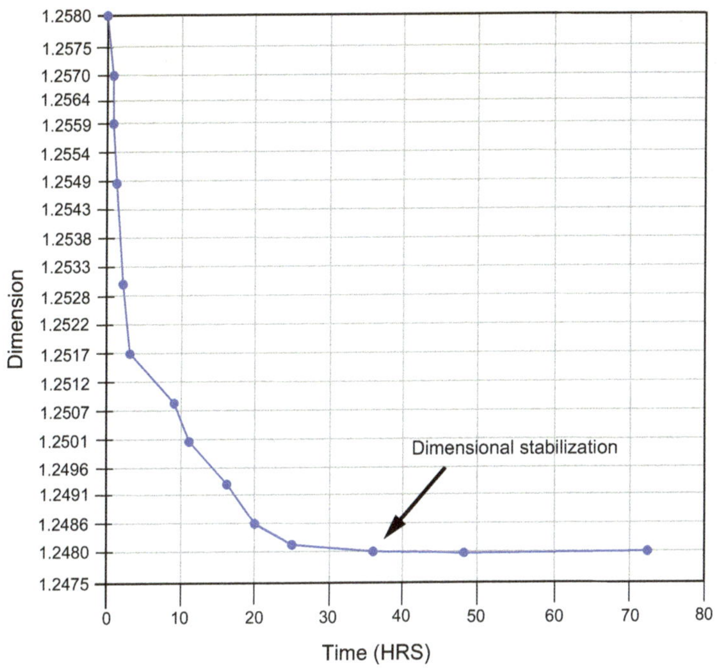

Figure 10.41 Post-mold shrinkage graph

10.4.1 Procedure to Measure Shrinkage

- Start molding with the established process parameters and let the process stabilize. The stabilization of the actual mold temperatures is an indication that the process has stabilized.
- Collect at least three shots as they are ejected from the mold and record the time each was ejected from the mold.
- After 15 minutes measure the parts from each shot and record the dimensions.
- Repeat after 30 minutes, 45 minutes, 1 hour, 2 hours, 4 hours, 8 hours, 24 hours, and so on until a stable reading is achieved. The time interval between measurements can increase as time after molding is getting longer. The goal is to collect the most measurements while the part is experiencing the highest rate of shrinkage, which is the timeframe shortly after molding. As the part ages, the rate of shrinkage decreases and the time interval between measurements can be increased.
- Generate a graph of dimension versus time.

If there are multiple cavities, it may be difficult to measure all parts in a short time during the initial part of the experiment. In this case, measure any one cavity. If the cavity fill is balanced and the mold cooling for each cavity is the same, one can assume that the parts will behave in a similar fashion. The parts from the other cavities must still be collected, with measurements being made at longer time intervals. Dimensional stability after longer relaxation times, e.g., at 2 hours or longer, is sufficient to measure the part dimensions.

10.4.2 How to Use This Information

Once the graph is generated, the time interval required for the part to become dimensionally stable will be obvious. Any secondary operations or direct part use must be done after the stabilization period. The stabilization time in Figure 10.41 is approximately 36 hours. This information can also be used to set what are called as 'hot specifications' during the production process. In the graph the dimension stabilizes after about 40 hours. However, it is not practical to wait for 40 hours to measure the dimensions. The dimension at a time such as 1 hour or 2 hours is recorded from the graph. In the graph, the dimension is 1.2570 inches at 2 hours after the part was molded. Based on the data one can conclude that if the dimension is 1.2570 inches at 2 hours, then the part will shrink to a final dimension of 1.2480 and stabilize.

10.5 Completion of Part 2 of the Recommended Process Development Procedure

The previous sections describe the science and the procedures to establish all process parameters. During each of the steps, the most robust areas of each process parameter must be recorded. These parameter settings will contribute to the make-up of the final process. Once the process has been established, parts must be molded using these process settings. The machine must be set to the established process and a sufficient amount of parts must be run to check the process robustness and the part quality.

Each procedure taken along the way may reveal an issue with the mold. For example, having a defect such as a short shot and flash at the same time is a clear indication of a mold problem. In this case, there is no process window and therefore it will be impossible to make an acceptable part. The mold must therefore be pulled out of the machine and fixed. This becomes a required mold trial iteration necessary for successful molding and production.

Likewise, other steps may reveal other issues, such as inadequate venting or improperly sized gates among others. An effort must be made to try to identify all issues with the fewest number of trials possible.

The procedure described is used to optimize the process and to make sure that the mold function is acceptable. It is therefore also called 'mold function qualification procedure.' Part dimensions were not the focus of this qualification, which does not mean that the dimensions should not be measured along the way. The dimensions could be measured and compared to the part drawing. If all the dimensions are acceptable and well within the quality requirements, the molding process can be recorded as final. This is the best-case scenario but the process capability or what to change when there is an issue is still not known. The next step should be Part 3 of the process development to evaluate the dimension and the process capability. There are several other benefits to performing Part 3 of the process development. The flow chart in Figure 10.42 shows the completion of Parts 1 and 2 of the recommended mold function qualification procedure.

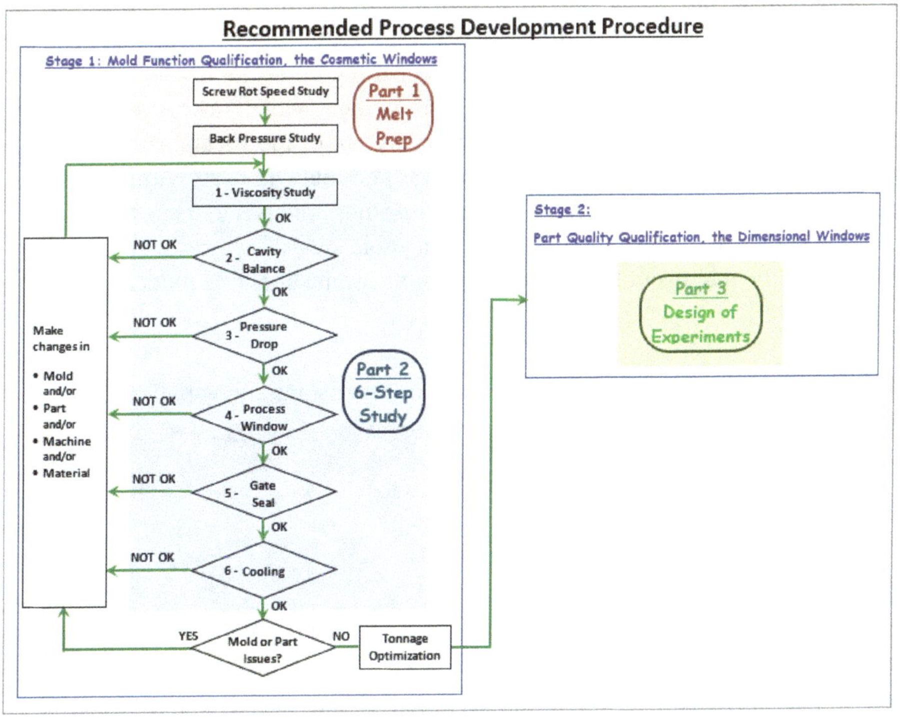

Figure 10.42 Parts 1 and 2 of the recommended process development procedure

10.6 Recommended Adjustments to Maintain Process Consistency and Robustness

Once the process has been established and if one does not want to perform Part 3, the Design of Experiments, it is time to consider the dimensions of the parts. Collection of the quality data at the established process is crucial; therefore, a good, statistically valid sample of parts must be measured. The minimum number of acceptable parts considered to provide reliable statistical data is 30. The data must be analyzed and a decision must then be made on how best to achieve the required quality standards. Table 10.9 describes the various scenarios and the recommended actions. In some cases, the natural variation can be larger than the difference between the lower and upper specification limits. For example, consider a part with the required nominal dimension to be 2.50 mm, the upper tolerance of 0.05 mm, and the lower tolerance of 0.05 mm. The total available tolerance is therefore 0.10 mm. If 30 parts were measured, and if the range (difference between the maximum and the minimum) was 0.12 mm, it will be impossible to mold all the parts within the required specifications, even if the average tolerance of the 30 parts was 2.50 mm. There will always be parts molded outside of specifications. In this case, opening up of the tolerances must be considered. Making any sort of steel adjustments will not help. If the range is less than the difference between the limits and the parts are out of specifications, then adjusting the mold steel will help to center the process and mold all parts within specifications. If the parts are functionally acceptable, the other option is to move the nominal dimension on the part print, and not modify the mold steel dimensions. This is a good option when the mold is not steel safe and the decision should be made by the part designer.

Table 10.9 Recommendations for Centering Part Dimensions and Variations

Location of average – acceptance of variation	Adjust steel	and/or	Move nominal	and/or	Open tolerance
At nominal – acceptable	NA		NA		NA
At nominal – not acceptable	No		No		Yes
Closer to the spec limits – acceptable	Will help	and/or	Will help		No
Closer to the spec limits – not acceptable	Will help	and/or	Will help	and	Yes
Out of spec limits – acceptable	Yes	and/or	Yes		No
Out of spec limits – not acceptable	Yes	and	Yes	and	Yes

10.7 Process Documentation

Injection molding is primarily a heat transfer process, involving variables related to speed, pressure, time, and temperature. It is important to record each of these process parameters in order to duplicate the process during the subsequent trials and final production. The mold qualification worksheets, machine setup sheet, water line diagrams, mold temperature maps, setup instructions, and operator instructions must all be documented and/or updated during each trial. Several sections in Chapter 11 describe this documentation.

Of particular relevance here is the mold qualification checklist. A master checklist must be developed for mold trials, depending on the requirements of every molding operation. During every mold trial, the checklist must be used to evaluate the mold and the process. This assures that every part of the mold or process is evaluated and all the problems and/or suggested improvements are documented. This documentation can then be passed on to the mold maker or those involved with project improvements. A sample mold qualification checklist is provided in Appendix F.

10.8 References

[1] Kulkarni, S. M., Hart, D., *SPE ANTEC Tech Papers* (2003) p. 736

[2] Mertes, S., Carlson, C., Bozzelli, J., Groleau, M., *SPE ANTEC Tech Papers* (1997)

Suggested Reading

Osswald, T. A., Turng, L., Gramann, P. J., *Injection Molding Handbook* (2007) Hanser, Munich

Beaumont, J. P., *Runner and Gating Design Handbook* (2007) Hanser, Munich

Beaumont, J. P., Nagel, R., Sherman, R., *Successful Injection Molding* (2002) Hanser, Munich

Rosato, D. V., Rosato, D. V., *Injection Molding Handbook* (2000) CBS, New Delhi, India

Kulkarni, S. M., *SPE ANTEC Tech Papers* (2003) p. 736

Cogswell, F., *Polymer Melt Rheology* (1981) John Wiley, NY

Dealy, J., Wissbun, K., *Melt Rheology and its Role in Plastic Processing Theory and Applications* (1990) Van Nostrand Reinhold

11 Process Development Part 3: Exploring the Dimensional Process via Design of Experiments

Chapter 9 introduced the first stage of the process development process where the boundaries of the process were explored based on the cosmetics of the part. No dimensions were considered. The bigger the cosmetic process window, the wider the availability settings ranges of the parameters, the higher the chances of a final robust process with dimensional consideration. The first stage is the basis for the second stage of process development, which deals with the dimensional requirements of the molded product. If the parts are not cosmetically acceptable, the dimensions would not matter. In any molding shop, the cosmetic visual check is carried out first. Once the parts are cosmetically acceptable, the parts are then checked for the dimensions. This second stage of dimensional analysis through experimentation is called Design of Experiments or DOE.

To understand what a DOE is, consider the relationship between part dimension and holding pressure as shown in Figure 11.1. If a given dimension of a part molded at two holding pressures is known and if one can generate a dimension relationship curve between the two pressures, then based on the equation of the curve, the dimension of the part can be predicted at a given set holding pressure. Conversely, to achieve a dimension the pressure can be predicted. This simple but powerful experiment gives the ability for us to understand the relationship between the dimension of the part and the holding pressure. This is the most basic DOE.

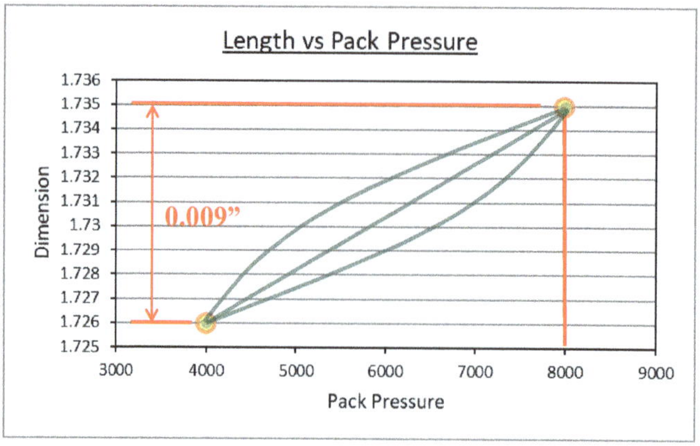

Figure 11.1 Relationship between holding pressure and part dimension

Planned experiments have been around for centuries. In the 17th century, a doctor planned some experiments to find a cure for an ailment of that day. He engaged various patients and various combinations of medicines to find the right cure. Planned experiments were used in agriculture to find the right combination of factors, such as soil type and fertilizer, to produce the highest yield of crops. Since the time involved in waiting for the results of these agricultural experiments was one complete season, which in some cases was one complete year, the technique of a planned experiment was very helpful. As planned experiments became popular as time-saving and efficient techniques, people with a background in mathematics and statistics became involved and developed them further. These planned and systematic experiments involving changes to the inputs and measuring the effect on the outputs became known as Design of Experiments or DOE. Each statistician came up with techniques that helped in analyzing different types of data in different scenarios. The people who have been most associated with the development of these techniques are G. Taguchi and Ronald Fischer. The field of designed experimentation is vast and considered a specialty of its own. Within the context of this book, we will focus on 'Factorial Experiments.' Factorial experiments work very well for injection molding. The use of any procedure (not just a DOE procedure) must be backed by a complete understanding of the underlying principles, which helps not only in the understanding of the analysis but also more importantly in the interpretation of the analysis. Factorial experiments, their analysis, and their interpretation are easy to comprehend and do not require a very strong mathematical background.

11.1 Why Perform DOEs?

If the process that was arrived at after performing Stage 1 molded acceptable parts not just cosmetically but also dimensionally, an obvious question is 'Why perform a DOE?' There are several advantages to performing DOEs. Although the parts are acceptable, the following questions have not been answered:

- If there is a dimensional issue, which out of the 11 + 2 should be changed?
- What about long-term variation?
- Are the dimensions that are in spec at the edge of a cliff of the molding process?
- If the process needs to be adjusted, how big are the process windows for each dimension?
- Are the alarm limits set on the inputs realistic or general default limits?
- Is there a more efficient process that the one established?

Performing DOEs helps in making data driven decisions and so should be performed. For example, if the dimension of the part needs to increased, then without performing a DOE the processor will need to rely on their past experience, which could tell them to reduce mold temperature and/or reduce melt temperature and/or increase cooling time and/or increase pack pressure and so on. A DOE will not only provide info to the most significant parameter but will also provide the magnitude of change.

11.2 The PVT Diagram and Injection Molding Relationship

Applying and using DOE in injection molding is relatively simple compared to its use in other manufacturing or production processes, because there are fewer parameters and types of responses. In contrast, a DOE in the pharmaceutical industry would be extremely complicated and laborious. In injection molding, there are some curvatures in the responses. These are based on the PVT diagrams and will be needed to be modeled into the equations and algorithms for analysis. The injection speed is nonlinear, for example. If large temperature differences between the high and low are considered then again there is a nonlinearity.

The phases of the complete injection molding cycle are related to the PVT diagram. An example of the PVT diagram for a PC-PBT blend is shown in Figure 11.2. Inside the mold, this is reflected in a cavity pressure curve obtained by placing a pressure sensor inside the cavity. Shrinkage is the change in volume as the plastic is being cooled. Based on the relationship between the volume and the temperature it can be concluded that the part dimensions also respond similarly to process parameters such as temperatures and pressures. Faster mold fill speeds will result in lower heat loss in the melt before it reaches the end of fill. Therefore, the plastic characteristics are in the top right hand side quadrant of the PVT graph. Decreasing the fill speeds will shift the curve to the left and result in a proportional change in dimension. The melt begins to solidify at one point and the volume continues to decrease. Every factor that becomes significant in the PVT diagram – cavity pressure curve will be a factor that will need to be considered to perform the DOE.

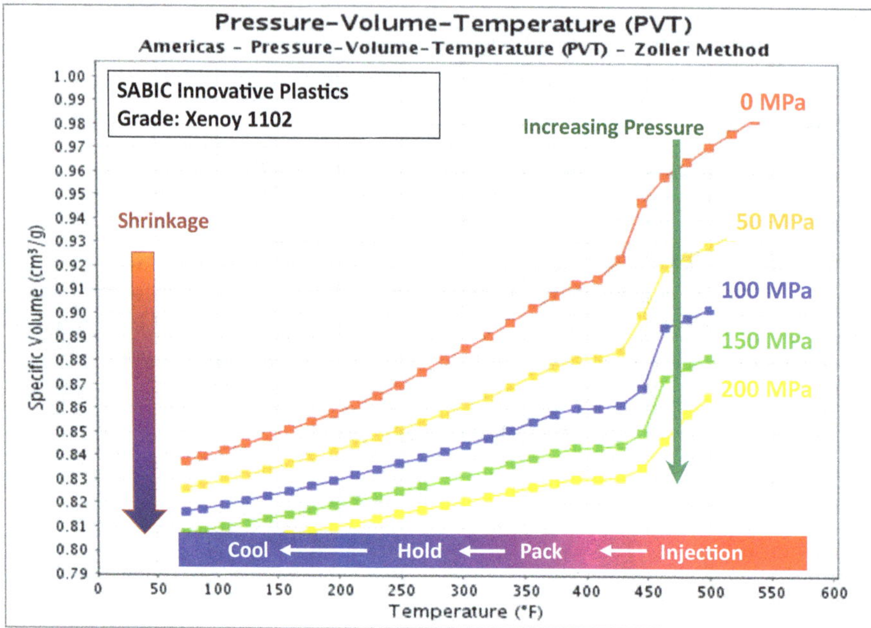

Figure 11.2 PVT diagram and the molding phases for a PBT-PC blend from Sabic Innovative Plastics

11.2.1 Design of Experiments: Definition

The simplest description of Design of Experiments (DOE) is a planned study. For example, studying the effect of holding pressure on the length of the part is a designed experiment. The length of the part at a low holding pressure and a high holding pressure is measured and is then plotted as a function of the holding pressure as shown in Figure 11.3. This is the most basic DOE that can be performed. It can be inferred from this experiment that an increase or decrease of pressure of 4000 psi will increase or decrease the length of the part by 0.009 in.

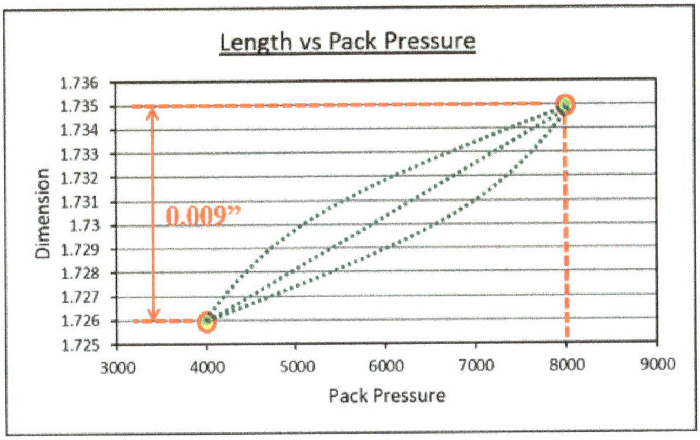

Number of Experiments = 2

No	Pack Press
1	Low (4000 psi)
2	High (8000 psi)

Dimension at 4000 psi = 1.726 in

Dimension at 8000 psi = 1.735 in

Change in dimension = 0.009 in

Figure 11.3 Effect of packing pressure on the dimension

When considering two parameters, holding pressure and melt temperature, and their influence on the length, we need to perform four experiments and determine the length at the following holding pressure/melt temperature combinations: low–low, low–high, high–low, and high–high, see Figure 11.4. Essentially, the experiment with the holding pressure discussed above is now repeated with the low and high melt temperature settings.

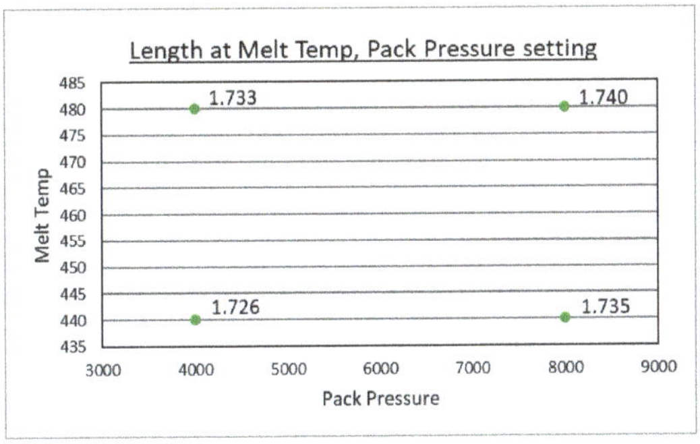

Number of Experiments = 4

No	Pack Press	Melt Temp	Dimension
1	Low (4000 psi)	Low (440 F)	1.726 in
2	High (8000 psi)	Low (440 F)	1.735 in
3	Low (4000 psi)	High (480 F)	1.733 in
4	High (8000 psi)	High (480 F)	1.740 in

Figure 11.4 Effect of holding pressure and melt temperature on part length

If we add one more parameter to this, for example mold temperature, we end up with eight necessary experiments. We are repeating the above four experiments at a low and a high value of the mold temperature, resulting in eight experiments, as shown in Figure 11.5. As the number of parameters to be studied increases, the number of experiments increases. With the addition of every parameter, the number of experiments will double and since there are 13 parameters this number multiplies very rapidly and it becomes impractical to run such a large number of experiments. This is where one of the powers of DOEs comes into play, reducing the number of experiments required to produce the required analysis and results.

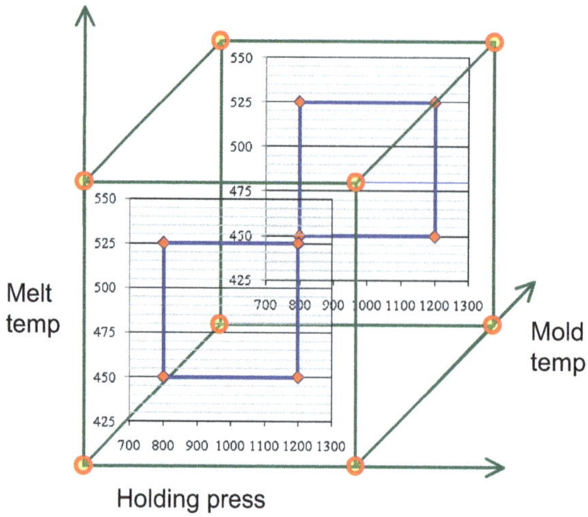

Figure 11.5 Effect of holding pressure, melt temperature, and mold temperature on part length

11.3 Terminology

When it comes to DOEs, the terminology used is different from that used in injection molding.

11.3.1 Factor

Any input to the process is a factor. Therefore, all processing parameters that are input into the molding machines are factors. Example: Holding pressure and melt temperature. Factors can be set to a desired value on the machine controller or can be selected from available options, such as a particular lot of material. Factors are classified as follows:

Control Factors: Can be controlled and changed when required, e.g., mold temperature

Noise Factors: Cannot be controlled, e.g., lot-to-lot variation

Constant Factors: Are not changed during the study, e.g., back pressure

Quantitative Factor: Can be continuously changed in increments, e.g., holding pressure

Qualitative Factor: Can be changed in discrete levels, e.g., material lots

11.3.2 Response

Any output from a process is a response. A response is the result obtained at the various level settings to which the factors are set to during an experiment. The value or attribute of a response depends on the setting of a factor. The response cannot be controlled directly. To get a required value of a response, the value of the factor must be changed. Part dimensions, fill times, cavity pressure, and the amount of splay are all examples of responses. Responses are classified as follows:

Quantitative Response: Are represented by numbers, e.g., length, weight

Qualitative Response: These are also called attributes; they are not represented by numbers and describe the condition of the response, e.g., splay, color

11.3.3 Level

A level is the number of points selected for the factor studied. For example, if we choose a low and a high value of holding pressure, there are two levels. If we choose a low, a center, and a high value, there are three levels. The number of levels is chosen based on the type of response to the factor. In injection molding, most responses to the factors are fairly linear. This means that when conducting a study at a high level and a low level, the response at a medium value of these two factors can be predicted to be the average of the responses. If the part length at 500 psi of holding pressure is 1.10 inches and at 1500 psi it is 1.20 inches, we can conclude that the length at 1000 psi will be close to 1.15 inches. There are some exceptions, such as cooling time, where a particular dimension may plateau off. In some cases, the holding pressure may also plateau off at very high values. Data with injection speeds are not linear. This is where a good amount of practical experience and engineering knowledge can help. In most cases, a two-level experiment followed by a confirmation study suffices most needs but in case of some factors such as injection speeds, a multilevel experiment may be required. Figure 11.6 shows some of the factors, levels, and responses in injection molding.

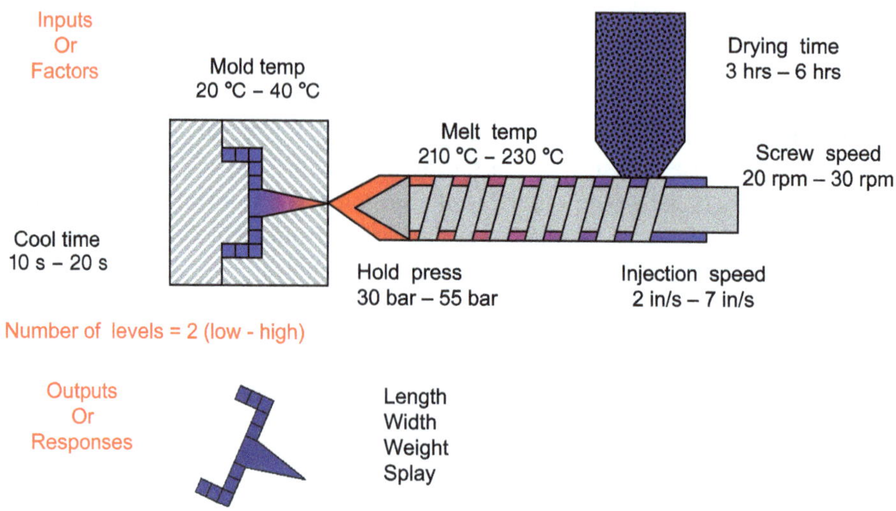

Figure 11.6 Factors, levels, and responses

11.3.4 Designed Experiment

A designed experiment is a study in which purposeful changes are made to the factors and the corresponding effects on the responses are recorded. The responses are analyzed and the results are used to optimize the process to make it robust. The experiments mentioned earlier are all examples of designed experiments. Table 11.1 shows a matrix for a three-factor, two-level, and two-response designed experiment.

Table 11.1 Matrix for a 3-Factor, 2-Level, and 2-Response Experiment

Run	Experimental run settings			Responses	
	Mold temp. (°C)	Cooling time (s)	Holding press (bar)	Length (mm)	Diameter (mm)
1	40	30	30	144.73	6.35
2	40	30	55	144.40	6.15
3	40	20	30	144.60	6.32
4	40	20	55	144.30	6.15
5	20	30	30	144.83	6.37
6	20	30	55	144.50	6.16
7	20	20	30	144.65	6.32
8	20	20	55	144.34	6.16

11.4 Relationships between the Number of Factors, Levels, and Experiments

The main goal of performing various experiments is to understand the effect of the factors on the final quality of the part. When the number of factors increases, the number of experiments also increase, see Table 11.2.

Table 11.2 Number of Experiments Based on the Number of Levels and Factors

Levels	Number of Factors												
	1	2	3	4	5	6	7	8	9	10	11	12	13
2	2	4	8	16	32	64	128	256	512	1024	2048	4096	8192
3	3	9	27	81	243	729	2187	6561	19683	59049	177147	531441	1594323

The relationship between the number of experiments, factors and levels is given by the following equation:

$$n = l^f \tag{11.1}$$

where

n = Number of experiments

l = Number of levels

f = Number of factors

For example, for 3 levels and 4 factors, we have 3 × 3 × 3 × 3 = 81 experiments. Running a large number of experiments can get expensive and be very time consuming. As discussed in Chapter 7, there are 13 primary factors that affect the part quality. If one was to perform a 2-level full factorial experiment, the number of experiments would be 8192; at 3 levels it would be 1,594,323 experiments! Both of these scenarios would be impractical to perform in any production environment. Even after the completion of the experiments, the data still needs to be collected. Inspection of all parts and collecting the data for the number of dimensions under consideration would be even more time consuming. In injection molding, it is common to work with multiple cavities and collect statistical data for several part dimensions. If this information needs to be collected, the required time and the effort involved can increase significantly. Through the use of statistical techniques, there are multiple designs of experiments available where the number of experiments can be reduced, yet reliable data can be obtained from them. These experiments are subsets of the maximum number of required experiments and this is where the advanced analysis techniques of designed experiments become a powerful tool.

Over the years, a number of researchers presented various ways to design fewer experiments, analyze the data, and provide reliable results. Each have their own advantages and disadvantages and are suitable to particular types of experiments. Some of the popular designs in injection molding are screening experiments, Taguchi designs, Plackett-Burman designs, and Box-Behnken designs. Although screening experiments are usually followed by full factorial experiments, in injection molding that is usually not necessary because there are only a handful of factors that truly make a difference in the quality of the part. Factorial experiments and the Taguchi methods are most commonly used. Analysis of the results does not require a lot of mathematical and/or technical background, which makes the Taguchi screening methodology very user-friendly. Once understood, common spreadsheets programs, such as MS Excel, can be used to analyze the data. However, because of the time involved in creating such a worksheet, it may be best to use software specially created for such purposes.

To understand DOEs further, we will introduce a number of concepts in the following.

11.5 Balanced Arrays

Table 11.3 represents a 2-factor experiment with 2 levels for each factor. Therefore, an experiment resulting in 4 test runs is required.

Table 11.3 Two-Factor, Two-Level Experiment

Experiment No.	Hold pressure	Melt temperature
1	Low	Low
2	High	Low
3	Low	High
4	High	High

An array is said to be orthogonal if it meets the following two conditions:

a) *For each factor, an equal number of highs and lows are tested.*

In this case, there are two factors, holding pressure and melt temperature.

Hold pressure – 2 highs and 2 lows are tested

Melt temperature – 2 highs and 2 lows are tested

b) *For each level within a factor, an equal number of highs and lows for every other factor are tested.*

In this case:

For low hold pressure – 1 low of melt temperature and 1 high of melt temperature

For high hold pressure – 1 low of melt temperature and 1 high of melt temperature

For low melt temperature – 1 low of hold pressure and 1 high of hold pressure

For high melt temperature – 1 low of hold pressure and 1 high of hold pressure

The matrix shown in Table 11.3 is an orthogonal array because it meets both conditions. There is a balance between the number of highs and lows of each factor of the experiment. If we consider the same concept of orthogonal arrays from a mathematical aspect, replacing the highs and lows with numerical values, low would be replaced with −1 and high would be replaced with +1. Replacing the names of the factors and calling them A and B will result in Table 11.4.

Table 11.4 Two-Factor, Two-Level Experiment

Experiment No.	A	B
1	−1	−1
2	+1	−1
3	−1	+1
4	+1	+1

Now, we will perform the following operations on the entries in this table:

Table 11.5 Orthogonal Arrays

Experiment No.	A	B	A × B
1	−1	−1	+1
2	+1	−1	−1
3	−1	+1	−1
4	+1	+1	+1
Sum	**0**	**0**	**0**

1. Create a new column called AB whose cells are the result of the product of the cells of A and B.
2. Add the cells in each of the columns in a new row.

11.6 Interactions

We will use the following example to explain what interactions within the concept of Design of Experiments represent: as the humidity and temperature increase, human comfort level decreases. For this experiment, the comfort level is graded on a scale from 0 to 10, where 10 is the most comfortable. When the humidity is low, a change in the temperature from 20 °C to 28 °C changes the comfort level from 9 to 8. In fact, most people do not even notice the change. However, when the humidity is high, the comfort level even at 20 °C is relatively low and with the same change in the temperature to 28 °C the comfort level falls to 2. People in coastal towns often face such discomfort, because of the high humidity even at low temperatures. If we plot these results, as shown in Figure 11.7, it is clear that the change in comfort between two temperatures at low humidity is different from the change in comfort at high humidity. There is a drop of 1 point at low humidity and a drop of 4 points at high humidity. Therefore, the amount of change in comfort due to temperature is dependent on another factor, the humidity. In technical terms, this means that there is an interaction between temperature and humidity when it comes to human comfort levels.

Expt No	Humidity %	Temp	Comfort Level
1	20	20	9
2	20	28	8
3	80	20	6
4	80	28	2

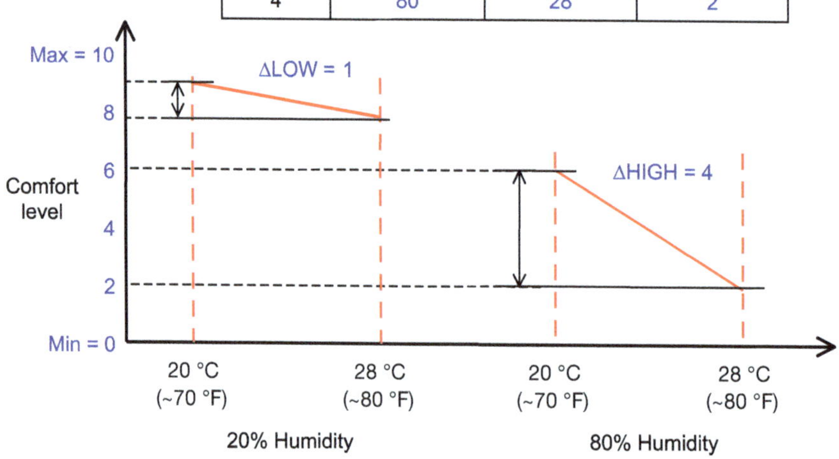

Figure 11.7 Interaction between humidity and ambient temperature regarding human comfort level

An example for no interaction between temperature and humidity is the pressure inside a container or an automotive tire. The change in tire pressure with an increase in temperature from 20 °C to 28 °C will be the same regardless of whether the humidity is 20% or 80%. Interactions can be non-existent, mild, or strong as shown in Figure 11.8.

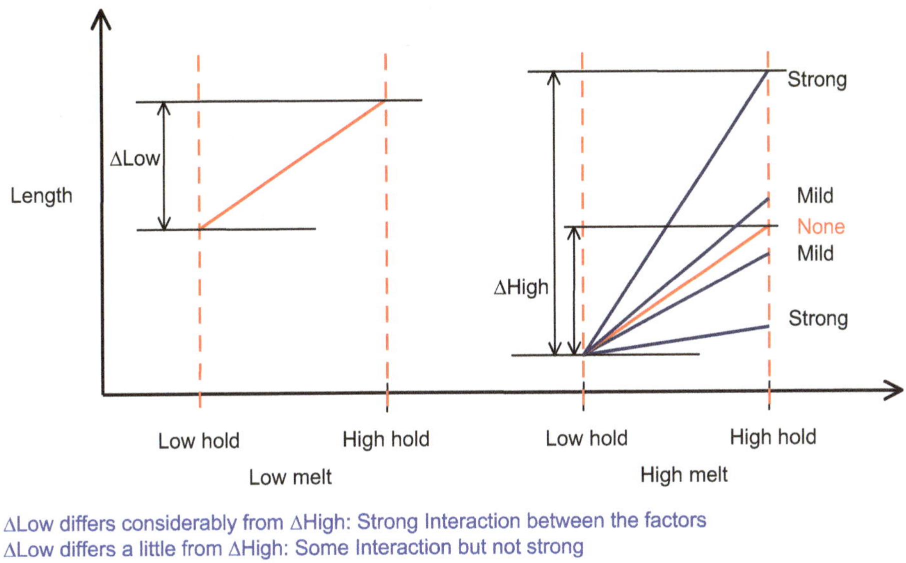

Figure 11.8 Types of interactions

Here, two factors are involved, and therefore we refer to two-factor interactions. Three-factor interactions have not been encountered in manufacturing processes, let alone injection molding. In injection molding one will always notice mild two-factor interactions. However, the effects are usually overpowered by the main factors. If strong interactions are seen, even between two factors, the measurement method must be questioned. The other possibility is that external effects may have caused interactions after the molding has taken place. For example, dimensions of the top of a deep draw box can be influenced during post-mold cooling because they may be largely unsupported. Mechanical stresses caused during the cooling of the part may be interpreted as interactions.

In the molding process, some mild interactions are usually observed between melt temperature, mold temperature, and any of the filling phases. Although the melt temperature is considered constant, the melt is cooling down as it fills the mold because filling is a non-isothermal process. Consider pressure and melt temperature as factors and the length of the part to be the response. If we measure the length of the part for the four experiments, we observe that the change in length between the low holding pressure and high holding pressure at the low melt temperature is slightly different from the change in length between the low holding pressure and high holding pressure at the high melt temperature. This means that the setting of the melt temperature has an effect on how the holding pressure influences the length of the part. At low temperatures, there is a greater influence and at higher temperature the influence is negligible. Therefore, we consider the interaction between the melt temperature and holding pressure to be present. If the changes in length at the low and high temperatures were similar, interactions would be considered absent. Interactions may be evident in cases where the levels of the factors are at extremes. Such an example is extremely low injection speeds, where the injection screw is barely moving at the low level, while at the high level, the screw is moving fast. At low fill speeds, the melt can cool down considerably before it reaches the end of fill, whereas in the case of fast injection speeds, the melt temperature can stay high through the filling stage. This would make a difference in the packing phase of the cycle and produce different results. Because the amount of heat that needs to be removed from the mold will also change, this can have a considerable effect on cooling. Mathematically, interactions are represented as the multiplication columns shown in the tables. Column AB is considered an interaction column between factor A and factor B.

11.7 Confounding or Aliasing

Consider the orthogonal array in Table 11.6, which shows a 3-factor experiment with two levels for each factor and all the interactions.

Table 11.6 Three-Factor, Two-Level Orthogonal Array with Interaction Columns

Experiment No.	A	B	C	A × B × C	A × B	A × C	B × C
1	−1	−1	−1	−1	+1	+1	+1
2	+1	−1	−1	+1	−1	−1	+1
3	−1	+1	−1	+1	−1	+1	−1
4	+1	+1	−1	−1	+1	−1	−1
5	−1	−1	+1	+1	+1	−1	−1
6	+1	−1	+1	−1	−1	+1	−1
7	−1	+1	+1	−1	−1	−1	+1
8	+1	+1	+1	+1	+1	+1	+1
Sum	**0**	**0**	**0**	**0**	**0**	**0**	**0**

The sum of each of the cells for each of the interaction columns is equal to zero. Therefore, this is an orthogonal array. In an orthogonal array, every column provides us with a unique set of information. Therefore, every row can be a unique experiment. Since we know that interactions in injection molding are either very mild or non-existent, we can take advantage of this fact. We can replace the names of each of the interactions columns by the names of any factors we would like to study. Therefore, in the above array, we can replace the four interaction columns by any other four factors we would like to study. Let us say that the original three factors were holding pressure, melt temperature, and mold temperature. Now we would like to study four more factors. Based on our engineering knowledge, we must first prioritize these four additional factors according to what we believe will provide the most experimental impact or information when studying these four factors. Let us say that after the first three main factors, the order of priority is cooling time, injection speed, holding time, and screw speed. We now rename columns 4 through 7 with these newly selected factors. This process of renaming an interaction parameter column with another factor is called *confounding* or *aliasing*; it is executed for our example in Table 11.7. Because the fourth column is a three-factor interaction and we know that three-factor interactions are almost non-existent, we can be almost certain that the interaction can be confounded or renamed as another factor. If the analysis shows that a confounded interaction has a more significant effect on the quality of the part than a non-confounded factor, the data must be checked and in some cases the experiments must be rerun. It is also possible that the non-confounded factors may have been wrongly selected. This is the reason why a molding background is required before performing DOEs. For example, holding pressure is a very significant factor and must always be a non-confounded factor. Consider a case where the non-confounded factors were back pressure, screw speed, and melt temperature, and one of the confounded factors was holding pressure. The holding pressure was confounded with the interaction between screw speed and melt temperature. If the experiment was run, the analysis would show that the holding pressure would be significant. But because it was confounded, it will show that the interaction between the screw speed and melt temperature would be significant, providing a false analysis.

Table 11.7 Introducing Confounded Factors in the Interaction Columns

Experiment No.	Non-confounded factors			Confounded factors			
	Holding pressure	Melt temperature	Mold temperature	Cooling time	Injection speed	Holding time	Screw speed
1	−1	−1	−1	−1	+1	+1	+1
2	+1	−1	−1	+1	−1	−1	+1
3	−1	+1	−1	+1	−1	+1	−1
4	+1	+1	−1	−1	+1	−1	−1
5	−1	−1	+1	+1	+1	−1	−1
6	+1	−1	+1	−1	−1	+1	−1
7	−1	+1	+1	−1	−1	−1	+1
8	+1	+1	+1	+1	+1	+1	+1
Sum	**0**	**0**	**0**	**0**	**0**	**0**	**0**

This shows that the biggest advantage of confounding is in the reduction in the number of experiments required to study the process. Based on Equation 11.1, repeated below,

$$n = l^f$$

a 7-factor, 2-level study would need 128 unique experiments, but with the process of confounding, these can be reduced to a mere 8 experiments. The general assumption when using confounding is that the interactions are minimum or non-existent. However, it must be stressed very strongly here that the above example of reducing the number of experiments from 128 to 8 is only for understanding purposes. In injection molding, one should not run experiments that are less than half the original number of experiments which would be a half factorial experiment.

11.8 Factorial Experiments

The experiments and tables used in the earlier discussion are all factorial experiments. The design with the maximum number of possible experiments without confounding is called a full factorial experiment. For example, a 2-level, 4-factor design would require 16 experiments in a full factorial design. Partial factorial experiments are designs that have fewer experiments using the technique of confounding. A half factorial experiment would mean half the number of runs of a full factorial experiment or 8 runs out of the 16 full factorial runs. Reliable data can be obtained from partial factorial experiments to analyze the effect of the factors. However, in injection molding, nothing less than a half factorial design should be run.

11.9 Data Analysis

Today, most analysis of DOE data is performed by computer programs that generate all the information and graphs in a matter of seconds. A typical analysis provides the following information:

- Factors that most influence the quality of the part
- The robustness of the quality of the part
- Prediction of the most optimized process
- Prediction of the capability of the process within the range of the parameters studied

To explain the results and the analysis it is best to consider an experiment. Consider the DOE results shown for the experiment in Table 11.8.

Table 11.8 Experimental Settings and Response Data for a 3-Factor, Full Factorial Experiment

Runs	Experimental run settings			Responses	
	Mold temp. (°C)	Cooling time (s)	Holding press (bar)	Length (mm)	Diameter (mm)
1	40	30	30	144.73	6.35
2	40	30	55	144.40	6.15
3	40	20	30	144.60	6.32
4	40	20	55	144.30	6.15
5	20	30	30	144.83	6.37
6	20	30	55	144.50	6.16
7	20	20	30	144.65	6.32
8	20	20	55	144.34	6.16

The three selected factors were mold temperature, cooling time, and holding pressure. There were two levels for each factor. The response was the length of the part and an internal diameter. Based on engineering knowledge and past experience, the order of importance for the factors was holding pressure–cooling time–mold temperature. However, if these three factors are arranged in the order of 'difficulty of change,' then they are rearranged in the order mold temperature–cooling time–holding pressure. 'Difficulty of change' simply means how soon the change is seen in the process. For example, once the mold temperature is changed, depending on the size of the mold, it may take anywhere from 15 minutes to an hour to see a change in the actual mold temperature. So this is a 'difficult' change. On the other hand, holding pressure is an easy change, because once the setting is changed, the change will be reflected in the next cycle. If the experiments are not going to be randomized, it is a good idea to have the order of the factors in the order of difficulty. This is an efficient way of getting all the experiments done by making the least amount of difficult changes during the experiments. So, in the above array, we will have to change the mold temperature (difficult change) only once compared to the holding pressure (easy change), which will have to be changed seven times after the first experiment.

With the help of this matrix, the experiments are performed. It is best to collect as many data as possible for accurate analysis. Typically, if statistical analysis is to be done, at least 30 parts from each cavity must be checked and the data recorded. However, as the number of cavities and recorded dimensions for a part increase, the amount of work to collect the data increases and this can sometimes become prohibitive. Performing full factorial experiments with a larger number of factors will again increase the number of experiments and therefore the number of measurements. In the experiment under consideration, we are going to measure the length of the part. We will take an average of five samples for each experiment.

The results are typically displayed in one of the following forms:

11.9.1 Tornado Charts

Tornado charts are similar to the commonly used Pareto charts. In DOE, a tornado chart is a bar chart in which the effect of each factor is plotted in descending order of magnitude. A factor that has a direct effect is displayed on the positive side of the y-axis, while a factor that has an inverse effect on the response is plotted on the negative side of the y-axis. A tornado chart is one of the most important and informative charts in DOE. It will display the factors that have the most effect on the quality of the part in descending order of importance. For the experiment under consideration, the tornado chart is shown in Figure 11.9.

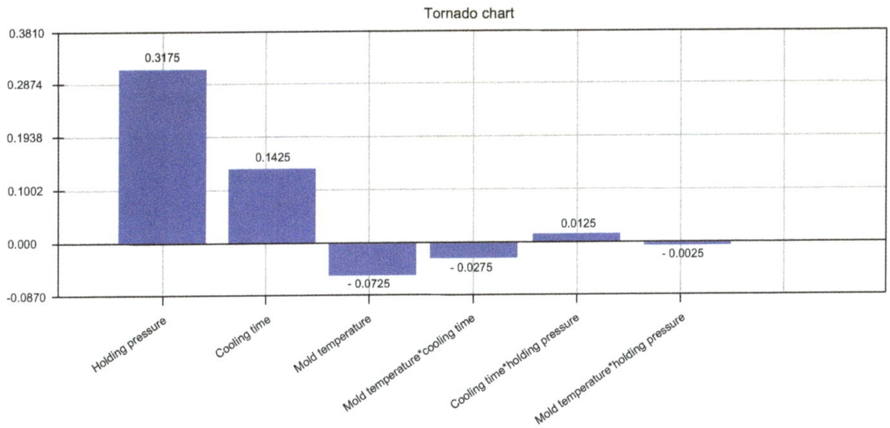

Figure 11.9 Tornado chart showing direct and inverse relations between factors and length response

Figure 11.9 shows that the factor with the most impact on the length of the part is the holding pressure. For the levels of the holding pressures tested (between 30 and 55 bar), the average change in length was 0.3175 mm, compared to the 0.1425 mm change observed between the cooling time levels of 20 and 30 seconds. Change in mold temperature had the least effect on the length of the part. The average change in length between 20 °C and 40 °C was only 0.0725 mm. Notice that the mold temperature bar is on the negative side of the axis. This means that when the mold temperature increases, the part length decreases. Positive values, as seen for holding pressure and cooling time, indicate that as these factors are increased, the length of the part also increases. The interactions are also plotted and are very insignificant. In Figure 11.9, holding pressure and cooling time have a significant effect and we can consider that the effect of the other factors is insignificant. The tornado chart for the diameter is shown in Figure 11.10.

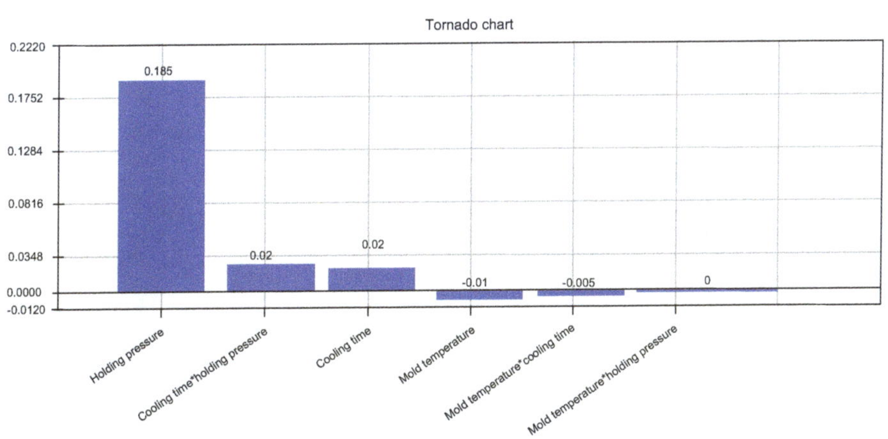

Figure 11.10 Tornado chart for diameter of the part

11.9.2 Contour Plots

Contour plots display contours that represent a constant value of a given response on a graph of two of the selected factors on the x and y axis. On any given point on a given contour, the value of the response will be constant, regardless of the values on the axes. Figure 11.11 shows a contour plot based on holding pressure and cooling as the factors.

Consider the highlighted contour that has a value of 144.65, which is also the nominal value. Anywhere on this contour, the value will always be 144.65. A combination of 29 s of cooling time and 42.5 bar of holding pressure (Point A) will produce a part with the same dimension as the combination 22 s and 52 bar (Point B). When the nominal value and the specifications are plotted on the contour plot, the process window inside of which dimensionally acceptable parts can be molded is visible. In Figure 11.11, the orange contours represent the upper and lower specifications. Contour plots help to determine the extent of dimensional process windows. The different types of process windows are explained in Chapter 9.

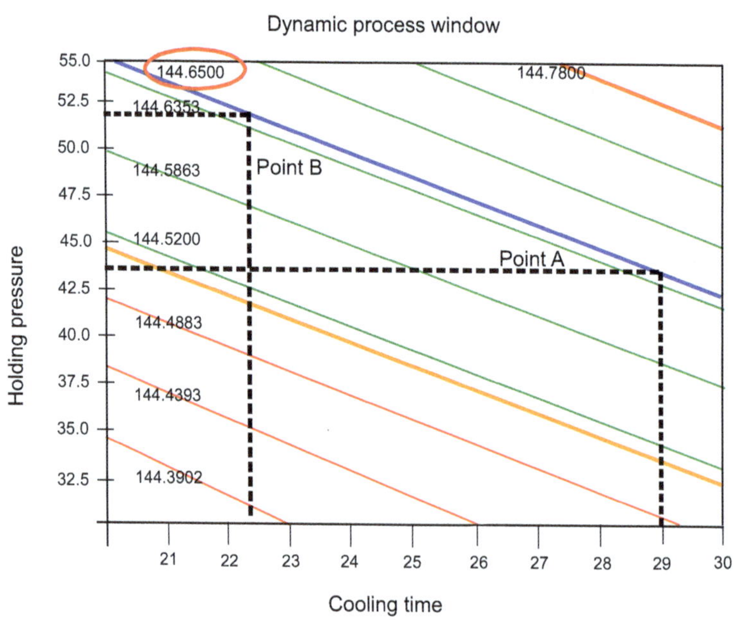

Figure 11.11 Contour plot for length of the part

11.9.3 Prediction Equation

A relation between the response and the factors can be established by mathematical means. Figure 11.1 showed the increase in the length of a part that resulted due to an increase in the holding pressure. A relationship based on the PVT and cavity pressure concepts must be established. Once the equation is established, the dimension at any value of holding pressure or the holding pressure required for a desired dimension can be predicted, see Figure 11.12. This is the basis of prediction equations that involve all factors and their interactions. Prediction equations are beneficial in selecting process parameters for robust processes and for hitting target dimensions. It is not always possible to hit target dimensions on all the responses. While one dimension is in tolerance, another may be out of tolerance. Overlaying contour plots or looking at composite and desirability functions is a good way to estimate the robustness and the process capabilities.

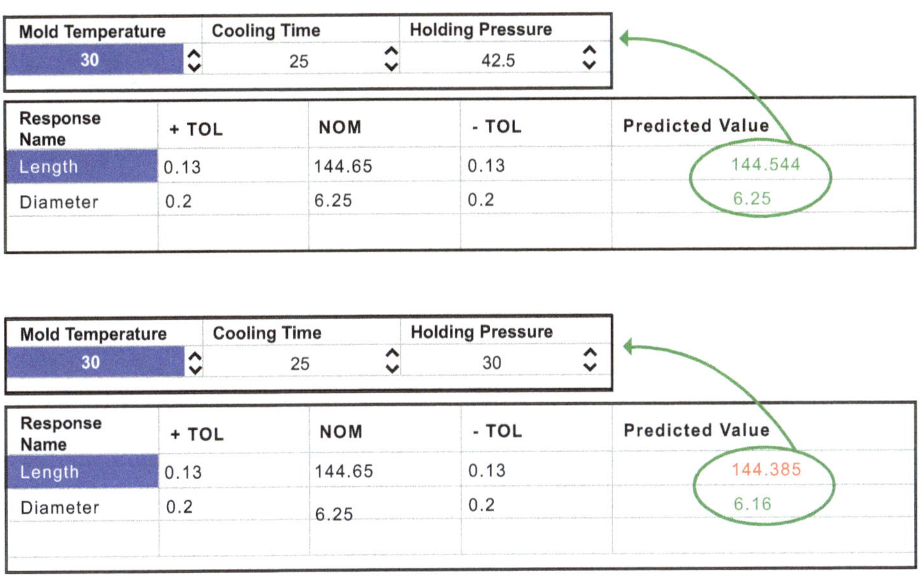

Figure 11.12 Prediction equation for length and diameter

11.9.4 Process Sensitivity and Outliers Charts

Process sensitivity charts are line graphs that provide a quick glance of the position of the response for each experiment. Such a chart can be considered as a visualization of the sensitivity of the response to a change in factors for each experiment. The upper specification limit (USL) and the lower dimensional specification limit (LSL) as well as the nominal are shown on the graph. If for all experiments the response is within the specification limits, the response is considered stable and not affected by the process. Such a dimension is shown in Figure 11.13. If this was considered a critical dimension to be checked regularly during production, a case can be made to eliminate this dimension for in-process inspection and it could instead be checked at startup only. Figure 11.14 shows a dimension that is easily affected by process changes. Composite process sensitivity charts display all responses on one screen to give a snapshot of all dimensions at one time.

Such a graph will also show if there are any outliers in the data. Any analytical software will analyze the data provided. If the data input is not reliable then the analysis is not reliable. A process sensitivity graph will show such outliers if any are present. If there are any such outliers or questionable outliers then the parts must be remeasured.

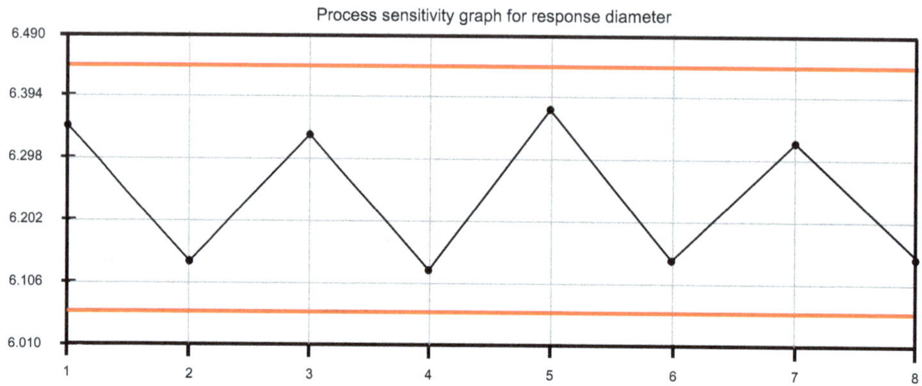

Figure 11.13 Process sensitivity chart for the diameter

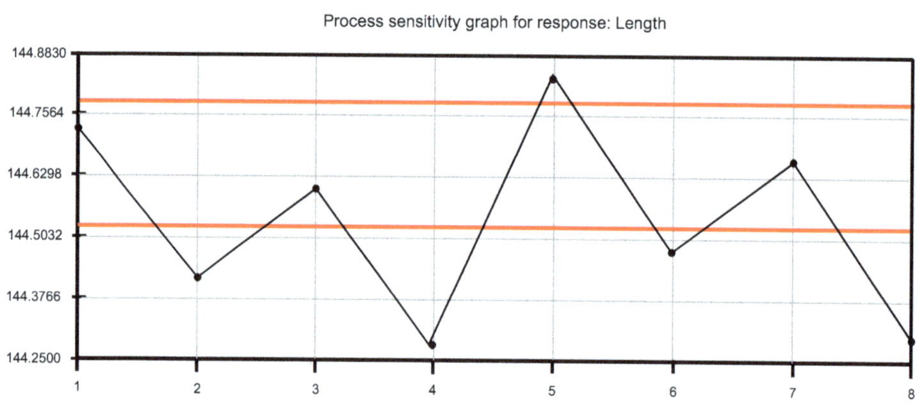

Figure 11.14 Process sensitivity chart or the length

11.10 Using the Results from DOE

The graphs and results from the above charts and equations can be used in a number of ways.

11.10.1 Process Selection

The contour plots will display the constant response curves together with the specification limits. The goal should be to achieve the nominal value of the response and be in a robust processing area, see Figure 11.11, where the LSL, USL, and the nominal are shown. To achieve the nominal, various combinations of the holding pressure and cooling time exist; for example, at 29 seconds and 42.5 bar or at 22 seconds and 52 bar. Any one of these combinations can be selected as process parameters because both are in a robust molding area. In injection molding, lower cycle times increase production efficiency. Therefore, the lower cooling time must be preferred and selected.

11.10.2 Cavity Steel Adjustment

If the nominal values and the specifications are located towards the corners or boundaries of a contour plot, it may be difficult to sustain production because the process may not be robust enough to mold the parts within specifications. For the part shown in Figure 11.15, the processing window between a short shot and out-of-specification is only about 50 psi (3.44 bar). The process must be kept within this 50 psi window to avoid shorts and parts out of specification. The process window is small and therefore not robust.

If the process variation is large, it is easy to produce parts that are either short or out of specifications. In such cases, without taking the dimensions into consideration, a process with a sufficient process window must be selected. Once it is determined that the process can produce parts with the required dimensional consistency (not the actual dimension), the steel must be adjusted to bring the parts within the required dimensions. For example, for the part in Figure 11.15, there is a danger of short shots because the holding pressure must be kept low to achieve the required dimension. If the pressure is increased, the parts will be out of specifications but the chances of short shots will be greatly reduced. Running at the center of the process window will produce parts that have the least or zero possibility of short shots. Parts must be molded at the center of the window and the dimensions measured. Depending on the dimensions, the mold steel must now be changed to bring the molded parts within the required tolerance. This will result in a robust process that will produce parts within the desired specifications.

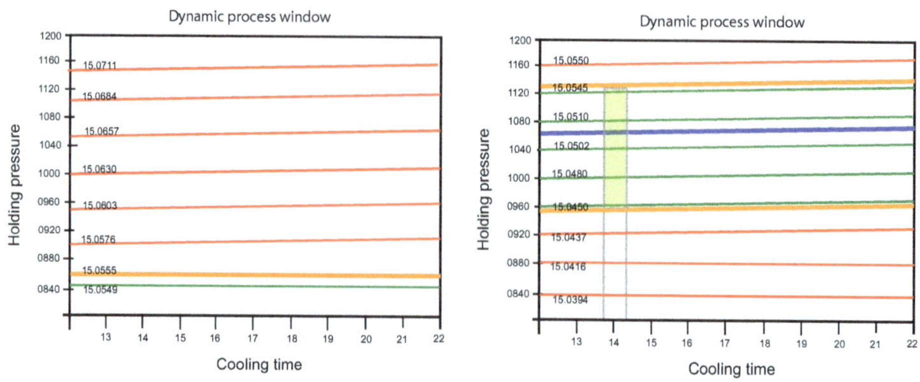

Figure 11.15 Using the DOE results to make steel changes and increase the processing window

11.10.3 Process Adjustment Tool

The factors used in the DOE were selected based on the assumption that they were the most influential to the part quality. The analysis results provided the quantitative effect of each of these factors. During a production run, if the part quality drifts, the analysis results can help in adjusting the process to get the part quality back to where it should be. For example, if the tornado diagram shows that the holding pressure and cooling time have the most significant effect on the length of the part, then a contour plot with holding pressure and cooling time as variables may help the processor adjust either one or both variables and continue making acceptable products. The guesswork regarding what parameter should be changed is eliminated. This will also keep the process sheet clean of changes to several parameters and limit the changes to the holding pressure and cooling time. Educating the workforce and providing access to this data at the molding machine is essential.

11.10.4 Setting Process Change Tolerances

The tornado charts and the contour plots provide information on the most significant process parameters and the extent to which they affect the part quality. Based on this data, the process change allowances and limits can be set. In the contour plot, a box must be set and the corresponding limits must be used as tolerance limits for allowable changes during production. Only these factors must be changed in case there is a quality problem.

11.10.5 Reducing Inspection

The results from a DOE can be very useful in reducing and even eliminating inspection. If the process sensitivity charts indicate that a dimension does not seem to be affected by the various process changes and provided that this dimension is within specifications, it can be checked at mold startup and then be assumed as within specification during the in-process inspection. The same holds true for other dimensions that are well within the process windows of the established process. For example, the dimension in Figure 11.13 does not go out of specification at all in the tested experiments. This dimension is stable and the measurement can be reduced in frequency or even eliminated.

11.11 The Dimensional Process Window (DPW)

The concept of the cosmetic process window was introduced in the previous chapter. This is the window or the boundaries of the process parameters within which the molded parts are cosmetically acceptable. Dimensions are not considered. Now that the topic of DOE has been introduced, we can consider the dimensional process window or DPW within which the molded parts would be within spec. Refer to Figure 11.16, which shows the DPW for one dimension for Cavity 1 of a two-cavity mold. A large DPW is a sign of a robust molding process. The process must be set in the center of the window.

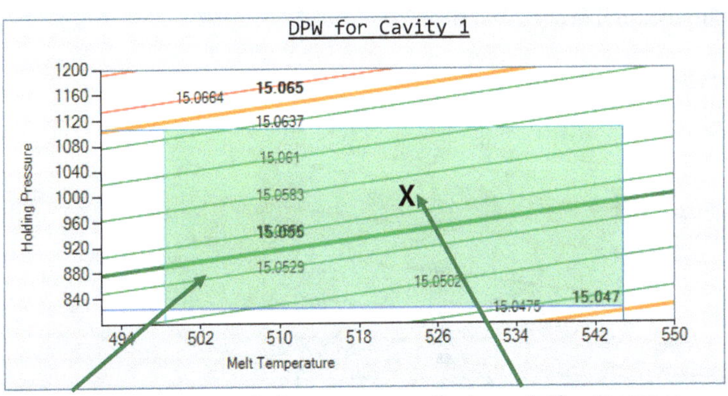

Dimensional Process Window:
Holding Pressure: 825 to 1100 psi
Melt Temp: 500 to 545 °F

Center of the Process:
Melt Temp: 525 °F,
Hold Pressure: 960 psi

Figure 11.16 The dimensional process window for Cavity 1

As the number of cavities and the dimensions increase, the window can quickly get smaller and smaller. Cavity balance plays a very important role in the size of the DPW. If all the cavities fill and pack at the same rate, the process windows for each cavity will overlap. If they do not fill and pack at the same rate, then each cavity will have its own DPW, but the overlap could be a small area that will reduce the overall process robustness and repeatability. Refer to Figure 11.17, which shows the DPW for Cavity 2. If any of the dimensions are cornered inside the CPW, then this is also a clear sign that the process is not going to be robust. The DPW for Cavity 2 is very small and is not acceptable. The window gets even smaller when the contours for both cavities overlap. Refer to Figure 11.18.

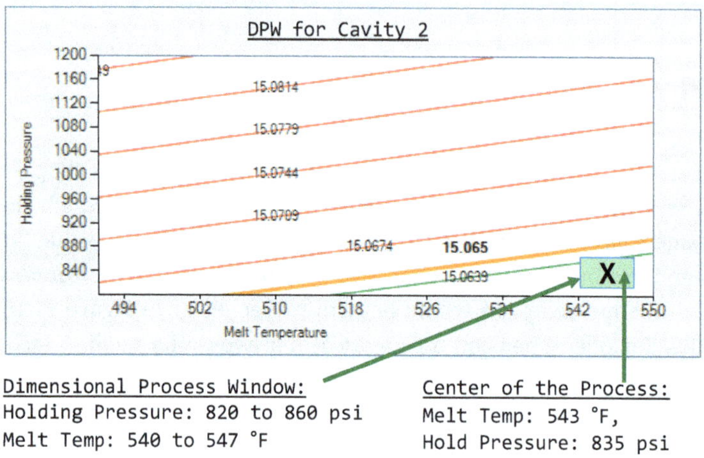

Figure 11.17 The dimensional process window for Cavity 2

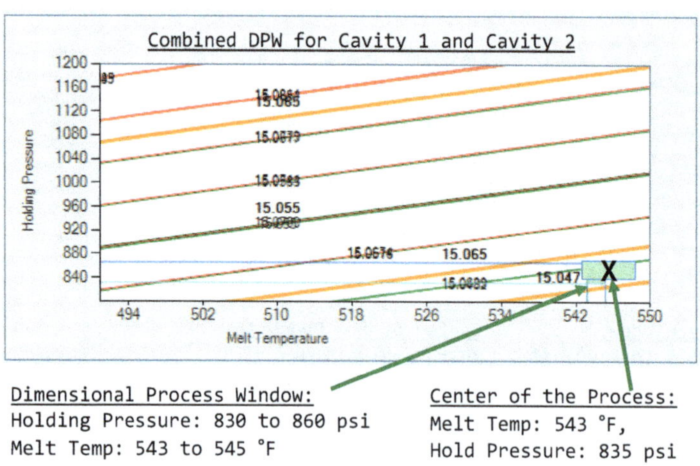

Figure 11.18 The dimensional process window for Cavity 1 and Cavity 2

A similar argument and explanation can be made considering dimensions. Refer to Figure 11.19, where two dimensions for Cavity 1 overlap, resulting in the reduction of the DPW. Each dimension should be considered individually first and the effort must be made to increase the DPW.

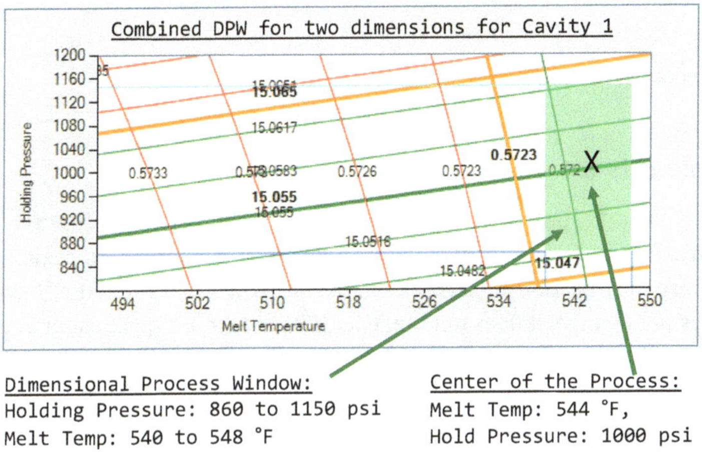

Figure 11.19 The dimensional process window for two dimensions for Cavity 1

The DPW is a subset of the CPW. Naturally, a wide CPW favors the possibility of a wide DPW, provided the nominal of the dimension is centered in the windows. If this is not the case, the data can be reviewed with the tooling engineer to support justifying a mold steel adjustment to center the part dimension within the specification limits.

11.12 Selection of Factors for DOEs

The final shrinkage of the plastic is easily influenced by a number of molding parameters. Molding process validations therefore call for a number of procedures to be followed to first find the most robust areas of each processing parameter, and then find the effect of each of these on part quality (DOE). The various processing parameters that are typically adjusted (inputs) are given below. These are the factors that directly influence the plastic and will therefore make a difference in the part quality. Factors such as mold open speed or ejection speed do not contribute directly to the plastic process and are therefore left out of the list.

1. Injection speed
2. Injection pressure
3. Packing and hold pressures

4. Packing and hold times
5. Melt temperatures
6. Mold temperatures
7. Shot size and transfer (switch over) position
8. Back pressure
9. Screw rotation (recovery) speed
10. Cooling time

Let us revisit the parameters above and discuss their optimization.

1. **Injection Speed** – Injection speeds are optimized using the data from the first step of the 6-Step Study. This is the viscosity study. The flat region (Newtonian region) of the curve is where the injection speeds are set. Because the viscosity in the Newtonian region does not change, it does not make a difference if the speed is set at a high or a low value within the Newtonian region. Therefore, injection speed is not a factor that should be typically included as a factor for DOE. Injection speeds typically affect the cosmetics of the part. If the difference between the speeds is large, then dimensions could be affected. Typically, injection speed should be set to 'as fast as you need' and not 'as fast as possible.' In most cases injection speed is not used as a factor for DOEs.

2. **Injection Pressure** – To achieve a set injection speed, a finite amount of pressure is required. The molding machine is set to a value that is higher than this number so that the required pressure is always made available as and when required. If the setting is below the required pressure, the process becomes pressure limited. As discussed, the injection pressure was set to about 2000 psi more than the peak pressure required. Setting to more than what is required will not have an effect on the injection speed because the machine will still use what it needs regardless of what is made available to it. Varying the pressure from a high to low value is not going to have any effect of the part quality. Therefore, the injection pressure is not used as a factor for DOEs.

3. **Pack and Hold Pressures** – As seen from the specific volume versus temperature graph (Figure 2.23), the specific volume is dependent on the pressure that is applied to the melt. The pack and hold pressures represent one of the most important factors and must always be considered as a factor for the DOE.

4. **Pack and Hold Times** – In cold runner molds, these are optimized using the gate seal study. This is Step 5 in the 6-Step Study. Using more time than required does not increase the amount of plastic inside the cavity and therefore does not affect the dimensions. Using less time than what is required to seal the gate will cause part weight variations, resulting in part dimensional variations and is therefore

not recommended. This parameter should therefore not be used as a factor for performing a DOE for cold runner molds. In case of hot runner molds, there is a land that needs to freeze off to stop the flow of the plastic into the cavity. However, the land area is small and the flow can easily be influenced by the hold time. The holding time should therefore be used as a factor for hot runner molds. The optimization technique is mentioned in Section 11.1.

5. **Melt Temperatures** – The melt temperature setting is responsible for achieving the required melt viscosity to fill the mold. In case of amorphous materials the recommended melt temperature range is wide and the so the melt temperature can affect the shrinkage. Therefore, melt temperature should always be a factor. In the case of crystalline materials, the melt temperature range is typically very narrow and does not significantly influence the shrinkage. Melt temperatures for crystalline materials should not be typically used as a factor for DOEs.

6. **Mold Temperatures** – Mold temperatures are responsible to maintain the flow of the plastic to the end of the fill of the part. The temperature of the plastic at the end of fill should be within the recommended range of the melt temperature. This is true for amorphous and crystalline materials. In the case of crystalline materials, the mold temperature has an additional function. Because the formation of the crystallites needs a certain amount of energy, this energy is supplied by the mold temperature. The optimum crystallite formation depends on the mold temperatures. The mold should therefore be maintained in the recommended temperatures and the mold temperature should be used as a factor for the DOEs. In case of amorphous materials, there is no requirement of crystallite formation and so mold temperatures are only critical for the flow and cosmetics. The lower end of the mold temperatures should be therefore used as long as the flow and cosmetics are achieved. The mold temperature is therefore not used as a factor for DOE for amorphous materials.

7. **Shot Size and Transfer (Switch Over) Position** – The volume between the shot size and the transfer position should be equal to the volume of the plastic injected in the injection phase. In practical molding, the mold is filled about 95–98%. Because this volume is fixed, both these process parameters should not be used as factors for DOE. When performing the experiments, the value of the transfer position should be changed to achieve the same 95–98% fill as and when required. For example, if the part is 98% full at low melt temperatures the part will more than likely be full at the high melt temperature. For the high-temperature set of experiments, the transfer position (or the shot size) must be readjusted to achieve the original 98% fill before collecting the parts for the experiment. The weight of the part in the injection phase should always be matched.

8. **Back Pressure** – Back pressure is used to help and achieve consistency in the amount of the plastic that is metered for the next cycle. It also helps in compressing the melt and squeezing off all the volatiles out of the melt. The back pressure is optimized by the back pressure study and therefore should not be used as a factor for the DOE.

9. **Screw Rotation (Recovery) Speed** – The screw rotation speeds were optimized using the screw rotation speed study and therefore should not be used at a factor for DOE.

10. **Cooling Time** – Cooling time is an important process parameter that must be used for the DOE for two reasons. First the as the amount of time the plastic remains in the mold changes, the heat transfer changes and therefore the ejection of the part on the specific volume versus temperature graph changes. The result is the change in the shrinkage value. Second, since injection molding is a business the processes must be as efficient as possible. Cooling time should be used as one of the factors for the DOE to find the optimum setting.

The factors mentioned above are the ones that are usually adjusted on the molding machine to mold parts to the required specifications. Refer to Table 11.9 for the summary of the factors to be used in the DOE. There are a maximum of five plastic factors that will truly make a significance difference in the part. All the factors are important, but to varying degrees. To make the DOE practical and efficient one must select the top hitters that will provide the 'low hanging fruit' and help to achieve the most optimum process. Understanding the molding parameters and the science behind their function is necessary. The same factor can have a different effect on the various dimensions of the part. For example, increasing cooling time can increase the length of a part but may have no effect on the diameter of the part. Each dimension must be dealt with on a case-to-case basis. The above discussion is a good guide that has been successfully used for several years by the author.

Table 11.9 Process Parameter Optimization Method and Selection for DOEs

No.	Factor	Method of optimization	To be used in DOE? Y/N
1	Injection speed	Viscosity curve	No (exceptions exist)
2	Injection pressure	Pressure drop study	No
3	Pack pressure	Cosmetic process window study	Yes
4	Hold pressure	Cosmetic process window study	Yes
5	Pack time	Gate seal study	No for cold runner Yes for hot tip, valve gate

No.	Factor	Method of optimization	To be used in DOE? Y/N
6	Hold time	Gate seal study	No for cold runner Yes for hot tip
7	Cooling time	Cooling time study	Yes
8	Melt temperature	Cosmetic process window	Yes for amorphous materials No for crystalline materials
9	Mold temperature	Cosmetic process window	No for amorphous materials Yes for crystalline materials
10	Screw rotation speed	Screw rotation speed study	No
11	Back pressure	Back pressure study	No
12	Shot size	Part fill of 95–98%	No
13	Transfer	Part fill of 95–98%	No

In most cases with a cold runner mold, a three factor, two level DOE is more than sufficient. These are the pack pressure, the cooling time, and the melt or mold temperature based on the morphology of the material. If the plastic material is a polyolefin, you may require a fourth factor, which would be the mold or melt temperature, which was not considered in the first three factors. In case of hot runner molds, the pack time must be used. Refer to Figure 11.20.

```
1. Hold Pressure .......YES
2. Cooling Time ........YES
3. Melt Temperature ┐
                    ├─ OR  ....... FOR POLYOLEFINS BOTH
4. Mold Temperature ┘
5. Pack Time ...........YES FOR HOT TIPS AND VALVE GATES
 MAX NUMBER OF FACTORS TO BE USED = 5, BUT IN MOST CASES = 3
```

Figure 11.20 Most commonly used factors for DOE

11.13 Analysis of Variance (ANOVA)

The topic of variation was discussed in Chapter 1. There is an inherent natural variation in every molding process. This is the reason process capability is calculated on some of the critical dimensions. Refer to Figure 11.21, which shows the variation and the spread of the dimensions for parts molded from two processes. The parts with

Process 1 (P1) are molded with a pack pressure of 55 bar, and the parts with Process 2 (P2) are molded with the pack pressure of 60 bar. P1 produces parts with an average of 15.25 mm and P2 with an average of 15.32 mm. The variation in both process is 0.10 mm. There is therefore a region where the spread of both processes overlap. Parts with a dimension between 15.27 mm and 15.30 mm can be molded from P1 and P2. ANOVA is the activity of comparing the inherent variation within a process to the difference in the averages of two processes. In the example, the inherent variation is 0.10 mm and the difference in the averages was 0.07 mm. There was a purposeful change of 5 bar in the packing pressure between P1 and P2. Detailed discussions are beyond the scope of this book. Most software programs will output this information.

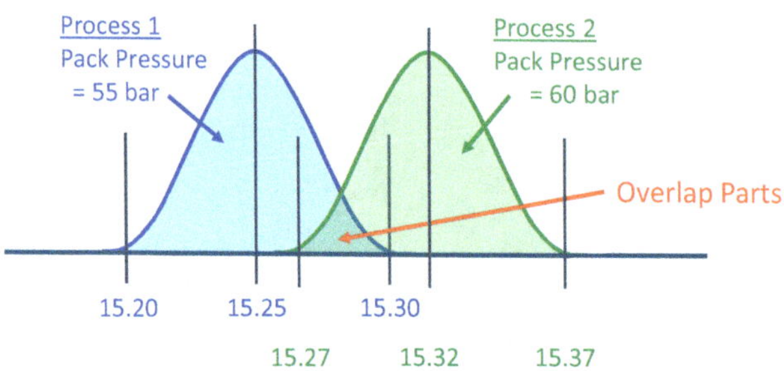

Figure 11.21 Parts in the overlap region that can be molded from two processes

While performing injection molding DOEs there should be a sufficient change in the molding process so as to make a difference in the quality of the parts (Figure 11.22). For example, an acceptable number for pack pressures should be about 275 to 350 bar plastic pressure (about 4000 to 5000 psi), for cooling times it should be about 8 to 10 seconds for a 2.5 mm thick part. Melt and mold temperatures should be taken from the data sheets.

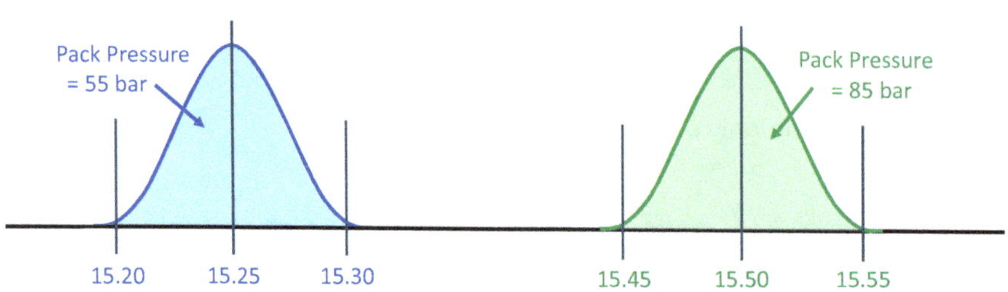

Figure 11.22 High and lows for DOE should be spread apart such that there are no overlapping parts

11.14 Collecting Shots for a Quality Check

A shot is defined as parts from one molding cycle. For a given experimental setting, at least five shots must be collected. Of these shots at least three shots should be measured. The reason for collecting five and measuring three is to eliminate any outliers if any in the three measured shots. All shots must be collected when the process is stable. Process changes fall under the categories of time, temperature, speed, pressure, and distance.

The following are the general guidelines about the number of shots between collections:

- Time: An example of time related settings are hold time, pack time, and cooling time. The number of shots in the barrel must be calculated. Once the times are changed, mold the number of shots equal to the number of shots in the barrel and discard them. Start collecting shots for measurement after the next shot. By doing this, the residence time in the barrel for all the shots is identical and what it would be if that setting was selected in production. Change in residence time will change resin properties. The mold temperature should also be checked for stability.
- Temperature: Examples of temperature related changes are melt and mold temperatures. In both these cases, it is advisable to stop the machine, make the change, and wait for the temperatures to reach the desired settings. Once the desired temperatures have been achieved, start molding, run the number of shots equal to the number of shots in the barrel, check the actual mold temperatures for stability, and collect parts. As a side note, start the DOE with the machine set at the lower value because it is easier to reach the increased settings than trying to drop temperatures down. Dropping the temperatures and reaching the set points is usually more time consuming.
- Speed: An example of speed related change is the injection speed. A change in the injection speed will change the fill time and therefore the rules of time mentioned above will apply. However, the change in the fill times from low to high are not high enough to be considered for waiting to empty the shots in the barrel. It is acceptable to wait only for a couple shots before the shots are collected. Screw rotation speeds are not a desired factor for DOEs. However, if these are changed, stability of melt temperature must be checked and then rules of temperature must be applied.
- Pressures: Examples of pressures are pack pressure and hold pressure. Machines will react to this change during the next molding cycle and in some cases instantaneously. For all practical reasons, these changes do not affect residence times or temperatures. Shots can be collected within two shots of making the change.
- Distances: Examples of distances are the shot size and the transfer position. These should not be used for a DOE, but if they are used, then the fill times will change and the rules discussed above for fill time can be used.

11.15 Choosing the Highs and Lows for DOEs from the Process Window

The cosmetic process window can be a parallelogram as shown in Figure 11.23. In a DOE, the values for the high and low pressures at the low melt or mold temperature should be identical to that at the high melt or mold temperature. To determine these values, one must therefore inscribe a rectangle within the parallelogram and use the high and low values of this rectangle. This is shown in Figure 11.23.

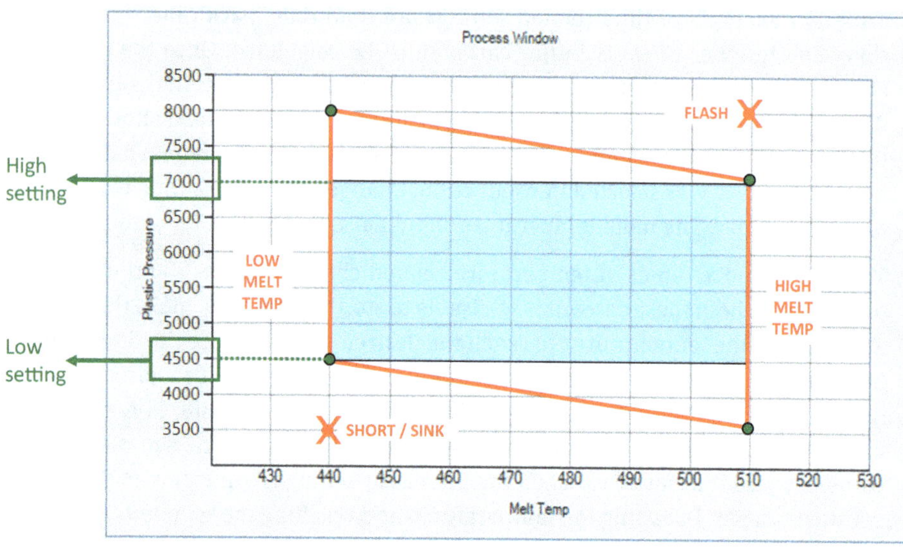

Figure 11.23 Selecting the high and low when the process window is a parallelogram

11.16 Randomization

The experiments in the set above are all arranged in a particular order. For example in Table 11.7, there are eight experiments, of which the first four are done with a low mold temperature and the next four are done with the high mold temperature. The rest of the factors also have a certain regular order. Randomization is not following any such order and completely selecting the experimental order at random. Each row is an experiment and randomization means running the experiments in a random order. Some experts advocate randomization in order to exclude any external effects that cannot be controlled. For example, if the first four of the above experiments are done in the morning, when the ambient temperature is low, and the remaining four are done in the afternoon, when it is hotter, the ambient temperature can have an effect on the response. The results will be compounded by the effect of the temperatures in the morning and the afternoon. This will make it difficult to separate the effect from morning and afternoon temperature change from the effect of the mold temperatures that were also different in the morning and the evening. There could be other factors related to ambient temperature that one may not be aware of. Tower water temperature, operator skills (not necessarily time efficiency), and material lot variation are examples of these factors. Therefore, mixing up or 'randomizing' the experiments will help to even out some of these effects, although not systematically. Randomization also helps to evaluate the robustness of a setting. For example, consider a knob with graduations being used to set the holding pressure, and a setting of 6 on the knob yielded a certain value of holding pressure. Setting it once in the beginning of the experiment and leaving it for a set of experiments will deliver the same consistent pressure. However, changing it to another value and then going back to 6 may yield another value of hold pressure. This demonstrates the robustness of the setting and going back and forth with the settings helps the evaluation of the repeatability of the equipment settings.

Randomization is not required in injection molding for several reasons. As described in Section 11.12, the number of factors used for the DOE is only five and in most cases only three. In most cases the number of runs is short so they can usually be completed within a couple hours. The effect of external factors during the short time is very small. Most machines have digital settings that are repeatable. If one changes the value of the settings and then goes back to the same value, the output is repeated. It is always good to make a note of the other factors that are constant. Noting the operator's name, material lot number, etc., is part of a good documentation procedure.

11.17 DOE Application to Optimize Pack and Hold Times

For cold runner molds, hold times are optimized by conducting a gate freeze study where the part weight is recorded as a function of the hold times. When the gate freezes, the part weight remains constant with increasing times. A second or so is added to the lowest value of time where the part weight stays constant, and this number is taken as the total time for the setting of the hold times. However, in hot runner systems or in valve gated systems, the gate area always has molten plastic and therefore the part weight curve never flattens out. Therefore, the above method does not produce acceptable results.

Excessive hold time and hold pressure can cause defects such as flash, stress in the part, or the part staying on the wrong side of the mold, whereas not enough pressure can cause shorts, sink, and dimensional issues. A combination of hold times and pressures must be found and then a DOE must be performed to optimize the parameters within the available limits.

Following is the procedure to optimize the hold times for hot runner and valve gated molds. The procedure has been explained with the help of an example for a two-cavity screen mold that is used in the irrigation industry.

A template to generate a cosmetic process window was prepared, see Figure 11.24. The *x*-axis corresponds to the holding time and the *y*-axis corresponds to the holding pressure. Such a template is called a visual inspection template (VIT). A filled out and completed VIT is shown in Figure 11.24. The red square markers represents a part with a defect at lower pressures and times (sink, short, etc.), a red triangular marker represents a part with a defect on the higher end of the pressures and times (flash, overpacked parts, etc.), and a green circular marker represents a part that is cosmetically acceptable.

Starting with a hold pressure value of 27.6 bar (400 psi), parts were molded from 27.6 bar to 96.6 bar (1400 psi) in steps of 13.8 bar (200 psi), with holding times from 4 seconds to 10 seconds in steps of 1 second. The data was collected and recorded in the VIT. The defect on the lower pressures and times was sink and on the higher side was flash in the screen area. One can now draw several cosmetic process windows (CPW) in the VIT.

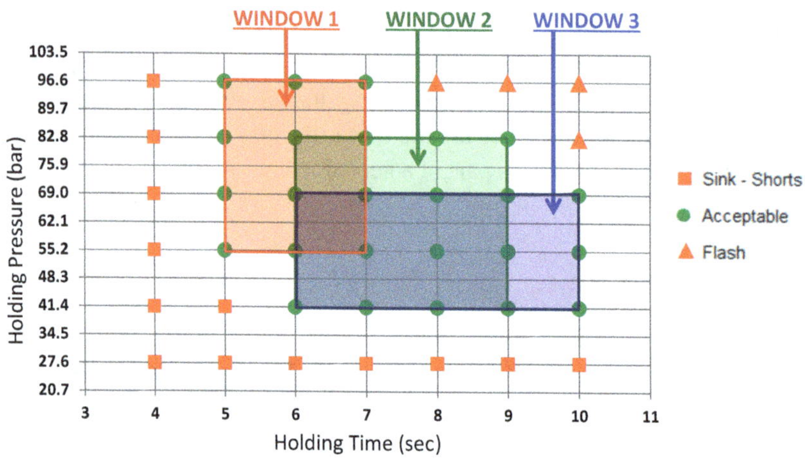

Figure 11.24 Visual inspection template with possible cosmetic process windows

The next step, once these windows are determined, is to look for dimensions. The technique of DOE is used for this. The process is varied between the limits of the windows and the dimensions are evaluated. After considering the three CPWs and other production requirements, it was decided to pick Window 3 for determining the limits of the DOE. The DOE matrix is shown in Table 11.10.

Table 11.10 DOE Matrix Based on the Cosmetic Process Window

Exp. No.	Hold Time (sec)	Hold Pressure (bar)
1	10	69.0
2	10	41.4
3	6	69.0
4	6	41.4

Parts were molded at the four settings in Table 11.10, and then the parts were measured for dimensions. The dimensional requirement on this part was on the length of the part and was specified at 126.80±0.07 mm. The analysis was carried out with the DOE module of the Nautilus Software. The results are shown in Figure 11.25(a) through (d).

Figure 11.25(a) and (b) show dimensional process windows for Cavity 1 and 2, respectively. The solid green contour lines represent Cavity 1, and the dotted contour lines represent Cavity 2. The green lines represent the settings where the parts are dimensionally acceptable, and the red lines represent the settings where the parts are dimensionally not acceptable.

Because there are two cavities here, we must then overlay the two contour plots on top of each other to determine a combined contour graph as shown in Figure 11.25(c). The area where the green contours for each cavity intersect represents the settings where both the cavities can be molded with acceptable dimensions.

One can now draw a composite dimensional process window inside which the parts will be dimensionally acceptable. Such a window is shown in Figure 11.25(d) for the above experimental data. Observing the window, one can notice that varying the holding time from 6.2 to 7.5 seconds and between 43.5 to 63.5 bar of holding pressure will produce dimensionally acceptable parts. Naturally, the bigger this window, the more robust the process is going to be.

Process robustness is the goal of every molder and therefore it is best that the center of the DPW in Figure 11.25(d) is taken at the process settings. A holding pressure of 55 bar and a holding time of 7.0 seconds were chosen as the process settings. Varying the holding time from 6.2 to 7.5 seconds and between 43.5 to 63.5 bar will still produce dimensionally acceptable parts. This is an indication of a robust process and will therefore produce parts not only to specifications but also with improved statistical process capability. Again, several dimensional windows are possible here. On molding machines, the extent of pressure variation is higher than time variation and therefore maximizing the pressure window is always preferred.

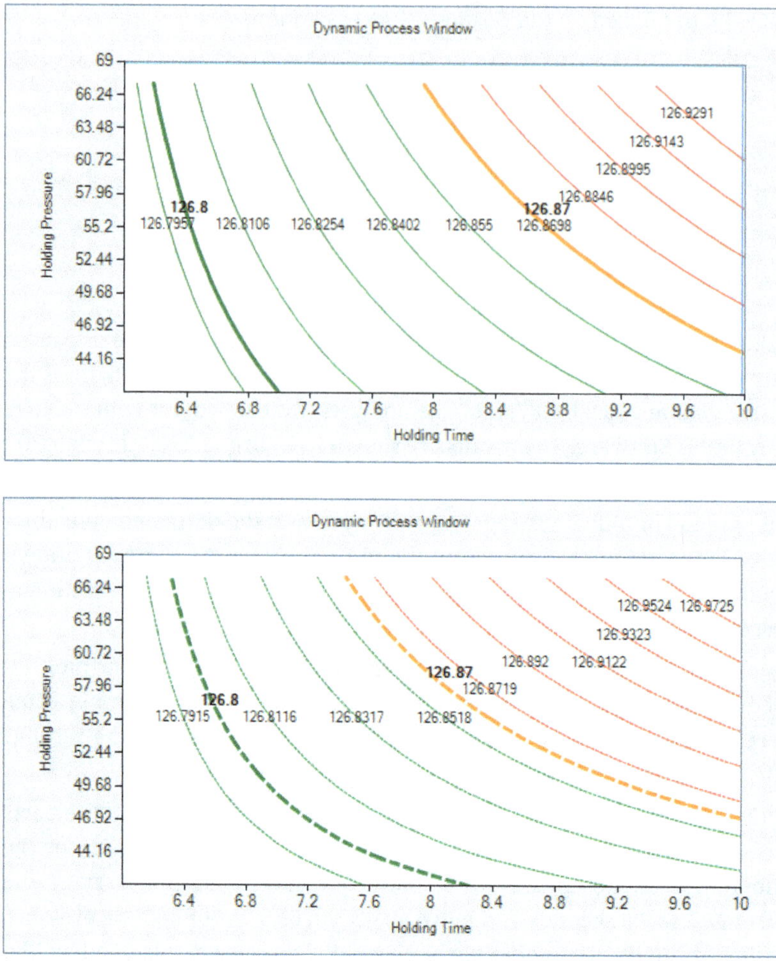

Figure 11.25 Dimension process window via DOE for hot runner molds; (a) dimensional process windows for Cavity 1, (b) Cavity 2, (c) both Cavities 1 and 2, and (d) a composite dimensional process window for both Cavities 1 and 2

The mold had gone through an iteration for steel adjustment resulting in an acceptable process window. During the first iteration, the mold was incapable to mold parts consistently to dimensional specification. The cosmetic process window was large but the dimensional process window was extremely small. Considering both cavities the molding process was not robust making it statistically not capable either. Remember that molding parts to spec does not mean that all the parts will be within spec. One must understand and measure the combined variation in the machine, material, process, and so on.

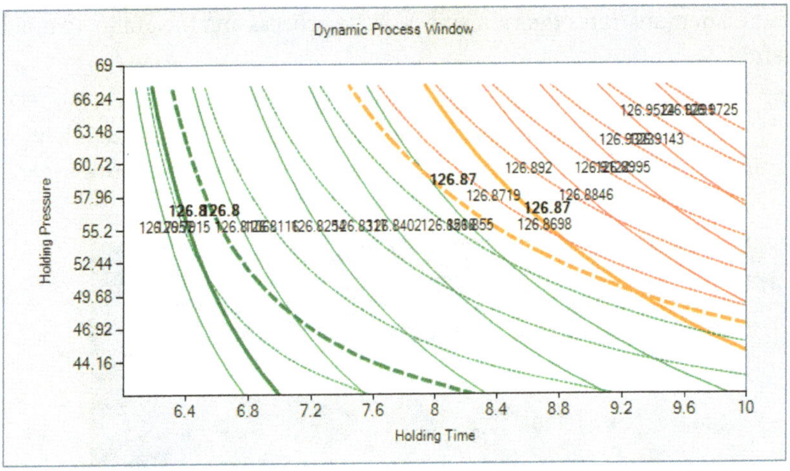

Dimensional Process Window:				**Selected Process Settings:**	
Holding Time	6.2	to	7.5 sec	Holding Time	7.0 sec
Holding Pressure	43.5	to	63.5 bar	Holding Pressure	55.0 bar

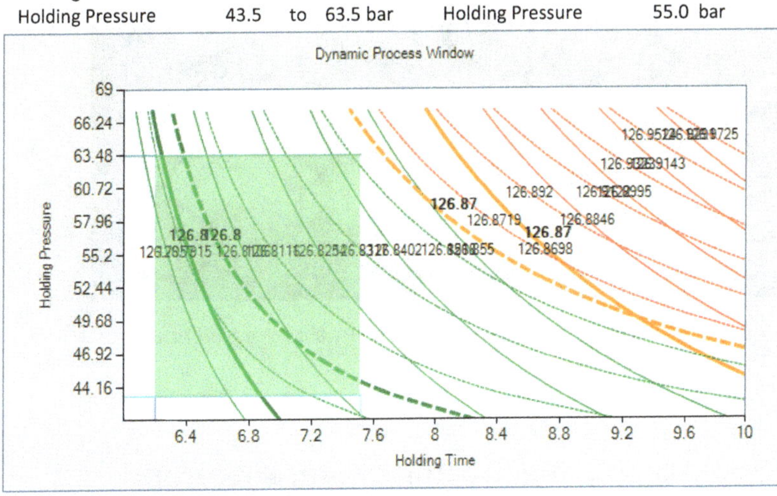

Figure 11.25 Dimension process window via DOE for hot runner molds; (a) dimensional process windows for Cavity 1, (b) Cavity 2, (c) both Cavities 1 and 2, and (d) a composite dimensional process window for both Cavities 1 and 2 *(continued)*

This procedure was described for hot runner molds but can easily be extended to molds with valve gated systems. The procedure remains the same and is in fact slightly simpler since the valve pin shuts off the gate eliminating the true hold phase from the original discussion of the pack and hold phase.

Optimization of holding pressure times for hot runner molds has always been an area of trial and error. The procedure described is a scientific way of determining the holding pressure and time in case of hot runners and valve gated systems. The procedure also evaluates and demonstrates the robustness of the process and the ability to mold parts consistently.

An interesting study was done using polarized light on an optical cover, see Figure 11.26. In this case, the sink in the gate area was visible under polarized light. A VIT was generated using these pictures followed by a DOE as described above.

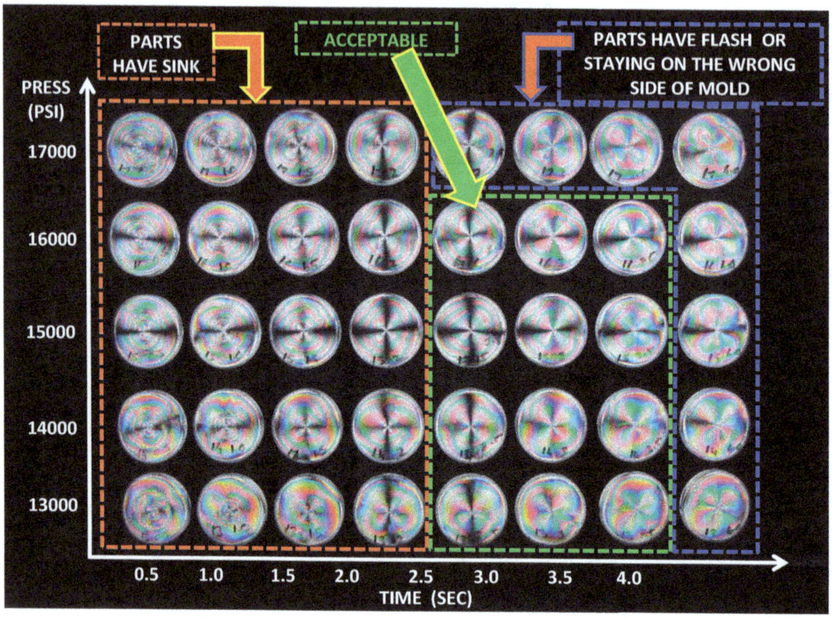

Figure 11.26 Visual inspection template and dimensional process window for an optical cover

11.18 Setting Acceptable Machine Tolerances and Alarms during Production

Tolerances are applied to the values of the inputs on the molding machine while alarms are applied on the outputs from the molding machines. For example:

Tolerance: Consider the tolerance on pack pressure is +/– 3.5 bar (50 psi). This means that if the set value is 70 bar, the processor can change the value anywhere from 66.5 bar to 73.5 bar.

Alarm: The limits on cushion is set to +/– 1.0 mm (0.04 in). This means that if the cushion value is 8 mm with an established process that is in production and if the cushion falls below 7 mm or goes above 9 mm, the machine will alarm.

Most companies have generic tolerance and alarms limit numbers for each of their processes. For example, melt temperature: +/− 10% or cushion value: +/− 2 mm. The problem in using a generic method to set the tolerances is twofold.

- Using a % is misleading. If the melt temperature is 200 °C, then +/− 10% is +/− 20 °C or a range of 40 °C. If the melt temperature is 370 °C, as is in case of a PEI, then a +/− 10% is +/− 37 °C or a range of 74 °C. So the higher the value, the wider is the tolerance range, which does not make any scientific sense.
- Process tolerances are applied to keep the product within the product quality requirements or within the LSL and the USL. If one puts a generic tolerance of +/− 10%, then one is assuming that if the parts are molded at the lower end (−10%), and the parts are molded at the high end (+10%), then both these sets of parts are acceptable. This may or may not be the case.

Machine tolerances and alarm limits must be set based on scientific reasons and should be determined from the experiments conducted during process development. The tolerances should be based on the results of the DOE. Output values during the DOE should be recorded and based on the final settings so that the alarm limits can be finalized. A contour plot and an example for setting alarm limits is shown in Figure 11.27.

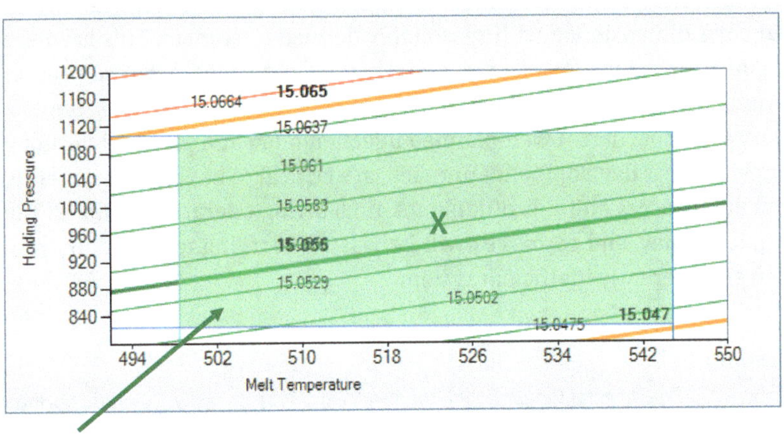

Dimensional Process Window:
Holding Pressure: 825 to 1100 psi
Melt Temp: 500 to 545 °F
} Resultant cushion: 0.015″ to 0.009″

Center of the Process:
Melt Temp: 525 °F, Hold Pressure: 960 psi

Process Tolerance:
Melt Temp: +/− 20 °F
Holding Pressure: +/− 135 psi

Alarm Limits:
Cushion: +/− 0.03″

Figure 11.27 Setting alarm limits

The above procedure is for one dimension and one cavity. One must consider all the cavities and all the dimensions. This can be a challenging task especially if the mold cavities are not balanced and/or the cosmetic process window is narrow. Most DOE programs will predict an optimal process. One can simulate the process using the software, and then set the tolerances. In case of tight tolerance parts, these process tolerances will be much narrower as compared to commodity molded products such as buckets, pails, forks, and spoons (where dimensions requirements are almost nonexistent).

11.19 Summary

Some would argue that the DPW and CPW get smaller and smaller as the number of dimensions and cavities increase. Unfortunately, that is the reality, and in fact, typically it is not acknowledged because of the effort involved in making the process windows larger. A small process window leads to a process with almost no room for any adjustment, resulting in a process sheet with very small limits for process changes. Tolerances on process parameters are usually established based on past molding experience. However, to be successful, process tolerances must be set based on the type of analysis described in this chapter. This further pleads the case for well-built injection molds and parts dimensioned with reasonably defined tolerances. This is where concurrent engineering principles and practices become important. A robust process requires less human intervention, which allows the potential for reducing the frequency of in-process inspection. A well-planned injection mold and part design, along with a disciplined process development approach, are critical to having an efficiently running production process and a profitable manufacturing operation. Without the above, much time, money, and resources will be wasted in trying to efficiently produce parts from an inherently inefficient system.

Suggested Reading

Lahey, J. P., Launsby, R. G., *Experimental Design for Injection Molding* (1998), Launsby, Colorado Springs, CO

Del Vecchio, *Understanding Design of Experiments* (1997), Hanser Publishers, Munich

12 Mold Qualification Flowchart, Production Release, and Troubleshooting

This chapter will discuss a flowchart that includes all necessary measurements and the documentation that must be carried out for successful release of a molding process into production. The tools required for mold qualifications and troubleshooting are also mentioned. Development of troubleshooting guides is a continuous process, and they need to be revised and updated during production runs.

12.1 Process Development Flowchart

Process developmental procedures described in the previous chapters are key to releasing the mold to production. A robust process needs little or no supervision during the production run. The required supervision usually involves for the processor to perform a visual check on the product, verify the process, answer any alarms, or perform scheduled or preventive maintenance such as cleaning the mold faces. If processes are not robust, a technician needs to constantly attend the machine to adjust the process parameters in order to yield acceptable parts. Robust processes yield consistent parts, shot to shot, cavity to cavity, and run to run.

The process development procedure can be split into two parts, the mold-function qualification and the mold-part qualification. The flowchart is shown in Figure 12.1.

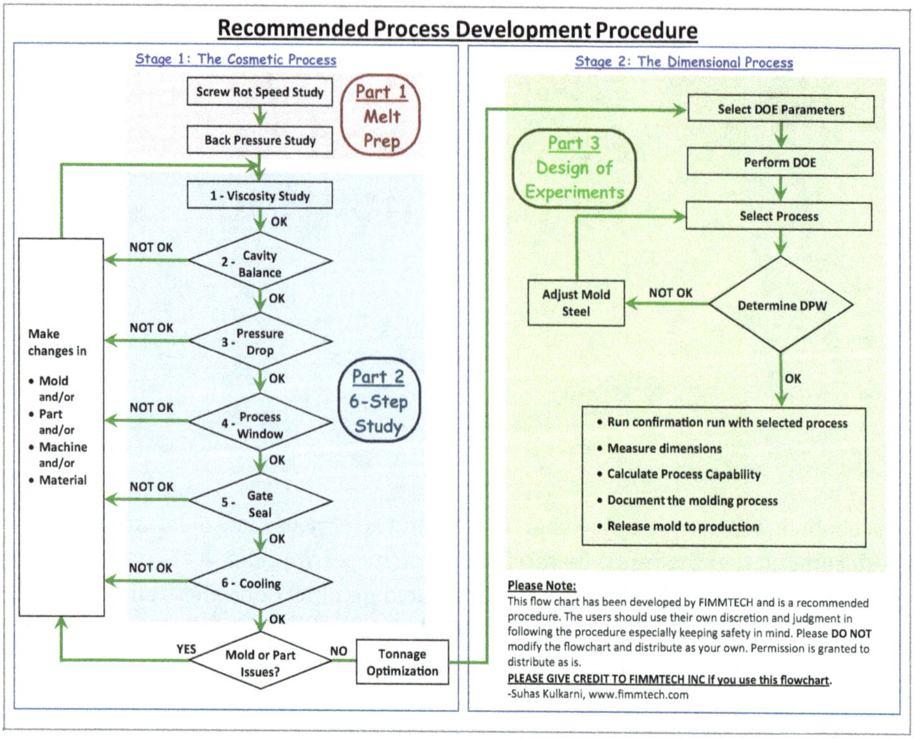

Figure 12.1 Recommended mold qualification procedure

12.1.1 Stage 1: Development of the Cosmetic Process

This stage is the mold function qualification stage and consists of two parts. The first part is the melt preparation study and the second is the 6-step study. During the melt preparation study, the melt is prepared to the required quality by optimizing the barrel heats, the screw rotation speeds and the back pressure. During the 6-step study the function of the mold and its components, the determination of some of the process parameters, and the size of the process windows are evaluated. During this study, dimensional analysis to measure process capability are typically not performed. The actual dimensions are not of major importance because the next stage is to adjust the mold steel dimensions to mold the parts within the required specifications. The mold function qualification step is to determine the cosmetic process window (CPW). Naturally, a wide window is desirable. Once the mold function and process windows are acceptable, the next step will evaluate the part quality.

12.1.2 Stage 2: Mold and Part Quality Qualification Procedure

This stage involves performing the DOE, selecting the process, and determining the size of the dimensional process window (DPW) and the control process window (CoPW). At this point, based on the knowledge from initial statistical dimensional analysis, there must be enough confidence that the process is stable and that the mold steel can be and must be adjusted to center the process in the DPW and thus in the CoPW. At this point, if the dimensional variation for a particular dimension is greater than the difference between the Upper specification limit (USL) and the lower specification limit (LSL), it becomes clear that it is impossible to consistently mold all the parts within specifications because some parts will always be out of specifications. The product designer must revisit the product specifications and the material selection for the parts. It is acceptable to stop the mold qualification process at the end of step 1 (the mold function qualification), if the dimensions are acceptable. However, the only way to know whether the process has been optimized is to perform a DOE. The benefit of performing a DOE outweighs the time and effort required to perform it. Process tolerances, alarm limits, and process robustness are some of the critical pieces of information available only through a DOE.

12.2 Mold Qualification Checklist

Having a mold qualification checklist makes sure that all features of the mold and the process have been assessed. A sample checklist is included in Appendix F. The checklist should be used during the mold trial and should be completely filled out with comments by the end of the trial. Any suggestions must be passed on to the mold maker or the concerned department, including reasons for any recommended or required changes. For example, if the mold is flashing, a sample part with flash must be given to the mold maker. An example of a completed mold qualification checklist is shown in Table 12.1.

Table 12.1 Mold Qualification Checklist

No.	Question	Comment
1	Are gate sizes acceptable?	Open gate size to 0.060″
2	Are runner sizes acceptable?	Yes
3	Is venting acceptable?	Vents on runners required
4	Is part fill acceptable?	Yes
5	Is the ejection acceptable?	Parts do not fall off the ejector pins
6	Has the rheology study been done?	Yes
7	Is the cavity balance acceptable?	No. Cavity 2 is shorter than the others
8	Is the process window acceptable	Process window is small. Part flashes soon after fill
9	Has the gate seal study been done?	Yes. Will be repeated after gate size change
10	Has the cooling study been done?	No

12.3 Process Documentation

Because there are a number of factors that affect the process, a detailed record must be kept of each of these factors. The factors include machine settings, actual process outputs, material drying parameters, machine setup instructions, mold setup details, operator work instructions, secondary processes, and any other factors involved in the journey of the plastic pellet until it is shipped out of the facility as a finished product.

12.3.1 Process Sheet

The process sheet is usually the first piece of documentation that is generated during the initial sampling of the mold. It primarily contains the machine settings for the process. It should also include other factor-setting information. A typical list of factors is included in the sheet shown in Table 12.2.

Table 12.2 Process Variables and Outputs to Be Recorded on a Process Sheet

Tempera-tures	Speeds	Pressure	Times	Other	Outputs
Drying	Injection	Injection	Drying	Shot Size	Part weight
Barrel	Mold open	Packing	Packing	Transfer position	Runner weight
Hot runner	Mold close	Holding	Holding	Material info	Injection time
Mold	Ejection	Back	Cooling	Nozzle length	Cushion
Fixtures	Screw	Tonnage		Nozzle orifice	Cycle time
Annealing				Coolant flow rate	Melt tempera-ture
				Core sequences	Screw recovery time

A process sheet must be generated and updated at the end of every mold trial iteration. Any new process changes that are made must be recorded over the old parameters, but a log of all the process changes with the corresponding dates must be maintained. This essential history is useful in debugging future issues. The history also provides information on the evolution of the process to the current state from "what it used to be." For example, over a few runs cycle times seem to drift away from the standard and the question always asked is "What changed?" A process change log is very useful in such cases and the changes over time can be evaluated. A process change log must also be maintained during production runs and should be a part of the process documentation.

12.3.2 Waterline Diagrams

Injection molding is a heat transfer process and the mold is the primary heat exchanger. The mold temperature plays a very important role in the part quality because it affects the rate of heat transfer. Mold temperatures must not only be kept constant during a run, but also must be maintained at the same values from run to run. This will help to achieve the run-to-run consistency goal. This leads to the necessity of hooking up cooling waterlines in the same manner for every run. The Ins, Outs, and the loops must be placed in the same place during every run to ensure the same heat transfer from the mold. A water line diagram must be maintained in the records, showing the hookup for every waterline. An example of a water line diagram is shown in Figure 12.2.

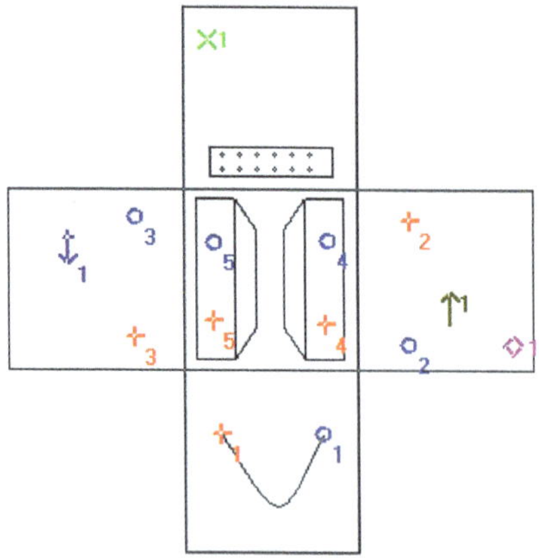

Line No.	Coolant	Temp In	Temp Out	Flow Rate
1	Water	140	145	0.5
2	Water	140	145	0.5
3	Water	140	145	0.5
4	Water	110	120	0.2
5	Water	110	120	0.2

Figure 12.2 Waterline diagram

Because the heat transfer also depends on the coolant temperature and flow rate, these values must also be recorded. The differential pressure or the pressure drop is useful in evaluating whether there is any scaling or other obstruction in the water lines. This information should typically be recorded by the tooling department.

12.3.3 Mold Temperature Maps

Even when the water lines are hooked up properly as documented in water line diagram, the actual mold temperatures are also important. For example, if the setting on the mold temperature control unit is 60 °C, but the water line is obstructed, the mold will not reach the required temperature. There is also a loss of heat between the temperature control unit and the mold. Water leaving the temperature control unit at 60 °C may cool down to a temperature of 50 °C by the time it reaches the mold. Therefore, the actual mold temperatures of the mold cavities must be recorded before startup. The temperature at each of the water fittings must also be recorded. Any plugged line will cause a different OUT line temperature than the rest of the lines. When the mold temperature is set close to ambient room temperature, a plugged line will not negatively affect the temperature distribution in the mold cavity, because the mold temperature will always find equilibrium in the cavity steel. It is therefore important to record and check the mold temperature when the mold has been running for some time. This time must be specified and recorded. Examples of mold temperature maps are shown in Figure 12.3.

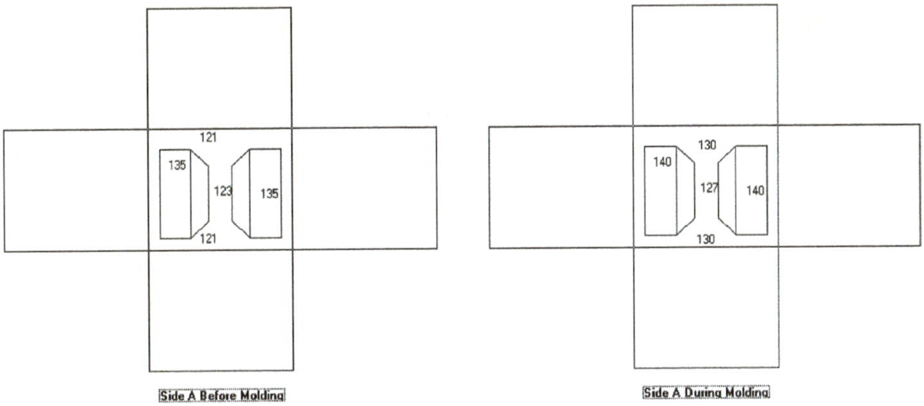

Figure 12.3 Mold temperature maps, before and during molding

12.3.4 Setup Instructions

The procedure and the tools required to set up the mold must be recorded to facilitate a quick and hassle-free set up. This is particularly important for molds that need special instructions, such as mold open and close sequences for a mold with cores. Molds can easily get damaged if the sequences are not correct, costing the company sometimes thousands of dollars for the components to be replaced, the manpower, machine time, loss of production, and so on. Mold start-up instructions must also be included. A mold must be started up by lowering both pressure and speeds. Mold shut-down procedures and material purging instructions must also be provided. Times for heat soaking of hot runner manifolds must also be provided for molds that utilize a hot runner system.

12.3.5 Operator Instructions

Packaging and any secondary operations performed by an operator must be clearly defined. The operator is usually the last person to handle a part before it goes into the box, making him/her the integral part of all quality control efforts. Operators should be given clear instructions regarding handling the product and performing any secondary operations, such as deflashing the product. Packaging is another important step, because parts not packaged correctly could become damaged during transportation and in some cases this could result in a change in quality, such as warpage.

12.4 Documentation Books

It is best to have two types of documentation records, the *mold qualification book* and the *production book*. The mold qualification book should record all details of the qualification process, and the production book must include all information required for production, including some of the qualification results to help the technician debug and fix any problems during production. Table 12.3 provides a list of documents that should be included in these books.

Table 12.3 Documents to Be Included in Qualification and Production Books

Document	Qualification book	Production book
Part drawing	Y	Y
Properties data sheet	Y	N
Processing data sheet	Y	Y
Mold setup instructions	Y	Y
Viscosity curve	Y	Y
Cavity balance	Y	Y
Pressure drop	Y	Y
Cosmetic process window	Y	N
Gate seal study	Y	Y
Cooling study	Y	N
DOE matrix	Y	N
DOE results – Pareto charts	Y	Y
DOE results – contour plots	Y	Y
DOE results – other	Y	N
Process sheet	Y	Y
Dimensional process window	Y	Y
Control process window	Y	Y
Process change log (during development)	Y	N
Process change log (during production)	N	Y
Operator instructions*	Y	Y

Document	Qualification book	Production book
Daily mold maintenance (during production)*	Y	Y
Start-up and shut-down instructions*	Y	Y

* These are living documents that should be started during mold development and must be constantly updated during production

12.5 Qualification Production Runs

Once part quality, process windows, and process capability are acceptable, a short qualification run must be performed. The length of this run should be determined based mainly on the production volumes and the criticality of part quality. For molds that would typically run for most of the year, a robust process and mold design will relieve production personnel of constant production management and will allow running a lean manufacturing operation. For such molds a qualification run of 24 hours should be considered. For critical parts, such as parts that require zero defect quality, the same procedure must be followed. It is extremely important that the process is closely monitored during this production run and process changes must be avoided. If process changes are required to maintain the quality, the whole process must be reevaluated, because this is an indication of a non-robust process. The parts molded from these runs must be initially quarantined and released to the customer once all the information and quality has been verified. Longer runs will provide a better understanding of the dimensional variations.

12.6 Mold Specific Troubleshooting Guide

Even for a commonly known defect, the solution to fix the problem may be different for each mold. For example, splay on one part could be a result of excessive injection speed, while for another mold running the same material, the splay could be the result of low mold temperature. Every mold has its own characteristics and therefore, although a general troubleshooting guideline is a good starting reference point, every production book must contain a living document that contains the record of the typical problems and the respective solutions. This record should be updated every time a new defect is seen and a solution for the defect has been found, saving effort the next time the defect is seen. The factors that must not be changed must also be mentioned. For example, if mold temperatures are important for a particular dimension, or if it was found not to be a significant contributor to the solution of the problem, then this must also be mentioned. Often, there is more than one solution to fix a defect and therefore the solutions must be prioritized. The results from the DOE must also be included here. They will give an indication of the most important factors that have an effect on the parts. An example of a troubleshooting guide is shown in Table 12.4.

Table 12.4 Mold Specific Troubleshooting Guide

No.	Defect	Recommendation	Do not change
1	Splay	Clean vents	Injection speeds
		Check material moisture	
		Reduce melt temperatures (not below 210 °C)	
2	Shorts on the sleeve	Clean vent pin	Holding pressure
3	15.055 dimension is undersized	Check water flow through the cores	Holding pressure

12.7 Molding Startup and Shutdown

12.7.1 Purging

The barrel must be purged off all the material present in it to eliminate any purging compound that was used and/or eliminate any degraded material present in the barrel. Injecting degraded materials in the mold can cause two potential problems. First, the degradation could have been so significant that all mechanical properties of the plastic have been lost and injecting this material in the mold can produce a brittle part, posing ejection problems. Parts getting stuck in the mold, parts breaking off during ejection, ejector pins piercing through the parts, breaking of ejection components due to stuck parts are all problems caused by degraded materials. The second problem that may arise is the excessive gassing that takes place due to the material degradation. The gasses build up in the barrel and if not purged, they get injected into the mold and again end up clogging the vents that are the last points to fill. The products of degradation are easy to transport through the mold and clog the vents. If the vents get clogged during the first shot and this problem is not noticed and fixed, the molded parts may develop cosmetic and dimensional problems immediately. This forces a process change further effecting parts and finally leading to a chain of process changes.

Hot runner systems must also be purged of possible degraded materials using fresh material. A large piece of cardboard must be placed on the moving side of the mold to avoid any degraded material from being sprayed on the mold, because if this happens, the mold often needs to be pulled out for cleaning, especially if there are complicated slide or lifter mechanisms. Care must also be taken to inject at lower speeds and pressures, because some hot runner manifold components rely on the clamping pressures to keep them from leaking under injection pressure. A good starting method is to use the screw speed and a high back pressure to extrude the melt out of the hot manifold. All molds are different and can pose unique problems with regard to purging. Each purging procedure must be evaluated for the particular mold before being carried out.

12.7.2 Startup of a Molding Machine

All molding operations involve high pressures, temperatures, and speeds. These can be extremely dangerous and accidents, from burns to fatalities, have been recorded. Safety should always be a priority and starting up the molding machine must be done following a careful and systematic procedure. From the process standpoint, it is preferable start up the molding process using a conservative approach, detailed in the following, before full pressures and speeds are set on the molding machine. This will prevent any accidental damage and over-packing of the parts in the mold. Over-packing of the parts causes flash and in some molds the flash can fill up internal vents. Stationary vent pins are examples of such internal vents. In addition, when the parts get over-packed, mold opening pressures increase and it becomes difficult to open the mold.

To better illustrate the startup procedure, here is an example with the following conditions: Holding pressure = 1000 psi, holding time = 8 seconds, and cooling time = 20 seconds.

Based on these conditions, a recommended startup procedure would be:

1. Verify the process with the documentation that was recorded during the qualification run
2. Check melt temperature
3. Check mold temperature and make sure water is flowing
4. Increase cooling time by an amount equal to the pack and hold time (cooling time = 28 s)
5. Set pack and hold, pressure, and time to zero (HP = 0, HT = 0)
6. Set the screw delay time equal to the set HT: Screw delay = 8 s
7. Purge the machine and take the first shot in a semi-automatic mode

8. Match the 'Injection Only' part with the recorded documentation or sample part
9. Take a couple more shots
10. Add half the hold and pack time and pressure and reduce the cooling time by half the set HT (HP = 500 psi, HT = 4 s, cooling time = 24 s)
11. Take about 5 shots
12. Set screw delay time equal to zero or the value in the recorded documentation
13. Set the process to match the recorded documentation

Note: Some parts cannot be molded short. In such molds, leave the holding time to the set value and gradually increase the holding pressure.

12.7.3 Shutdown of a Molding Machine

The shutdown procedure is equally important. If a heat sensitive material or an easily degradable material is molded, it must purged from the barrel before shutting the machine down. Molding must be stopped and an appropriate purging compound must be used to purge the barrel. If the mold has a hot runner system, follow the recommendations of the hot runner supplier to purge the hot runner system before shutdown. All material in the barrel must be emptied out and the screw must be left in the front position. If the material is not emptied out and the screw is not left in the forward position before being shut down, a solid cylinder of plastic will be formed once the machine has cooled, which will pose a problem during start-up.

Because plastic is a bad heat conductor, the center of the cylinder takes a long time to melt; therefore, the solid cylinder prevents any purging or injection from taking place. When chilled water is run through the molds, the water is shut off and a few shots are run to prevent water condensation in the cavities.

The last shots with the runners must always be saved with the mold. No secondary operations such as degating or deflashing should be done on these shots because they are evidence of the condition of the mold and process before shut down.

12.8 Troubleshooting

Troubleshooting is probably the most important job function in any production environment. As equipment and processes have become more and more complex, the need for highly knowledgeable workers with good troubleshooting skills has become increasingly important. The molding process is complex in terms of plastic flow characteristics and the speeds, pressures, times, and temperatures involved with the process.

Each of these factors can affect the quality of the part. For example, increasing melt temperature can increase flow rate, or increasing injection speed can also increase flow rate. Increasing cooling time can increase the dimension of the part, while decreasing the mold temperature can also increase the dimension of the part. Because multiple actions can have similar effects on part quality, different people have their own preferred method of solving a problem. The downside of this fact is that the process sheet is often updated with the latest changes and after a few months of running the mold, the set process ends up being completely different from when it was first created. This illuminates two problems: first, the process was not established scientifically and second, the personnel were not trained to troubleshoot the process in a systematic manner with the right tools. But even with a robust process, often production problems may still occur. When a robust process is established, documenting the complete process with all inputs and outputs must be done; however, it is sometimes impossible to capture the status of each related process or equipment. For example, recording the response time of a hydraulic valve controlling the injection or a pneumatic air valve controlling the gate is difficult and probably never done, although this may have an impact on the part quality. If the conditions of these components change, so will the molded part quality requiring troubleshooting skills.

There are no hard and fast rules for troubleshooting an injection molding process. First, the problem with the parts must be understood and then the possible reasons for the issue must be considered. No process parameter must be changed until the complete set of process parameters is compared with the original recorded process parameters and the outputs are scrutinized. Every other factor that is not recorded on the process sheet must also be looked into.

The following are some guidelines that can be followed for general troubleshooting.

1. Compare the set process to the original documented process. If there are any changes, do not change the process back to the original (because they may have been made for a reason).
2. Observe the whole process for ten shots and record all the outputs for these ten shots. These include: injection fill time, cushion, screw recovery time, and cycle time. Compare the above actual outputs to the documented ones.
3. Stop the machine and compare the actual mold temperatures to those that were documented in the book.
4. Record the melt temperature and compare with the documented measurements.
5. Check the temperature of the hoses to check for water flow as described earlier. Remove any obstructions in the lines and let the water circulate for some time to allow the mold temperature to stabilize.
6. If there have been any changes compared to the original process sheet, refer to the results from the DOE and see if those changed parameters would have any effect

on the quality of the part. If not, set the process back to the original. If the changes are influential to part quality, check if the out-of-specification part was a result of this change made. If so, change the process parameter back to the original. If not, make the appropriate change and bring the parts back into specifications. Any process changes must be recorded.

Once the parts are running meeting quality requirements, the process and quality must be constantly monitored for about a shift's worth of production. Any unusual trends, such as increasing injection pressures, decreasing fill times, or any other variations must be noted. This change may sooner or later cause the parts to go out of specifications and one must be proactive to find the root cause of the trend or variation. A good example is inadequate cooling of the mold because of lack of cooling lines. In its role as a heat exchanger the mold must remove all heat that is put into it to maintain steady state. If the heat is not removed, the mold temperature will rise and affect the quality of the part. This change will be gradual and will be manifested in a certain trend in the dimensions of the part.

A process change log sheet must be maintained with the production book. The original process must always be present in the book, together with the updated process sheet. Every time a change is required, the old and new process parameters must be recorded on the process change log sheet. This disciplined record keeping will ensure that a history of all changes made to the process sheet is maintained for review.

Checking water flow through the mold is not a very easy task. When the mold temperature control water unit is turned on, it is assumed that the water is flowing through each line in the mold. One of the most common reasons for dimensional issues in production is that a water line (or oil line) becomes obstructed. These issues can be detected only by physically checking the temperature of the hose with a pyrometer, if the water temperature is high. For temperatures between approximately 30 to 45 °C (85 to 115 °F), the most efficient way is to hold a hose by hand and feel the heat of the water flowing through the lines. For water temperatures below 30 °C, this technique can be confusing because the temperature is not very different from room temperature and holding the hose by hand and trying to feel for a warm hose may not indicate water flow. Machine vibration can easily be confused for water flow and therefore trying to 'feel' the water flowing through the mold must be avoided. It is good practice to raise the temperature to approximately 35 °C and check for the water flow. Safety must be practiced. The practice of checking for water flow must also be followed at the start of every job. The water must be set to approximately 35 °C, the water flow must be checked, and then set to the required process temperature. Certainly for molds that use oil as a coolant a pyrometer must be used because the mold temperatures can be very high depending on the material. For example, PEEK is processed at mold temperatures between 175 °C and 205 °C (350 °F to 400 °F). The importance of safety cannot be stressed enough here because of the high temperatures of the coolant oil that is used in processing such materials.

Below are some of the common problems encountered on the production floor:

1. Pinched or kinked waterlines
2. Water line control valve not turned on
3. IN and OUT reversed
4. Inlet volume of water to the manifold is less than the required output to the machine
5. When using a T-junction for water lines, the IN must be such that it is the vertical line of the letter and the OUTs are the horizontal lines. If the water IN is one of the horizontal lines, the OUT on the vertical line can get starved of water. A venturi-meter works on the principle that vacuum is generated in a vertical line of a T-junction if the internal geometries and sizes of the T-junction and the fluid velocities satisfy Bernoulli's equation. Although the generation of vacuum is unlikely, the effect of reduction of water flow is highly possible.
6. Water lines from the controller to the mold are too long, causing loss in temperature and increasing pressure drop
7. Feed throat temperature is not regulated

12.9 Important Equipment and Tools for Qualifications and Troubleshooting

Although this seems like a topic that does not need discussion, it is included here more as a checklist and to emphasize the importance of this equipment.

1. Melt pyrometer
2. Surface pyrometer
3. Weigh scale with the following accuracy (preferred)
   - 1 g for parts weighing over 250–300 g
   - 0.1 g for parts weighing between 50–250 g
   - 0.01 g for parts weighing less than 50 g
   - 0.001 g for micro-molded parts
4. Flow meter
5. Magnifying glass
6. Flash light and mirror
7. Brass tools – rods of various sizes (diameter and length), pliers.

8. Flame torch
9. Heat gloves – thin and thick
10. Processing data sheets
11. Camera – still and video
12. Calculator
13. Notepad, forms (with copies) and procedures of mold qualification

The use and benefits of most of the equipment is self-explanatory. The actual melt and mold temperatures must be measured with the pyrometers, rather than recording the settings. Accurate weigh scales are required to perform gate seal tests or to perform statistical analysis on part weights. Part weight is a valuable output that directly correlates to the quality of the part. If no variation is seen in the part weights, the weigh scale resolution is not small enough. An accurate scale with sufficient resolution to detect small changes in part weight is a necessity. A flow meter need not be installed on every single water line, but it should be used to check the water flow during the preventive maintenance schedules. Those numbers should be used to compare the flow rates during production, if there is a problem with part quality. When dealing with plastic or parts stuck in the mold, brass tools or those made of soft materials must be used to prevent damage to the mold. For highly polished surfaces, it is best that only experienced technicians handle any stuck plastic or any issues with the mold. The polished surfaces can easily get damaged and repairs are not only expensive but also time consuming. Video cameras can help with infrequent problem occurrence. In such cases, setting up and then reviewing the recording is an easy solution. The recording can be paused or played back in slow motion to pinpoint the issue. Because documentation is a large part of the whole qualification process, documents, such as data sheets and forms, should be readily accessible to make the process efficient. Laptops and workstations are becoming commonplace on the production floor and the need for paper documents is slowly decreasing.

12.10 Common Defects, Their Causes, and Prevention

Defects in injection molded parts could result from any one of the five factors that contribute to part quality. These are part design, material selected, the mold design and the build, the choice of the molding machine, and the molding process. The quickest and easiest change that can be done is always in the molding process and, therefore, it seems like processing could take care of all defects. In most cases, it is quite the opposite. See Figure 12.4, which is a representation of the part that had the gate in the wrong location. In the process of packing out the thick section at the end of the fill, the rib area near the gate would get overpacked and stick in the mold. The thick section at the end of fill had sink. The problem was clearly with the mold design. The actual part is not shown for anonymity of the customer.

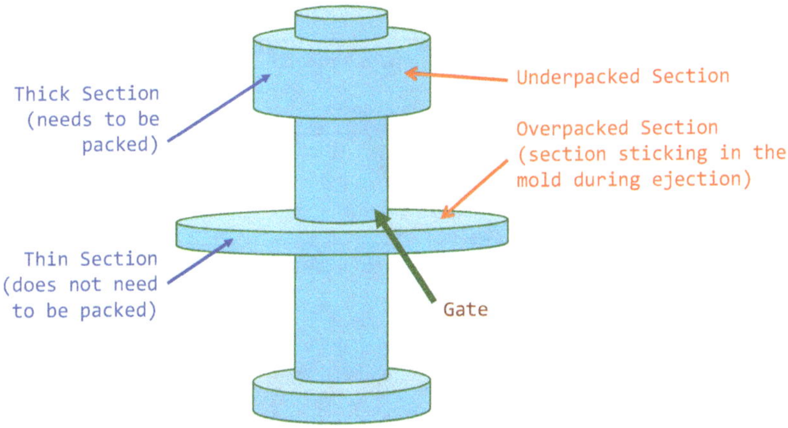

Figure 12.4 Mold design related issue

12.10.1 Splay: What Is It? How to Get Rid of It?

In injection molded parts, the most common cosmetic defect is splay. Almost every molding operation rejects parts for this reason.

Injection molding is the process of injecting molten plastic into a mold. The cavity steel has a desired texture that gets picked up by the molten plastic and is replicated on the part. Depending on the base polymer being molded, the plastic melt temperature is anywhere between 175 °C and up to even 400 °C, as is in case of some PEI resins. At these temperatures, water turns to steam, and some of the low molecular additives can burn to produce volatiles. The speed of injection of the plastic into the mold generates shear. Excessive shear can degrade the molecules. Steam and volatiles (collectively grouped as volatiles) from degradation flow with the plastic into the mold. Because of the fountain flow of the plastic into the mold cavities, the volatiles get to the surface and prevent the molten plastic from coming in contact with the mold steel, at the same time, smearing the volatile on the interface of the melt and the mold steel. This shows up as streaks and is called splay. Splay is also called silver streaks. See Figure 12.5.

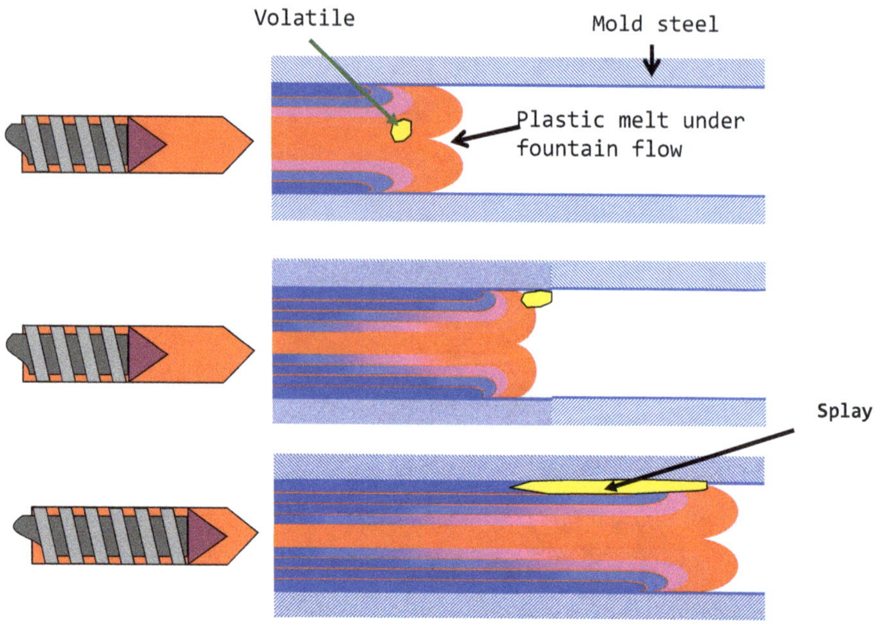

Figure 12.5 Formulation of splay

Sources of splay:

- Moisture in the material.
- Gate sizes are smaller than required. If the gates are too small or have sharp edges, the plastic can get excessively sheared and produce volatiles. In that case, mold design is the culprit.
- Sometimes a worn out screw and/or barrel can also cause excessive shear resulting in splay.
- Regrind is another source of splay especially if the regrind has excessive fines. During the regrind process some of the plastic gets ground up into very fine and almost powder like material called as fines. The fines do not convey well because they can get stuck to the screw and finally degrade causing volatiles.
- The vents in the mold provide an outlet for the volatiles. If the vents are plugged up and/or are not deep enough and/or are not sufficient in number, then the volatiles have nowhere to go resulting in splay.

There will always be some volatiles and definitely air that gets trapped in the melt stream. This air is the air that is present between the pellets as the pellets go from the feed section to the compression section of the screw. Back pressure applied during the screw recovery process helps get rid of this air and volatiles. Back pressure should always be kept to a minimum because excessive back pressure can also increase shear and result in splay.

12.10.2 Defects in Molding

All troubleshooting must be done during the mold qualification stage. As discussed, it is important that a mold have a large cosmetic window, which will then increase the chances of a robust molding process. Following are some of the defects in molded parts. The solutions listed are a living, dynamic list and are only possible solutions. A solution to a problem in one part may not be the solution to the identical problem in another part. So judgement must be used.

1. Short Shot

Possible Reason:

- Molten plastic is not reaching the mold cavity section.

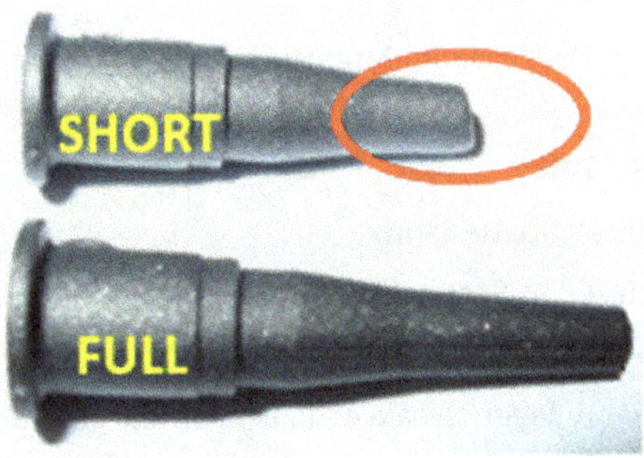

Short Shot

Possible Solutions:

- Increase melt temperature
- Increase mold temperature
- Increase injection speed
- Increase injection pressure if process is pressure limited
- Check if the mold is vented in the area of short shots
- Increase gate and runner sizes

2. Flash

Possible Reason:

- Molten plastic flowing into unwanted sections of the mold cavity.

Possible Solutions:

- Check for mold shut-off and mold damage
- Decrease melt temperature
- Decrease mold temperature
- Decrease injection speed

3. Sink

Possible Reason:

- Plastic is shrinking as it cools but additional plastic cannot be further compensated for the shrinkage.

Possible Solutions:

- Increase pack and hold pressures
- Increase pack and hold times
- Decrease mold temperature
- Decrease melt temperature

4. Splay

Possible Reason:

Refer to Section 12.10.1 for a detailed explanation.

- A layer/streak of a unwanted gaseous byproduct from the melt or moisture in the material, which comes in between the melt flow and the cavity walls, preventing the texture from being picked up and, in addition, eventually leaving a residue.

Possible Solutions:

- Dry plastic to suggested moisture levels
- Decrease injection speeds
- Decrease melt temperature
- Decrease screw rotation speeds
- Decrease back pressure
- Increase mold temperature
- Increase venting
- Increase gate sizes

5. Warpage

Possible Reason:

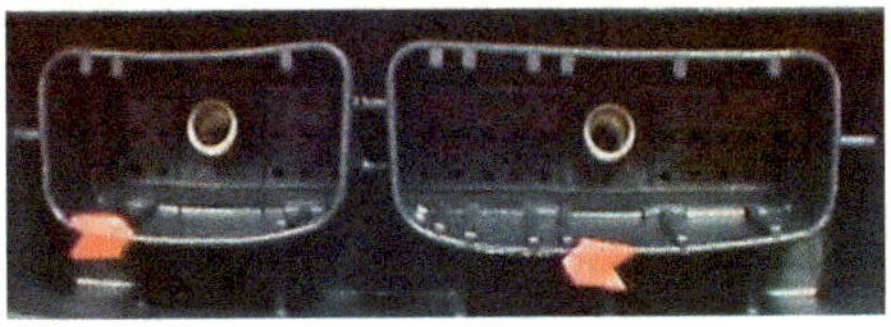

- A differential cooling rate of the melt in two sections of the molded product.

Possible Solutions:

- Decrease melt temperature
- Experiment with differential mold temperatures on the fixed and movable sides
- Increase pack and hold pressures
- Increase pack and hold times
- Increase cooling time

6. Burn Marks

Possible Reason:

- When air and gasses get trapped inside the mold cavity during plastic injection, the high pressure results in the dieseling of the of the plastic resulting in the burning of the plastic.

Possible Solutions:

- Increase venting in the mold
- Decrease injection speed
- Decrease melt temperature
- Decrease screw rotation speeds

7. Contamination or Black Specks and Streaks

Possible Reason:

- This can be caused by degraded plastic and/or foreign material that can get mixed with the plastic to be molded.

Possible Solutions:

- Reduce melt temps
- Reduce injection speeds
- Reduce screw speeds
- Find source of foreign material

8. Voids

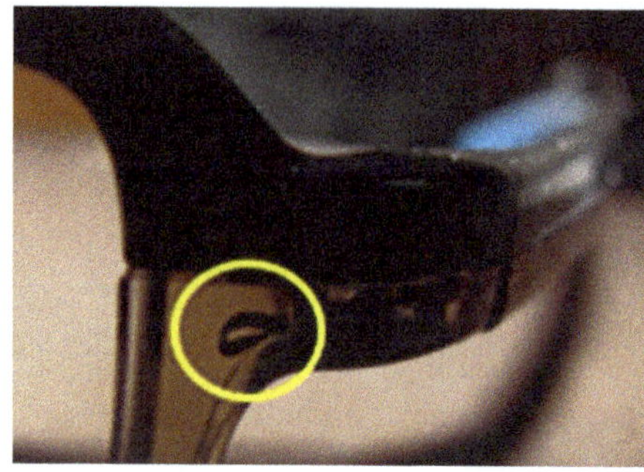

Possible Reason:

- Usually occur when the parts are thick; the walls solidify and the plastic melt shrinks towards the wall and moving away from the center. This creates a vacuum void inside the part.

Possible Solutions:

- Reduce melt temperature
- Reduce mold temperature
- Reduce injection speed
- Increase pack and hold pressures
- Increase pack and hold times

9. Bubbles

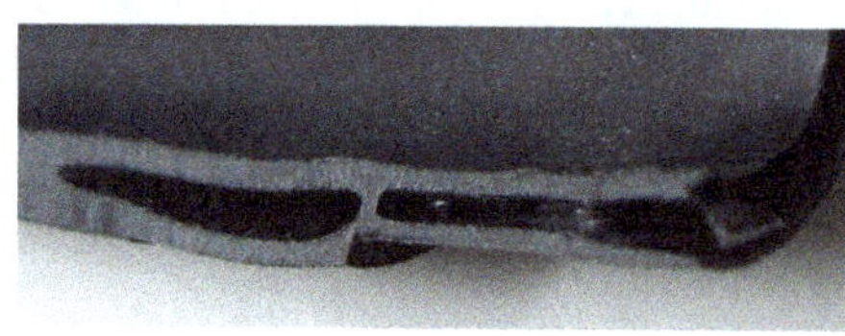

Possible Reason:

- When moisture and/or a gaseous by-product gets mixed with the melt and injected into the mold cavity, this moisture or gas, if embedded inside the melt, can show up as bubbles.

Possible Solutions:

- Dry material to suggested moisture levels
- Increase back pressure
- Reduce melt temperature

10. Gate Blush

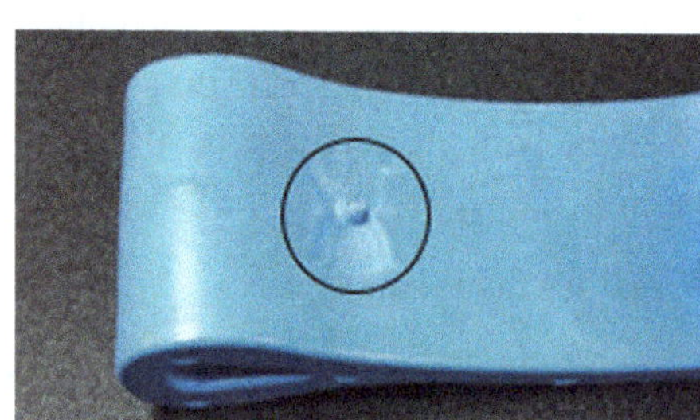

Possible Reason:

- Shows up at the gate when the material is sheared differently as compared to the rest of the part.

Possible Solutions:

- Slow down the injection speeds in the gate area
- Profile the injection speeds if necessary

- Experiment with increasing and decreasing melt temperatures or hot tip temperatures, in case of hot runner molds

11. Jetting

Possible Reason:

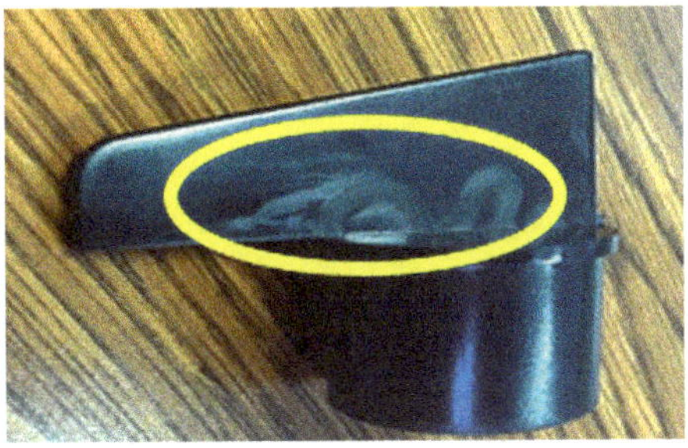

- Seen when the part is thick, and causing the injected plastic 'snake fall' on to the cold mold surface. The incoming plastic does not blend with the first material, causing the jetting marks.

Possible Solutions:

- Reduce injection speed
- Increase melt temperature
- Change gate location

12. Weld Lines

Possible Reason:

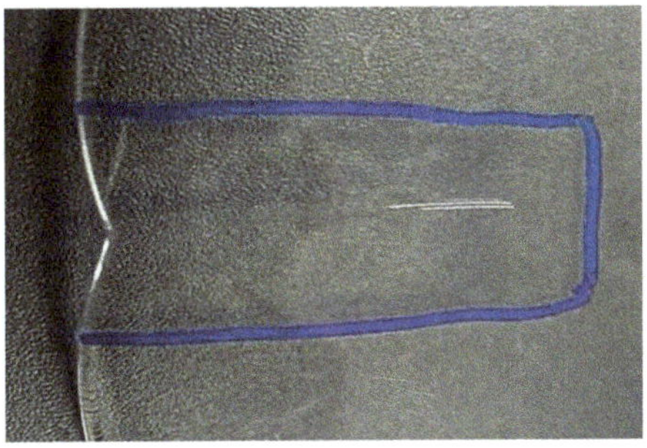

- The melt flow front is usually cold due to exposure to the cold cavity. When two flow fronts meet, as in a flow around a mold pin, they do not fuse uniformly causing a weld line. Sometimes air can also get trapped inside to form the defect.

Possible Solutions:

- Increase melt temperature
- Increase mold temperature
- Increase injection speeds
- Increase venting

13 Role of Mold Cooling, Venting, and Regrind in Process Development

13.1 Mold Cooling

The mold is basically a heat transfer unit. The hot plastic melt is injected into the cold mold cavity where it is cooled down until it is ejected out of the mold. The mold is kept at a lower temperature compared to the melt, which is sufficient to bring the melt down to the ejection temperature of the plastic. The temperature of the mold is very important because it controls the rate of heat transfer. The aim of scientific processing is to achieve three different consistencies: shot-to-shot consistency, run-to-run consistency, and cavity-to cavity consistency. Consistent heat transfer is important to achieve this goal. The rate of heat transfer between the melt and the mold is directly proportional to the difference in the temperature between the two. If melt and mold temperatures are held constant throughout the molding run, it is safe to assume that the rate of heat transfer was consistent during every shot. Melt temperatures are achieved by setting the barrel temperatures on the molding machine and typically they are not a major source of variation. It is always good practice to set alarms on the actual barrel temperatures to detect any changes and out-of-limit conditions. Although it is not common practice to have a melt temperature sensor inside a mold, this is an excellent way to measure the consistency of the melt temperature.

Maintaining a constant mold temperature is much more challenging compared to maintaining constant melt temperatures. This is where the mold design is critical. Cooling circuits must be designed for efficient heat transfer. The choice of the coolant must also be considered. The word *cooling* is used because it is necessary for the mold to be at a lower temperature than the melt in thermoplastic injection molding. Mold temperatures can be as high as 162 °C (325 °F) for materials such as polyimides, with melt processing temperatures as high as 400 °C (750 °F). Therefore, a few important considerations regarding the design of cooling channels for molds are discussed in the following.

13.1.1 Number of Cooling Channels

Uniform heat transfer between every section of the melt and the mold steel will reduce the melt temperature evenly throughout the part. This will lead to consistent shrinkage, eliminating any warpage or built-in stresses in the molded part. The most ideal scenario is to have a constant steel temperature at every point in the cavity, which means that the coolant needs to envelope the cavity steel. Naturally, this is not possible and therefore an attempt must be made to place the maximum amount of cooling lines around the part. This is not an easy task because with an increasing number of cooling lines there is always a risk of mold damage and weakened mold cavity steel. The plastic is injected at very high pressures and therefore the mold steel needs to be capable of withstanding these pressures. The presence of ejector pins, core pins, or other mold components can also weaken the mold plate. The need for these mold components often presents the problem of how to fit them in without interfering with desired flow channels. On the other hand, too few water lines will cause insufficient heat transfer, leading to hot and cold spots that result in inconsistent cooling. Over time the overall mold temperature can also increase, changing the heat transfer rate and therefore the part quality. The goal is always to find a balance between mold integrity and under-cooling.

The other factor is the total length of the flow channel, including the hoses that connect to the mold temperature unit. Long flow lengths have high pressure drops, reducing flow rates and thereby reducing heat transfer. There are no easy formulas to calculate or estimate the number of cooling lines required because each part is different, with different surface areas, thickness, orientation in the mold, and so on. Computer programs are available to predict mold and part temperatures during molding. These should be used as guidelines when possible. A recent development called contour cooling is gaining popularity. Here, the cavity is built in laminates or plates in which the cooling channels are machined following the contours of the part surface. When the mold is assembled, the cooling channels form a circuit that resembles the part contours. This is an effective way to achieve uniform cooling, but the price can become prohibitive.

13.1.2 Reynolds Number of the Coolant Flow

The Reynolds number gives an indication whether the coolant flow is laminar, transitional, or turbulent. For maximum heat transfer the flow must be turbulent and never laminar. In laminar flow, the coolant flows in layers. As it passes through the mold and begins to pick up heat from the cavity, the temperature of the coolant layers close to the steel increases. Because the heat transfer rate is directly proportional to the difference in temperatures between the steel and the coolant, an increase in temperature of the coolant layer causes a drop in heat transfer rate. A change in heat transfer rate in turn will cause a change in part quality and affect consistency. In turbulent flow, the coolant molecules do not move in layers but are constantly mixed around in the flow channel, absorbing the heat from the steel and distributing it evenly in the coolant. This helps to maintain water temperature and therefore the heat transfer rate remains constant. Figure 13.1 shows the difference between laminar and turbulent flow. Even with turbulent flow conditions, the water can slowly begin to heat up, indicating that there is more heat to be transferred. In this case, the diameter of the water lines or the number of water lines need to be increased. If the diameter of the water line is increased, the flow rate of the water must also be increased. A Reynolds number higher than 4000 will ensure turbulent flow (some texts recommend a Reynolds Number of approximately 3500 for turbulent flow).

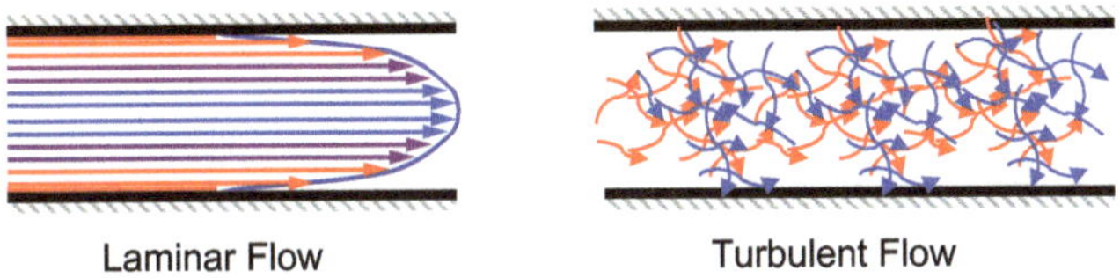

Figure 13.1 Laminar and turbulent flow

The calculation for the Reynolds number Re is given in Equation 13.1.

$$\mathrm{Re} = \frac{\rho V D}{\mu} \tag{13.1}$$

ρ = the coolant density

V = the coolant velocity

D =the diameter of the pipe

μ =the dynamic viscosity of the coolant

Calculations of the Reynolds number for water with a viscosity table are given in Appendix C.

13.1.3 Type of Coolant

As described in Section 13.1, the word 'coolant' is used because it is cooling the melt. The coolant temperature can be as high as 165 °C, where it is essentially heating the mold but still cooling the melt and maintaining the mold at a desired temperature. Water and oil are typically used to maintain the mold temperature. Water is inexpensive, easier to work with, can easily achieve turbulent flows, and is therefore most widely used. If an additive, such as ethylene glycol, is added, the properties of the additive must be considered regarding heat transfer and type of flow calculations. The disadvantage of water is that it can only be used up to temperatures of approx. 95 °C (~205 °F), which is close to its boiling point.

For temperatures above 95 °C oil is frequently used. Many high performance materials, such as polyimides, need mold temperatures in the range of 162 °C (325 °F). Oil must be used in such cases. The disadvantage here is that oil flow is rarely turbulent, making heat transfer more difficult. From a safety standpoint, oil heaters are equipped with features that shut their pumps off in case of a sudden rupture of an oil line. In some cases, electric heaters are also used but these are less common. There is no mechanism to take the heat added by the plastic out of the mold. An injection mold relies on convection and conduction to transfer the heat. The temperature in the cavity steel area is therefore not reliable and cannot be maintained accurately. Some mold designs do call for a hybrid system and use a combination of electric and oil heaters. In this case the mold temperature is more reliable.

13.1.4 Series and Parallel Cooling

When the number of water lines in a mold increases, there may not be sufficient outlets on a water manifold to hook up each individual water line separately. In this case, some of the water circuits through the mold are looped or ganged together and supplied by a single water line. There can be two types of such arrangements, as shown in Figure 13.2, and each has its own advantages and disadvantages. The important factor is to make sure there is turbulent flow in the lines.

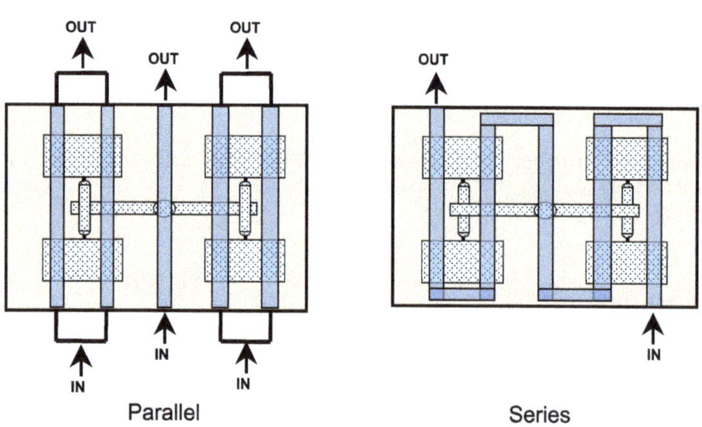

Figure 13.2 Parallel and series configurations for coolant lines

Parallel cooling lines exhibit less pressure drop allowing the coolant to flow at a higher velocity and therefore they pose less risk for laminar flow of the coolant through them. However, the flow through the lines may not be uniform. Because the coolant will take the path of least resistance, it is possible that a cooling channel with any restriction or blockage may not get sufficient coolant flow. It will be difficult to observe this without having a flow meter on every individual line. In case of a series arrangement, any restriction or blockage in a water line can be easily located; here, one flow meter is sufficient. If all molded parts suddenly start having quality problems, this could indicate a problem in a water line. Water lines hooked in series can exhibit large pressure drops, if the overall length of the line is too long.

13.2 Venting

The air inside the cavity needs to be evacuated to ensure plastic filling the cavity. Therefore, vents are added to the cavity blocks in the mold. Air that is not evacuated gets pressurized and super-heats, resulting in a dieseling effect on the plastic parts. This will cause the plastic to burn and/or create an unfilled area, causing a short shot. Over time, the mold steel can get damaged because of the excessive air pressures in a local area at the end of fill or in corners where the air and plastic tend to get pressurized. Figure 13.3 shows a part with burn marks before venting and without burn marks after venting. Figure 13.4 shows a rib that was short for lack of venting. Internal voids are another common defect that is seen with insufficient venting. Depending on the flow pattern of the plastic, the air gets trapped inside the part, forming voids. An example of a part with internal voids is shown in Figure 13.5. Once the vents were added to the mold, the voids disappeared. Lack of vents can also create excessive pressures in the cavity, causing the mold to open sufficiently enough to cause flash on the parting line. Some machines have an option for mold 'breathing' before the start of the holding phase. The mold is allowed to open slightly to let the air out and then is clamped before the start of the pack and hold phase.

Figure 13.3
Elimination of burn marks after the addition of vents

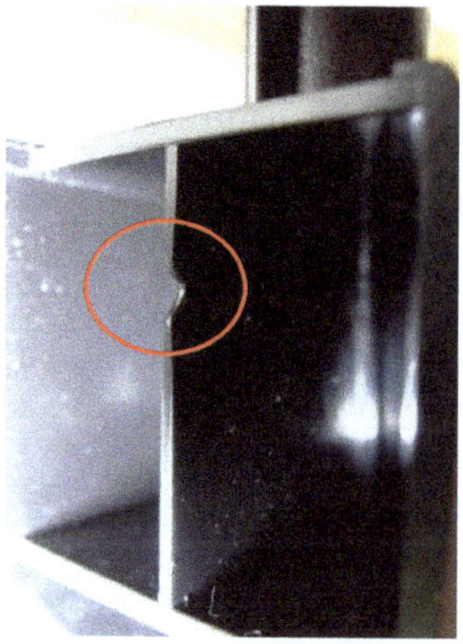

Figure 13.4
Rib not filled completely for lack of venting

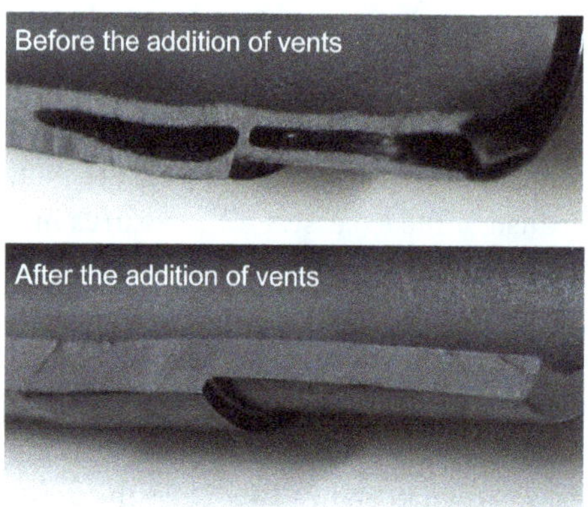

Figure 13.5
Elimination of voids after the addition of vents

13.2.1 Dimensions of the Vent

The vents connect the inside of the mold to the outside the mold. The viscosity of the plastic must be high enough to prevent it from flowing out of the mold through the vent. Figure 13.6 shows the cross section of the mold with the vent area. The relieved section that is closest to the cavity steel is the primary vent. The dimensions of the primary vent are the most critical. First, the vent depth should be such that it provides evacuation of air but prevents leaking of plastic material through the vent. Vent depths are discussed in detail in the next section. Second, the length of the vent land should not cause a pressure drop, preventing air from being pushed out, nor should it be so short that the plastic finds its way out to the secondary vent. Typical land lengths should range from 1.2 to 1.5 mm (0.060 to 0.080 in), assuming the vent depth is designed correctly. The vent width should be at least between 5 and 8 mm (0.200 in to 0.320 in) in size. On the high end, it can be as wide as desired and in some cases it can run around the entire perimeter of the part (ring vents).

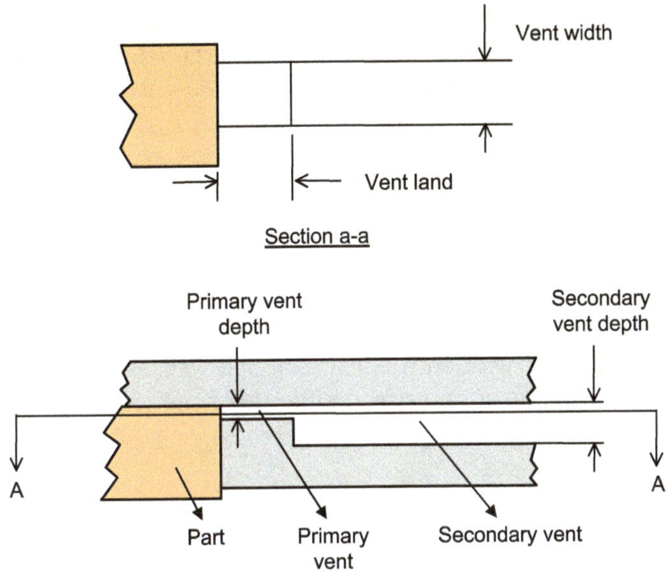

Figure 13.6 Details of the vent section

The secondary vents are also called vent reliefs. The dimensions of the secondary vents are larger than the primary vent and therefore provide easy transport of the air inside the mold out to the atmosphere. Secondary vents should be approx. 0.25 mm (0.010 in). In all cases, the vents must be well polished to avoid any build-up of residue caused by the gases. The vents must also be draw polished in the direction of air flow. If the path of the air to the atmosphere is long, another step in the vents must be considered. The additional step could be as deep as 0.6 mm (0.025 in). These tertiary vents help in reducing the pressure drop and should also be draw polished in the direction of air flow.

13.2.2 Primary Vent Depths

Vent depths depend on the viscosity of the plastic at processing temperatures and therefore differ from plastic to plastic. Typically, the material manufacturer will provide recommendations regarding vent depths. For example, the recommended vent size for ABS is 0.05 mm (0.002 in). Mold makers typically follow these recommendations when building an injection mold and tend to stay at the lower end of the recommendation to avoid flashing of the mold. If the plastic is able to enter the vent, this indicates that the vent is too deep and that it must be corrected. This involves welding and repairing the surface. To avoid this, mold makers are conservative and leave the vent depth steel safe. It has always been thought that it is the viscosity of the plastic that is the only factor important when determining the size of the vent. In a recent study it was found that the size of the vent depends not only on the viscosity of the plastic but also on the thickness of the plastic section behind the vent. To conduct this study, a special mold was constructed (see Figure 13.7). The test part produced by this mold was center-gated and had 18 tabs of varying combinations of tab thicknesses and vent sizes, shown in Figure 13.8. The center gate delivered the melt to each of the tabs at the same time. There were three tab thicknesses: 0.125 in (3.175 mm), 0.0625 in (1.587 mm), and 0.0312 in (0.792 mm) and six vent sizes: from 0.0005 in (0.0127 mm) to 0.0030 in (0.0762 mm) in steps of 0.0005 in (0.0127 mm). Different materials were molded and for each tab the minimum vent size that produced flash was recorded. The test results indicated that it was the thickness of the tab that also played an important role in determining the vent size. The thicker tab was able to accept larger vents without flashing. For example, with an ABS material that was tested with a 0.125 in tab, the vents flashed at 0.0030 in vent size. On the other hand, the vents for the 0.0312 in tab, flashed at 0.0020 in vent size, see also Figure 13.9.

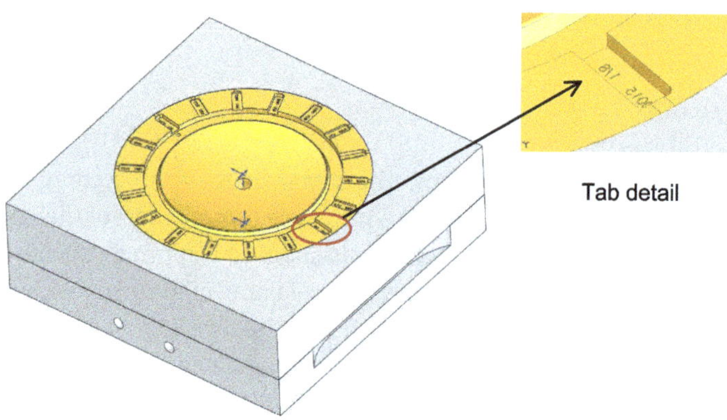

Figure 13.7 Vented mold for studying vent sizes

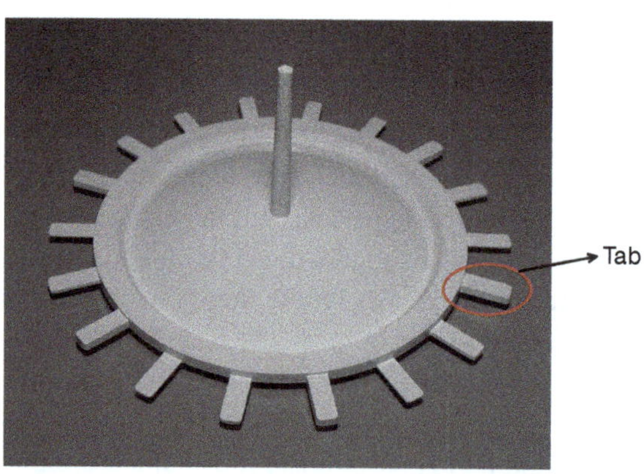

Figure 13.8 Test part molded with the vent test mold

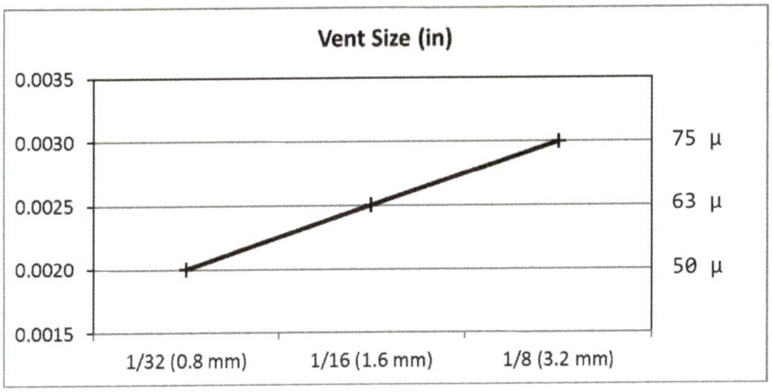

Figure 13.9 Graph showing the combination of tab thickness and vent depth where flash begins to occur for ABS (medium flow)

As the plastic is forced into thinner sections, the plastic pressure increases, thus forcing the plastic out of the primary vent. In thicker sections, the overall plastic pressure is lower and therefore the vent sizes can be larger. The published vent size value for ABS is 0.002 in (0.0508 mm), but the test results from the vented mold showed that a vent size of 0.003 in (0.0762 mm) is acceptable. This value was used on several molds successfully. Figure 13.10 shows the various tab/vent size combinations that defined the start of noticeable flash in the vent. Tests on nylons also showed surprising results. The published vent size value for nylons is 0.0005 to 0.0007 in (0.0127 to 0.017 mm), but for thicker sections, a vent size of close to 0.0015 in (0.038 mm) could be used. In all cases it is important that proper processing techniques were followed. This again makes a case for following scientific molding procedures to establish robust processes.

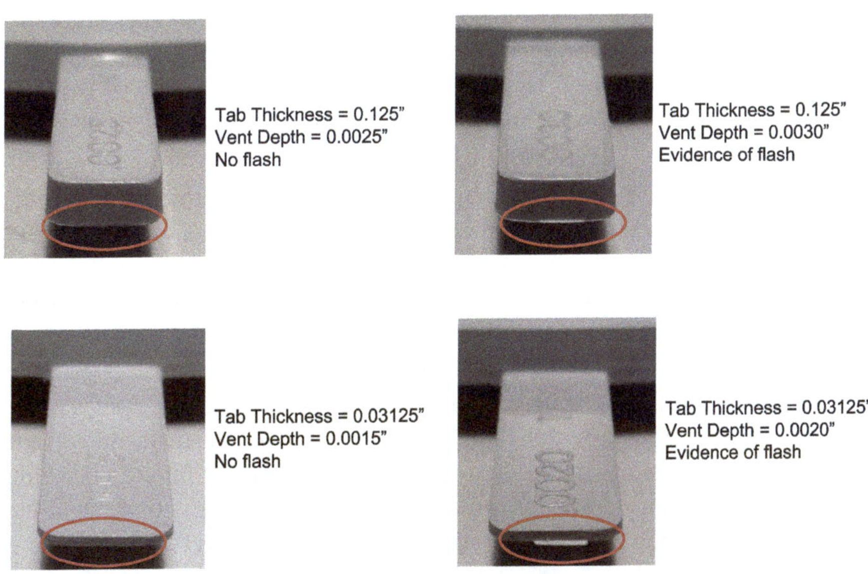

Figure 13.10 Tab/vent combinations showing no flash and evidence of flash

13.2.3 Location of Vents

The most efficient location of a vent is at the end of fill of the part. As the plastic flows into the cavity, it begins to push the air out and continues to displace the air until the cavity is full of plastic. This means that the flow rate of the air leaving the mold is equal to the flow rate of the plastic entering the mold. If there was only one vent at the end of fill, the size of the vent would have to be equal to the size of the gate. But if this was the case, the plastic would easily get out of the mold through the vent during the pack and hold phase of the molding cycle. Because the vent depths have to be substantially smaller, the required amount of cross sectional vent area must be distributed across the mold. The area most accessible in the flow path of the plastic is at the parting line of the mold. Therefore, the parting line must be vented as much as possible. Once the cavity in the area of the parting line is filled, the plastic now pushes the air to other areas of the mold. Any ejector pins, core pins, cavity inserts, or cavity blocks in these areas should be used to help vent the air out of the mold. For each of these components, vents must be added based on the recommended vent depths. Ejector pins offer the most advantage, because they are self-cleaning. The residue from the gasses released through the ejector pin vents will get cleaned as the ejectors cycle back and forth during every cycle. Using ejector pins for venting does build up residue elsewhere in the ejector box and increases the frequency of maintenance. The vents on stationary components, such as the core pins, can get plugged easily and may need frequent cleaning. Sometimes a positive air blast during the ejection can help clean the vents. In recent years, porous steel has become increasingly popular. For example, in deep draw parts, where placing an ejector pin is not possible or not acceptable, the end of fill areas are made of porous steel inserts that allow pressurized air to escape from the cavity. These inserts do require frequent cleaning and therefore the molds are typically built such that these inserts can be taken out of the mold while the mold is still in the machine. The inserts are removed, cleaned, and put back into the mold while it is still bolted into the molding machine. Using porous steels does not allow for mirror finishes or high polish surfaces.

One area of the mold that is often not given much consideration with respect to venting is the runner. If the runner is not vented, the complete volume of air in the sprue and runner enters the cavity through the gate of the part. This air burdens the cavity vents with additional work. Runners must therefore be vented as much as possible and even up to a depth where some flash can be seen. It does not really matter if the runners exhibit some flash because the molded parts are of primary interest. The sprue puller pin must also be vented. Often, at the end of the screw recovery, the screw is sucked back without rotation to relieve any pressure on the melt and prevent it from drooling from the nozzle tip. This causes air to be sucked into the machine through the nozzle. Because the air is present in the heated nozzle with some of the plastic, there can be some build-up of gasses in the time after screw recovery and before injection. During the start of injection, this air and gas is injected into the mold. This is another reason that the runners must be vented. Figure 13.11 shows a mold with possible vent locations on the parting line.

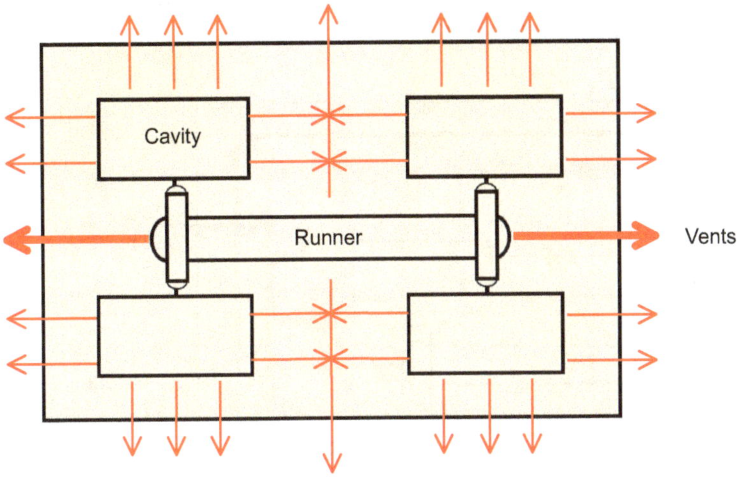

Figure 13.11 Possible location of vents on the parting line

13.2.4 Forced Venting or Vacuum Venting

In every mold there will be corners where air will get trapped. Because the volume of the air may not be large, defects are not easily seen. Close examination will show a rounded corner rather than a crisp corner. In other molded parts, shorts may be easily seen with the naked eye. Placing ejector pins in these areas to provide venting may not always be possible. If the corner detail is important to the function of the part, the only solution to eliminate the defect is to install vacuum in the mold. The vacuum evacuates all air from the system before injection, so that all cavity details fill with plastic, thus producing an acceptable part. Molds that need to be fitted with vacuum have parting line vents in the cavity blocks that are relieved into an outlet to which the vacuum line is attached. There is a seal around the cavity inserts to prevent outside air from being sucked in. This seal is formed when the mold is closed. Molds with slides and other contours are not good candidates for adding vacuum vents, because their mold faces are not always flat. Adding the seal may require additional real estate, increasing the size of the mold. The ejector pins may also need seals to hold the vacuum. Cold runner molds have an open sprue and will leak the vacuum. Although vacuum can be used successfully on cold runner molds (by slight modification of the process), valve gate molds and hot runner molds are better candidates for vacuum assist. In many cases, vacuum is added after several attempts of adding conventional vents have failed to produce acceptable parts. By then, all the ejector pins and/or core pins are vented and the parting lines, cavity blocks, and slides have been relived to the atmosphere. Additional work may now be required to fix these so-called leaks in the molds. Venting must therefore be considered at the mold design stage and the product designer must clearly specify the requirements on the part print. Figure 13.12 shows a mold configured for of vacuum venting.

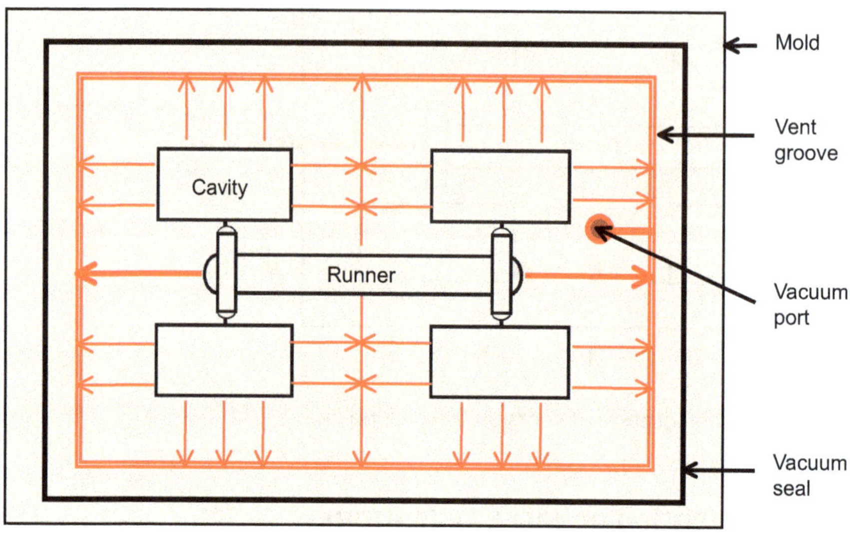

Figure 13.12 Venting using vacuum

13.3 Regrind

In injection molding manufacturing processes, runners that fall out of the mold or parts that are found defective can be ground up for reuse. This material is called regrind. Regrind material is generated from plastic that comes with at least one heat history cycle at the processing facility. Materials supplied by the resin supplier are classified as virgin materials, even though they may have gone through a melt processing process for compounding of additives such as fillers or colors. Regrind material is traded either directly or through third parties between companies that cannot use regrind material, such as companies manufacturing medical applications and those companies that produce certain consumer products, such as garden equipment or trash carrying equipment, who can use some amount of regrind and still manufacture products within their product requirements. Park benches are an example for a product where 100% regrind material is used.

13.3.1 Effect of the Molding Process on the Part Properties

The molding process has been redefined as the collection of all the processes the plastic is subjected to from the moment it enters the manufacturing facility as loose pellets to the point when it leaves the facility as a finished molded product. The process can be broken down into the drying, melting, mold filling, cooling, and the final packaging of the part. Each phase of the process can affect the plastic, the additives in the plastic, and the fillers. During the drying process there is no breakdown in the molecular weight of the plastic. However, as discussed in Chapter 4, exposure of the material to excessive drying times can cause a loss or breakdown of the additives. Drying times in some resins are cumulative. If the lost additives happen to be those that contribute to the heat stability of the melt, then the plastic molecules will break down during melt processing. Fillers are not affected by excessive drying times.

During the melting phase, the plastic is transformed from pellet to melt in the injection barrel. The plastic is subjected to heat from the heater bands and through shear from the rotating screw. The plastic material may contain fibrous or non-fibrous fillers. There are other additives, such as plasticizers, heat stabilizers, and a variety of other materials that are compounded into the material as it moves through the barrel. The heat from the heater bands and the shear from the rotating screw will cause the polymer molecules to break down and it is the extent of this breakdown that must be considered. Processing the plastic at the high end of the recommended melt temperature can be increasingly more detrimental to the molecular weight of the plastic because the plastic is now subjected to higher temperatures where the allowable residence time may not be large. Having the screw speed set at a high value can also increase the shear rate, again contributing to the breakdown of the molecules. High screw speeds can create large mechanical forces breaking down the fibers, while non-fibrous fillers may not be as affected.

During the mold filling phase, the material passes through the nozzle, the sprue, the runners, the gates, and into the cavity and finally reaches the end of fill. During this filling phase, the material is subjected to high shear forces, causing a breakdown in the plastic molecules. The fillers may not be impacted, unless the gate sizes are small. Shear is higher in smaller diameter runners and therefore smaller runners cause larger drops in molecular weight. Large tab gates offer the least amount of molecular weight decrease through the gates. The part geometry can also play a role in the degradation of the plastic. Thinner cross sections increase the shear and therefore increase the chance of plastic degradation. Insufficient venting in the mold can cause a dieseling effect and burn the plastic which in turn lowers the molecular weight. This may be difficult to observe on dark color parts.

The discussion above shows that each phase of the process has an effect on the molecular weight of the plastic and the integrity of the fillers. If any one of these processes is not controlled properly, it is easy to degrade the plastic. In many cases, plastic degradation or filler breakdown may not be cosmetically noticeable, making it easy to assume that the process was acceptable.

The generation number for a quantity of regrind is defined as the number of times the regrind has been generated from the same batch of plastic. For example, if a runner was ground up to be used back into the process, it is considered as first generation regrind. When the regrind is reprocessed back into the machine and regrind is generated from the molded runner, then this regrind is called second generation regrind. With each progressive generation, there is a deterioration of the plastic properties.

When regrinding, the runner or the parts are reduced to a size close to the virgin plastic pellets with the help of a granulator (grinder). Because of the random mechanical breakdown of the plastic, the particle size distribution ranges from fines to coarse particles. Low rpm grinders are best suited to achieve a narrow particle size distribution. A large amount of fines can cause two problems: First, due to a lack of mass, the fines melt immediately and stick to the feed section of the screw, causing inconsistent screw recovery. Second, these fines tend to degrade much faster, causing defects in the part. Fines must therefore be removed before they enter the feed throat. This is done with fine separators also called cyclone separators. In addition, regrind pellet size must be close to the virgin pellet size to achieve consistent mixing and feeding of material that contains regrind.

13.3.2 Using Regrind

It is clear that using regrind can cause a varying degree of loss in material properties. Therefore, the amount of regrind that can be used in a product will depend on how this loss affects the performance of the final product. It is impossible to provide general hard and fast rules or equations to calculate the acceptable amount of regrind that can be used in a part. The only reliable method is to experiment with varying percentages of regrind and different generations of regrind. A good starting point may be to run the product with 100% first generation regrind and perform functional testing on the parts. Surprisingly, on some products this may be completely acceptable. If the initial process was carefully controlled and if the molding process was developed using scientific molding principles with large process windows, the material properties may stay well within the required specifications. If the 100% product is unacceptable, various ratios of virgin to regrind must be evaluated. When testing the product, the properties that are important to the product must be evaluated. Testing of the product in its final assembly and service environment is critical. Lab results from tensile or impact testing can only provide a comparison of the property, but product testing results will provide the actual value. The testing procedures must be standardized and should be used in regular production. An acceptable part can also be used to create a standard which all products can be compared to or checked against to determine their acceptability.

13.3.3 Batch and Continuous Processes of Incorporating Regrind

There are two methods of incorporating regrind into the product. In the batch process, the regrind is generated offline, away from the molding machine. A predetermined amount of regrind is either blended with the virgin material and loaded into the machine hopper, or the regrind and virgin material are blended together at the molding machine hopper, using blending equipment. In this system, the regrind will have to go through the drying process together with the virgin plastic. Blending equipment based on weight (gravimetric blenders) is most reliable. Some systems rely on loading time of virgin plastic and regrind to achieve the right percentage. For example, to achieve an 80:20 blend the virgin plastic is fed into the system for 80 seconds and the regrind for 20 seconds. This system is not reliable because time does not equate to weight considering the possible differences in the bulk densities of the virgin plastic and regrind. If the plastic gaylord or container, which feeds the hopper, runs out of material, it may only be loading either the virgin or the possible worst case, only regrind. Gravimetric feeders look for a particular amount by weight of the virgin and regrind before delivering the mixture to the molding machine. If one of the components is not available, either the blending equipment or the molding machine will set off an alarm for lack of material. Blending equipment is also physically mixing virgin material and regrind before delivering it into the hopper of the molding machine. Systems that rely on time may end up having layers of virgin and regrind in the hopper because they alternately deliver each component for the specified length of time to the hopper of the molding machine.

In a batch process, the chances of contamination are very high and this is probably the main reason why regrind projects often are not successful. It is easy for two runners of the same color but of different materials to be thrown into the same grinder. Contamination can show up right away when the barrel is being purged for molding. If the contaminant melts at a relatively lower temperature but is now subjected to a higher temperature, it can easily start gassing or spurting. It may also show up as splay or other cosmetic defects. If the contaminant is a higher melting plastic, there is a possibility of it getting stuck in gates and hot tips or it may show up as unmelted pellets in the part if the gate sizes are large. In both cases, delamination, structural integrity or other issues can be the result and the part quality will suffer.

With a continuous process of regrind incorporation, the runners are picked up with the help of a sprue picker or robot and directly dumped into a granulator; the grinder grinds up the runner and delivers the regrind to the hopper immediately. The runners are not separated in a separate area to be mixed later, such as in the batch process. The continuous process is a cleaner way of incorporating the regrind with almost no chances of contamination. The disadvantage here is that some molecules have gone through the process multiple times because this is a continuous process and the runner is always recycled. It is impossible to figure out the generation of the regrind, because it is all one homogeneous mixture. The smaller the ratio of part to runner weight, the greater is the percentage of older generation regrind present in the system. Continuous processes are suitable for parts that are able to accept a large percentage of regrind and that have a larger ratio of part to runner weight. The smaller the runner, the lower is the amount of regrind and therefore the fewer are the chances of processing problems or product failure. If the runner is directly fed back to the feed throat, in most cases drying is not required, eliminating the risk of additive loss caused by the drying process.

13.3.4 Estimating the Amount of Regrind from Different Generations

Consider a part to runner weight ratio of 80:20. Assuming the whole runner is incorporated back into the plastic, the percentage of each generation of regrind present at each pass of molding is shown in Table 13.1. The concept of generation at each pass may seem confusing. First generation regrind during pass four will be the regrind generated from the virgin material from pass three. It will not be regrind generated during pass one. The regrind generated during pass one becomes third generation regrind. Because part and runner are molded in the same shot, the percentage of regrind will be the same in the runner and in the part.

Table 13.1 Percentage of *g* Generation Regrind after *p* Number of Passes for a Part to Runner, Ratio of 80:20

Regrind generation number (*g*)	Pass (*p*)				
	1	2	3	4	5
0 (Virgin)	100	80.00	80.00	80.00	80.00
1	–	20.00	16.00	16.00	16.00
2	–	–	4.00	3.20	3.20
3	–	–	–	0.80	0.64
4	–	–	–	–	0.16
Total	**100**	**100**	**100**	**100**	**100**

It is a common perception that as the number of passes increases, the material starts to deteriorate very fast. This is true in cases where the part to runner weight ratio is high; for example the runner weight is 75 grams and the part weight is 25 grams. In the above example, where the part to runner ratio is 80:20, Table 13.1 shows that there is always 64 g or 80% of virgin material in the part, 12.8 grams of 1st generation regrind, 2.56 grams of 2nd generation regrind and so on. If a disciplined approach has been taken to dry the plastic and process the material, the properties of the plastic may well be within the required specifications. This will be true with filled or unfilled plastics and in most cases this should be acceptable. If the part design was not robust, it is possible that the small decrease in the properties would result in the failure of the part.

Table 13.2 Weight of *g* Generation of Regrind after *p* Number of Passes in an 80 g part with a 20 g runner

Regrind generation (*g*)	Pass (*p*)				
	1	2	3	4	5
0 (Virgin)	80	64.00	64.00	64.00	64.00
1	–	16.00	12.80	12.80	12.80
2	–	–	3.20	2.56	2.56
3	–	–	–	0.64	0.51
4	–	–	–	–	0.13
Total	**80**	**80**	**80**	**80**	**80**

Table 13.3 Weight of *g* Generation of Regrind after *p* Number of Passes in a 20 g Runner of a 80 g Part

Regrind generation (*g*)	Pass (*p*)				
	1	2	3	4	5
0 (Virgin)	20.00	16.00	16.00	16.00	16.00
1	–	4.00	3.20	3.20	3.20
2	–	–	0.80	0.64	0.64
3	–	–	–	0.16	0.13
4	–	–	–	–	0.03
Total	**20**	**20**	**20**	**20**	**20**

The following formula can be used to estimate the percentage of regrind from each generation in a part:

$$\begin{aligned}&(1)\text{If}(p-g)<1,\ \text{then}\ R=0\\&(2)\ \text{If}\ (p-g)=1,\ \text{then}\ R=\left(\left(\tfrac{x}{100}\right)^{g}\right)100\\&(3)\ \text{If}\ (p-g)>1,\ \text{then}\ R=\left(\left(\tfrac{x}{100}\right)^{g}\left(1-\tfrac{x}{100}\right)\right)100\end{aligned} \qquad (13.2)$$

with

x = percentage weight of the runner

g = generation of the regrind

p = molding pass number

R = percentage of regrind

Example

Part weight = 35 g, runner weight = 7 g

therefore, $x = \left(\frac{7}{7+35}\right) \times 100 = 16.67$.

In the 4th pass (p = 5), the amount of 3rd generation regrind (g = 3) can be determined with $(p - g) = (5 - 3) = 2 > 1$ as

$R = ((16.67/100)^3\,(1 - (16.67/100))100 = 0.39\%$

Regrind tables for various runner ratios are given in Appendix G.

13.3.5 Effect of Regrind on Processing

Regrind can have lower molecular weight, degraded additives and/or their byproducts, damaged or destroyed fillers, and contaminants. Depending on the extent of these changes it may be necessary to change the process to compensate for some of the changes in the outputs and make acceptable parts. A decrease in molecular weight will cause a drop in viscosity of the plastic, affecting all the outputs related to viscosity. Fill times and cushion values will be lower. If part of a processing aid or a viscosity reducer was lost, it is possible that the fill times and cushion values will be higher. If the filler was fibrous, the breakdown of the fiber length will cause the plastic to flow easier, resulting in shorter fill times and lower cushion values. Lower viscosity can also lead to an increase in plastic cavity pressure, although the pressure behind the screw remains unchanged. Excessive amounts of decrease in molecular weight can cause the plastic to degrade easier, causing excessive gassing and plugging of the vents. Cosmetic defects such as splay can occur. These effects can be compensated for by changing some of the process parameters. For example, reducing melt temperature will increase the viscosity of the plastic or reduce the gassing that occurs. In all cases, it is important that the process be robust and more importantly be redeveloped based on the presence of regrind.

13.3.6 Closing Remarks

Using regrind is a good and effective way to save money and at the same time protecting the environment from waste plastic being dumped into landfills. However, many processors tend to switch back to virgin resin as soon as they detect a quality problem. Often the quality issue is fixed and the processors are hesitant to go back to using regrind because they identify the cause of the problem being the regrind. Although this could be true, the reason why the regrind was causing the problem must be investigated. Sometimes the regrind is not properly dried or there may be too many fines from the regrind process that could have caused the problem. A proper analysis of the amount of regrind and a procedure to incorporate it into the molding process are two very important steps that must be followed. The implementation plan can be pursued as far back as in the part design stage. A part designed to be molded from virgin plastic can be overdesigned to compensate for the loss of properties when regrind is used. Using regrind should be part of the production plan and as seen in the above 80:20 example, the part will always contain 92.8% of virgin and first generation regrind resin and should therefore be able to function satisfactorily. Most companies that are not successful in implementing a regrind program fail because of lack of discipline and training of the shop floor personnel. For example, the operators who assist in grinding the runners need to be educated about the different types of plastics and that every clear runner in the molding facility is not necessarily the same material. Instituting a regrind program requires discipline and is a culture change.

Suggested Reading

Osswald, T. A., Turng, L., Gramann, P. J., *Injection Molding Handbook* (2007) Hanser, Munich

Beaumont, J. P., Nagel, R., Sherman R., *Successful Injection Molding* (2002) Hanser, Munich

Rosato, D. V., Rosato D. V, *Injection Molding Handbook* (2000) CBS, New Delhi, India

14 Related Technologies and Topics

This chapter will introduce a variety of technologies and techniques that can be applied to enhance the robustness of a process and/or to speed up the complete project from conceptualizing the part to its release as a molded product. Cavity pressure technology, which takes process monitoring and control to a completely new level, increasing product quality and plant efficiency many fold, will be introduced. Companies can build a knowledge base of their experiences to apply to future projects based on their past experiences. Concurrent engineering is an extension of the teamwork concept, keeping the whole team involved with the decisions and progress of the project.

14.1 Cavity Pressure Sensing Technology

When we discussed processing in the previous chapters, everything has all been related to the molding machine. For example, process optimization dealt with optimizing the processing parameters, such as injection speed and holding pressure. However, the part is finally made inside the mold, and therefore knowledge of what happens inside the mold can give the most valuable information about the quality of the part. The molten plastic will follow the specific volume–temperature graph (Figure 2.10) that dictates the part quality. Tracing this information for every shot will reflect the quality of the part. Needless to say, if every shot follows the same curve each time, all shot will be identical. Although it is not easy to output a specific volume–temperature graph for the mold, there are other indirect methods. Placing a pressure transducer inside a mold can provide information about the melt pressure in the mold and as the melt starts to cool, the pressure decrease can be plotted. Temperature transducers inside the mold will provide information on the temperature of the plastic. Although the "transducer" is technically the right term, the term sensor is most commonly used and we shall adopt that terminology in this text.

14.1.1 Sensors and Output Graphs

Figure 14.1 shows the placement of a 'Near the Gate Sensor' and an 'End of Fill Sensor' in a mold. A typical cavity pressure graph is shown in Figure 14.2. The graph also identifies the injection, holding, and cooling phases. There are three pressure traces in the figure that will be described in the following:

Hydraulic Pressure Curve: To measure hydraulic pressure, the sensor is placed in the hydraulic line. Since the injection phase is a fast dynamic phase, the pressure increases very rapidly until the start of the holding phase is reached. Once the switch-over from injection to holding takes place, the pressure drops to the holding pressure value and stays constant until the end of the holding time. After the holding time is complete, the hydraulic pressure drops to zero. If the screw rotation starts immediately, the back pressure shows up right after the end of the holding phase. If there is a delay, the trace picks back up indicating the back pressure for the screw recovery time.

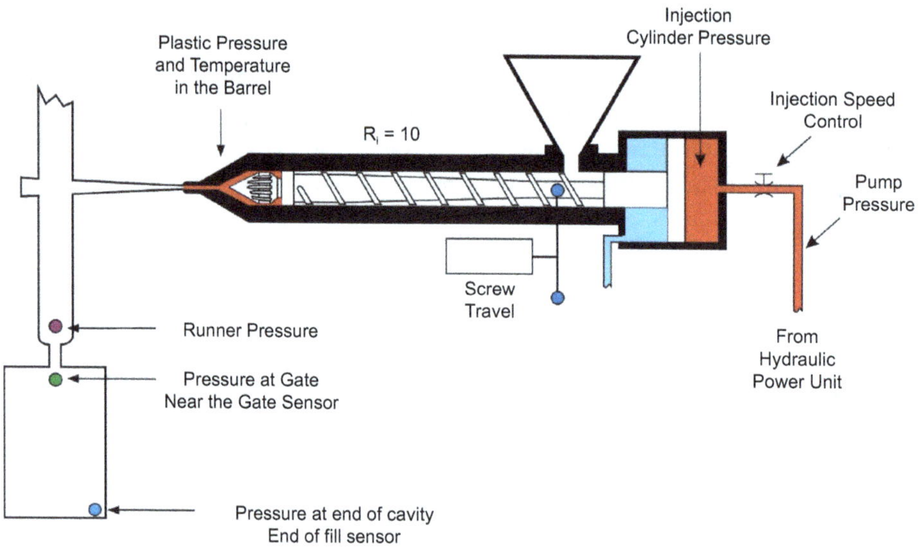

Figure 14.1 Schematic of placement of sensors inside a mold (Courtesy: RJG Inc.)

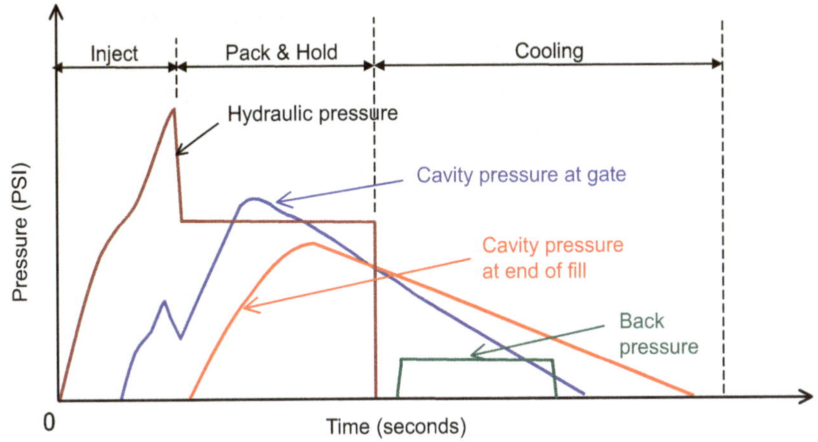

Figure 14.2 Typical representation of a cavity pressure graph

Cavity Pressure Curve Near the Gate: As the plastic begins to enter the mold and the cavity, it is in contact with the sensor placed just behind the gate. The pressure increases until the injection phase is complete, but does not drop down like the hydraulic pressure, because the cavity is pressurized and the gate is frozen or closed. The decrease in melt pressure is caused by the shrinkage taking place inside the mold and it causes the plastic to move away from the mold wall and the sensor. The start of the curve has a delay which is equal to the time the plastic takes to reach the sensor from the start of injection.

The Cavity Pressure Curve at the End of Fill: This curve looks similar to the curve described above, the only difference being the starting point of the curve. Here, the starting point represents the time the plastic requires to reach the end of fill. If the process is decoupled (see Section 7.5.4), this time is almost equal to the injection fill time.

14.1.2 Types and Classification of Pressure Sensors

There are various types of sensors used in plastics processing. Their classification is based on the type of technology used to collect information.

Strain Gage Sensors: The underlying principle of a strain gage is based on the Wheatstone bridge that has a network of resistance elements. The electric current flowing through this network is measured. When any external force is applied to the strain gage, the amount of current changes. This change in the amount of the current is proportional to the applied force and hence the amount of force can be determined.

Quartz-Based Sensors (Piezoelectric Sensors): Some materials, such as quartz, generate an electrical potential when an external stress is applied. The electrical potential is directly proportional to the applied force. A quartz-based sensor can therefore measure the force applied on the face of the sensor.

Another type of sensor classification is based on the location of the sensor in the mold.

Direct Sensor: These are also called flush mount sensors, see Figure 14.3. Here, the sensor face forms part of the cavity wall. The sensor is fitted such that it comes in contact with the molten plastic and measures the pressure. Such sensors are not commonly used because it is not always possible to get the sensor inside the mold cavity. It can also be a risky task to work inside the cavity to fit the sensor. The cavity wall would need rework if damaged. Some molds are subject to high temperatures and this could also pose a problem because pressure sensors may not withstand high temperatures.

Figure 14.3 Flush mount sensor (Courtesy: RJG Inc.)

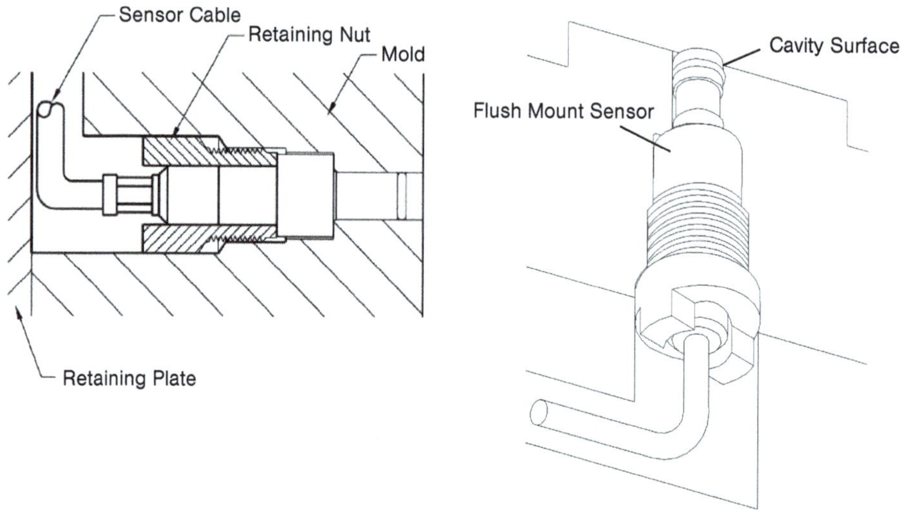

Figure 14.4 Flush mount sensor (Courtesy: RJG Inc.)

Indirect Sensors: When a sensor is placed underneath an ejector pin or a core pin, it is called an indirect sensor. The force is applied on the face of the pin and transferred to the base of the pin where the sensor is located, see Figure 14.5. These sensors are more common because there is room to work with inside the ejector box where the sensor has to be fitted.

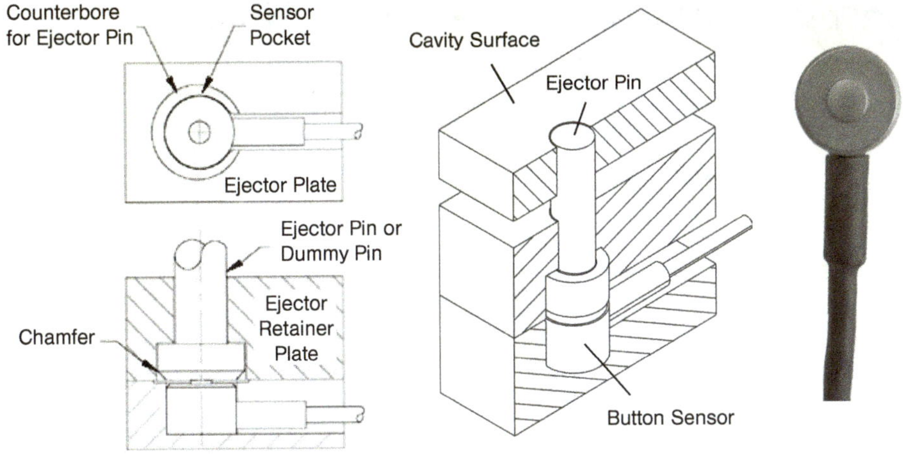

Figure 14.5 Button sensor behind an ejector pin (Courtesy: RJG)

Yet another type of classification of the sensors is based on the function the sensor will perform in the mold.

Control Sensors help control the process (which will be described in the following).

Monitor Sensors are used for monitoring the process. Process alarm limits can be set based on the information acquired from this sensor.

14.1.3 Use of Information from the Pressure Graphs

The information obtained from the pressure graphs is valuable in a number of ways.

- The pressure graphs that are traced provide information about the cavity pressure and therefore the part quality based on the specific volume-temperature relationship.
- Inconsistencies from shot to shot can be observed. If the pressure traces do not repeat, the part quality is not repeated, see Figure 14.6. It can be observed that the hydraulic curve overlays itself every shot, proving the hydraulic pressure is consistent. However, the plastic pressure curves are not being duplicated. If there was no cavity pressure hookup, one would automatically assume that the process is consistent and that the part quality should also be consistent from shot to shot. A leaking check ring can cause such a variation and can only be noticed with a cavity pressure sensor.

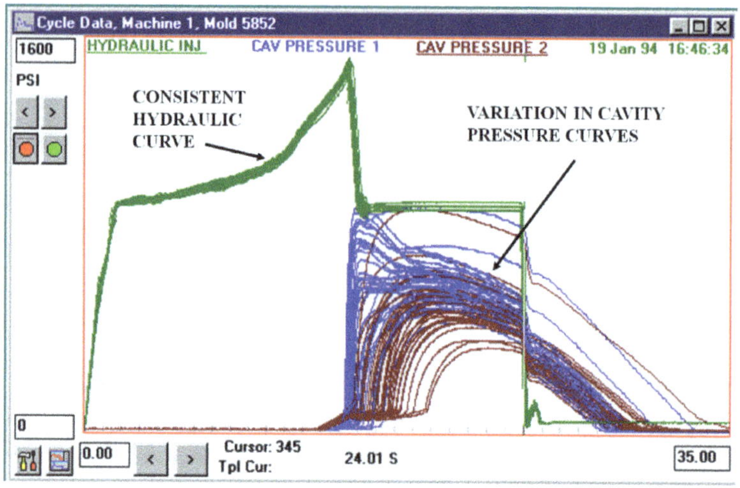

Figure 14.6 Repeatable hydraulic curve with varying cavity pressure showing inconsistency in part quality (Courtesy: RJG Inc.)

- If there are sensors present in each cavity, the quality of each cavity can be compared, see Figure 14.7. In this example, there are four cavities and the traces for each cavity are different. Based on the specific volume–temperature curve, each cavity will produce different parts. Although there is consistency within a cavity, there is cavity-to-cavity inconsistency.

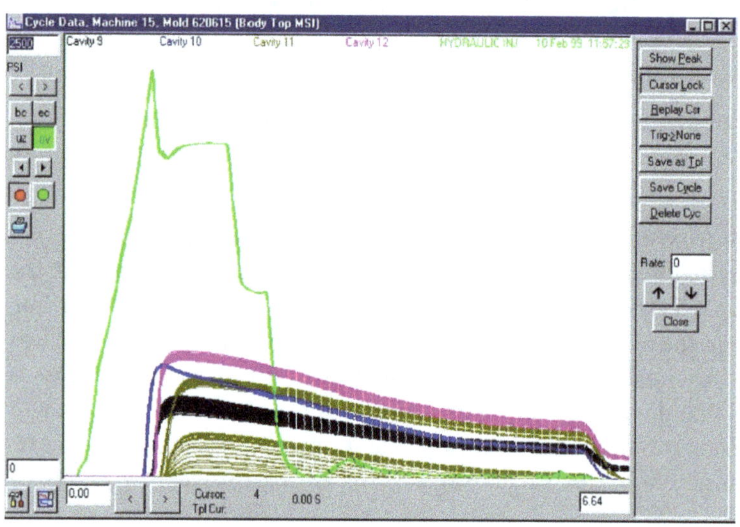

Figure 14.7 Cavity pressure variations between cavities showing cavity-to-cavity inconsistency, note that cavity 11 has the greatest variation, ranging from well-packed to nearly short (Courtesy: RJG Inc.)

- The cavity pressure graph can also provide gate seal information. If the gate is not sealed, the plastic will leak out of the cavity, causing a sudden drop in cavity pressure. The curve will therefore show a sudden change in its slope. A frozen gate will show as a gradual drop in cavity pressure rather than a sudden change in slope, see Figure 14.8.
- If the sensor is at the end of fill, a short shot will result in a 'zero' value for the pressure. This can be used to detect short shots and contain the parts. Usually, a signal is sent to a robot to reject the parts or sent to a conveyor to reverse and reject the parts.

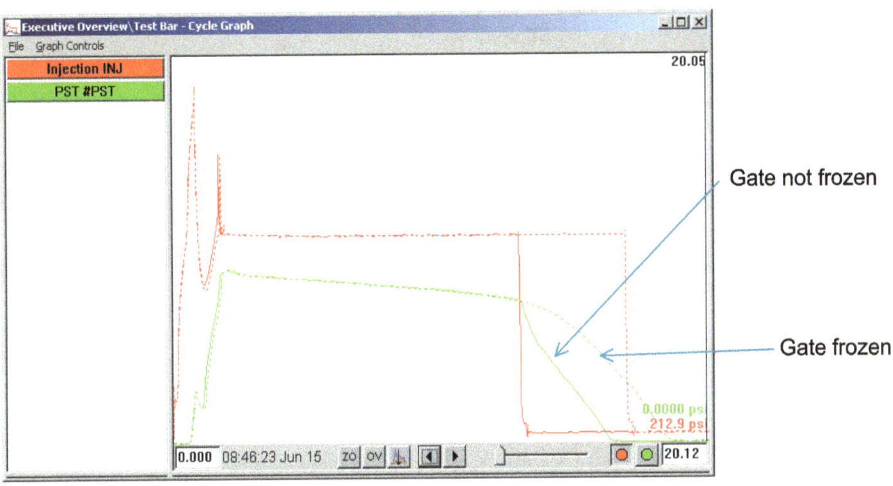

Figure 14.8 Cavity pressure curve showing the status of gate freeze (Courtesy: RJG Inc.)

14.1.4 Controlling the Process with Cavity Pressure Sensors

A decoupled process is the most efficient and consistent method of molding. The switch-over from the injection phase to the holding phase can be initiated by monitoring one of the following parameters: hydraulic pressure, time, or screw position. The most common and consistent method used is the screw position. When the screw reaches a preset position, it transfers to the holding phase. The screw position controls the percentage of the volume filled and, when using decoupled molding (see Section 7.5.4), this percentage should be between 95 and 98%. In response to the injection phase there is an increase in the cavity pressure and then a slow decrease. Figures 12.9 through 12.12 show the various cycle integrals.

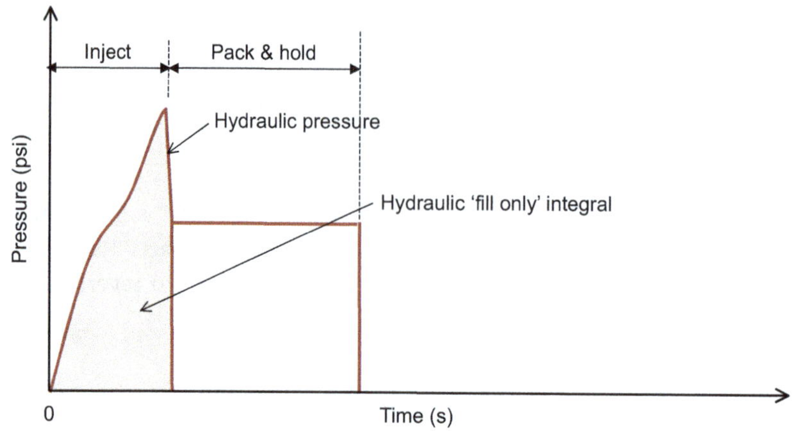

Figure 14.9 Shaded area representing the hydraulic 'fill only' integral

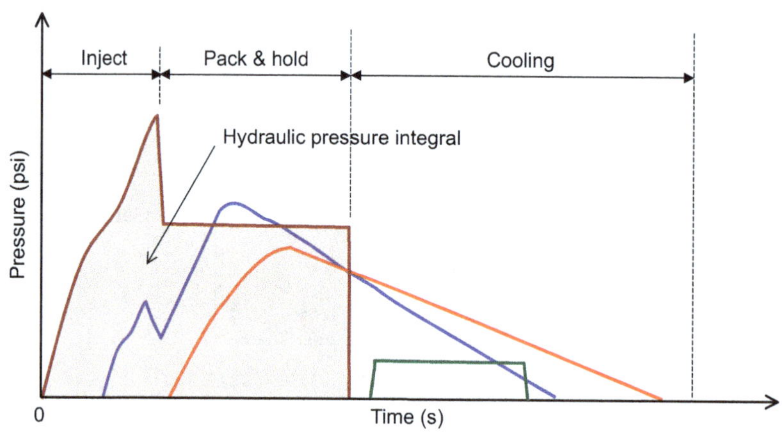

Figure 14.10 Shaded area representing the hydraulic integral

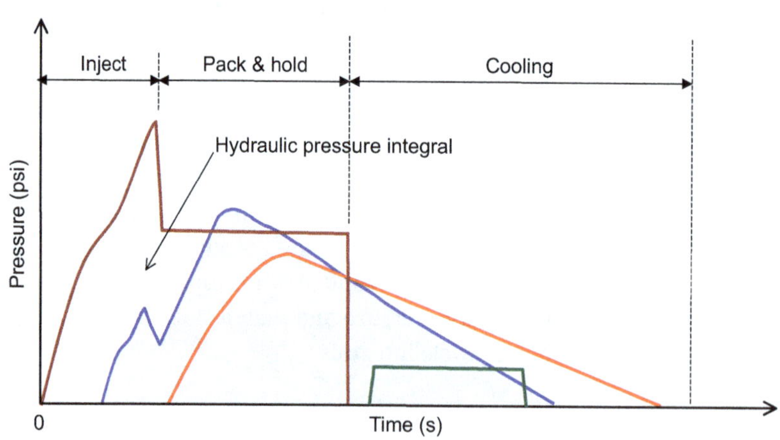

Figure 14.11 Shaded area representing the post-gate cavity pressure integral

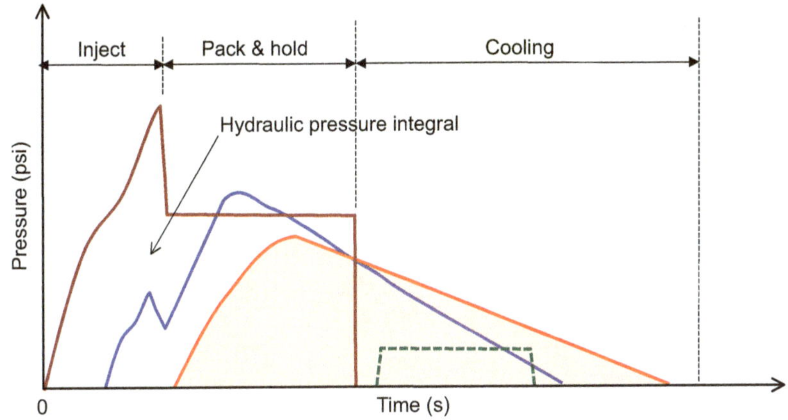

Figure 14.12 Shaded area representing the end-of-fill cavity pressure integral

Peak pressure, time to the peak pressure, and the area under the curve before the peak pressure (shaded area) represent the injection phase. Therefore, if this is duplicated every shot, the fill is duplicated every shot. If the information from the sensor can be fed back to the machine and if the switch-over is now controlled using a value on the graph (usually the peak pressure), the switch-over for every shot will take place at the precise fill amount of the cavity, resulting in shot-to-shot consistency. Injection will always be ended when the desired pressure is reached in the cavity. This is called process control using cavity pressure sensors. Other than process consistency, the other advantage lies in the fact that the process is now machine-independent. If the same curve is repeated on another machine, the part quality will be the same. Controlling the process using pressure sensors leads to the most consistent process.

14.1.5 Sensor Locations

Sensor locations are important because the location will dictate the amount and quality of the information it will acquire. Preferably, sensors to control the process must be located near the start of the fill (in the first third of the part) and the sensors to monitor the process should be near the end of the fill, in the last third of the part. For extremely small parts, a sensor can be located in the runner. The compromise here is that real gate seal information is not available because it can be obtained only from the pressure traces in the cavity. Table 14.1 shows the benefits of certain types of sensors and their locations (direct or indirect).

Table 14.1 Comparative Benefits of Types of Sensors and Technology (Courtesy: RJG Inc.)

	Types of sensors and technology			
	Flush mount (direct)	Under the pin (indirect)	Piezoelectric	Strain gauge
Accuracy of data	Yes	Yes	Yes	Yes
Ease of calibration/ configuration	Yes			Yes
Location flexibility	Yes			
Lower installation cost		Yes		
Easier replacement		Yes		
Removable cables			Yes	
Limited mold space	Yes		Yes*	
High temperature			Yes	
Easy troubleshooting		Yes**		Yes
Use with long cycles				Yes
Lower costs				Yes

* Smaller connector
** Easier removal

14.2 Building a Knowledge Base

Many molding companies try to find and dominate a niche in the market. For example, a molder may be known to be an expert in molding gears with high precision while another may be an expert in two-shot molding. Companies reach these levels of expertise having perfected the technologies over a number of years. The collected knowledge comes from a pool of experience, trial and errors, and applied science. This knowledge and solutions to the problems should be recorded to make future projects efficient during project development, the release to production process, and finally in production. For example, consider a company that molds connectors of various types, primarily from nylons and polyesters. Over the years they may have built hundreds of tools for similar types of connectors. There may be a commonality in the shape of the connectors, the gate locations and sizes, cooling channel locations, wall thicknesses, and so on. If the scientific processing principles are followed, the processing parameters would also end up being very similar. For example, the mold and melt temperatures or the plastic pressures in the cavities can be very similar for parts being molded from the same materials. Similar processing parameters will lead to similar shrinkage values. This can be verified by taking the steel dimension and comparing it with the part dimension. If the cavity steel for a number of molds and their respective molded parts are measured, a database can be built and a simple formula or a trend graph can be developed. When a new similar project comes in, this database can then provide a better estimate of the shrink values that can be used when designing the mold. This technique can be very useful when designing components that are difficult to make and when mold dimensions cannot be kept steel-safe. A threaded insert is a good example of this type of component. A 16-cavity threaded core insert would be very expensive to rework, if the final thread size on the molded part was not within specifications and not within the required capability. Here, prior knowledge of a similar mold would come in very useful. If the plastic flow is similar in an existing mold compared to a new mold, the molecular orientations can be similar, leading to similar shrinkage values. If the process using the existing mold was robust, using similar parameters will lead to a robust process in the new mold. Again, the part quality becomes predictable. In referring to the threaded part mentioned earlier, determining the size of the cores can be done with a higher confidence level. Although shrinkage can be estimated with the help of simulation programs, the data acquired by modeling is rarely accurate, because of the basic assumptions that go into building the software algorithms. For example, a software program will assume perfect venting and perfect packing of the parts, although in practical molding this is never the case. Building a knowledge database for similar products eliminates the need for assumptions and at the same time takes into account all common factors, making the estimation more reliable. It is important to keep the steel safe and to start the mold dimensional modifications only after the first mold trial where a robust process was developed.

Building a knowledge base helps taking the guess work out and reducing the number of iterations to the final product release to manufacturing. This saves time and money, making the overall process efficient. An example of a summary sheet is shown in Figure 14.13. The data can also be used to build a second mold for an identical part, especially when in the first mold some of the required part dimensions were unattainable, possibly because of the cost or time of modification or of making new mold components. In such cases, the mold design can be reverse-engineered and the parts in the new mold can be molded within the required specifications.

Summary of Mold Dimension, Part Dimension and Process Parameters for Part No.: 84736-2

Part Description : Cover Connector
Material ID : PBT
Grade : Valox 420 (Sabic IP)
% Filler : 30% glass

Runner Size feeding the part : 0.3175 inch
Type of Gate : Sub D-gate
Number of Gates : 1
Gate Size : 0.060 inches
Number of Cavities : 4

Melt Temperature : 480 deg F
Mold Temperature : 160 deg F
Fill Time : 0.98 sec
Cavity Pressure-Post Gate (Max) : 8500 psi

Machine Shot Size : 150 gms
Barrel Usage : 65.75 %
Machine Tonnage : 175 tons

Critical Dimensions	*Steel*	*Part*	*% Shrinkage*
D1 - Length	2.540	2.502	1.5
D2 - Width	1.740	1.705	2
D3 - Shroud Height (@C)	0.507	0.497	2
D4 - Shroud Height (@B)	0.507	0.492	3

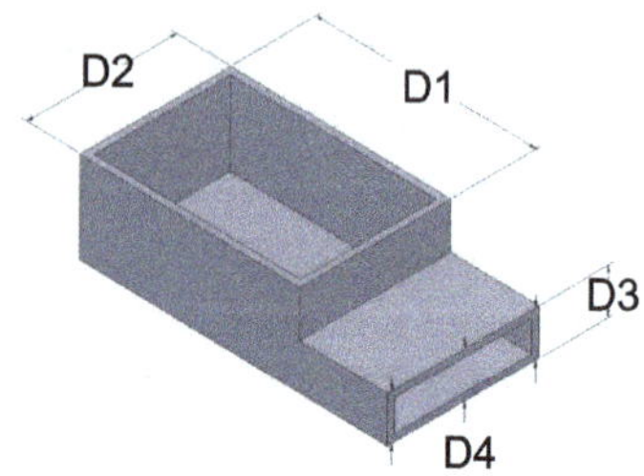

Notes:
Venting affects the fill and the dimensions of the shroud. Build shroud for easy access of vents. Vents must be cleaned at the start of every shift.

Figure 14.13 Knowledge worksheet

14.3 Concurrent Engineering in Injection Molding

A molding company usually receives a new project request through the sales department. The sales department then asks the engineering department about the feasibility of the molding. Once the job is accepted, the tooling department gets involved. They design the mold and help building the mold through the mold maker. The mold is then mounted on the machine and the expectation is that the process engineer can mold acceptable parts in the first attempt. Unfortunately, molding the parts successfully in the first attempt is not common, because the process, although having followed a logical sequence, did not involve all departments in the appropriate sequence of the decision making process, leading to certain failures. Many mistakes are often discovered during the first trial of the mold. 'They should have done it this way' or 'The parts will never fall off the mold' are typical comments during the first trial of the mold. This could have been avoided, if all departments were made aware of the project and its requirements before the mold was built. 'Over the wall engineering' is a term used to describe the project activity where each department is disconnected from each other. Each department performs the activities requested by the previous one and passes the project on to the next (Figure 14.14).

Figure 14.14 Over the wall engineering – departments are disconnected from each other

'Concurrent Engineering' involves representatives from every department in the project at the earliest stage (Figure 14.15). This should be followed by subsequent meetings, where the project details and any changes should be reviewed by the team to evaluate the impact a change may have on their own individual departments. Therefore, when the product designer mentions that the material is, for example, a long fiber thermoplastic (LFT), the process engineer knows, he will need a different screw and nozzle tip, or the scheduler knows that he will need to schedule time on the one specific machine in the shop with the required special screw. Raising a red flag at this stage saves time and money and delivers the product on time. Although conventional 'Over the Wall' approaches are commonplace in most companies, disengaging each department is a common cause of failures, inefficiencies, and delayed launch of the product.

Figure 14.15 Concurrent engineering

The concept is further explained in the matrix in Table 14.2. The first column is a typical list of departments involved in the design and manufacture of a product. The first row describes the activities involved. A 'Yes' at the intersection of a row and a column means that the particular activity has an impact on that department. For example, for the process engineer, the mold design is very important, while for the quality department, the mold design has no direct impact. Please note that this is a universal matrix and there will be exceptions. This matrix should be used as a guideline to set up individual company matrices.

Table 14.2 Suggested Personnel – Activity Chart for Implementing Concurrent Engineering

Job function	Activity						
	Product design	Material selection	Mold design	Mold construction	Machine selection	Process development	Quality requirements
Product designer	Yes	Yes	No	No	No	Yes	Yes
Tooling engineer	Yes	No	Yes	Yes	No	Yes	Yes
Mold maker	No	No	Yes	Yes	No	Yes	No
Material supplier	Yes	Yes	No	No	Yes	Yes	No
Process engineer	Yes	Yes	Yes	No	Yes	Yes	Yes
Quality engineer	No	Yes	No	No	No	Yes	Yes
Sales team	No	Yes	No	Yes	No	No	Yes

14.3.1 The Product Designer

Conventional Involvement: Product design is where the concept first takes shape through a CAD model or a prototype. At this time, there is a general idea of what the requirements for the plastic material are and therefore a selection may or may not have taken place. The machine, the mold design, the molding process, or any of the other factors are of no concern to the designer, who is focused on a functional concept.

Required Involvement: A product designer should understand the manufacturing process, especially the processing part of it. Design for manufacturability principles must be implemented. For example, the designer must be aware of the fact that a thick section in a part must be cored out to reduce sink or sufficient draft must be provided on the part to release the part from the mold. An explanation of the function of the part to the molder and his team would be beneficial to the molder. This way, the process engineer knows the particular material he needs to process or can raise a flag when tolerances are impractical for a specific material.

14.3.2 The Tooling Engineer

Once the product designer presented an acceptable design, the tooling engineer gets involved in the design and the economics of building the mold. Of all job functions, the tooling engineer is probably the one who is most involved in the project because his direct involvement is required at the various stages.

Conventional Involvement: The tooling engineer is usually the liaison between the mold maker and the molder. His job is to deliver the mold to the molder and fix any issues the molder discovers during processing. Any design changes to the parts – features and/or dimensions – need to be reflected in the mold and the tooling engineer is responsible for these changes.

Concurrent Requirement: A tooling engineer must understand the concepts of developing a robust process. The techniques and the benefits of Scientific Processing should be familiar to him. The tools must be qualified using these techniques and changes to the mold must be made to support a robust process. For example, identifying a process window is extremely important. The tooling engineer must closely work with the process engineer, the quality engineer, and the mold maker.

14.3.3 The Mold Designer and Mold Maker

Most mold makers work with their own mold designers. They are given the part model and design the mold accordingly.

Conventional Involvement: The mold designer receives the design from the part designer, designs the mold for the chosen number of cavities and the mold maker fabricates the mold. He knows the type of material that will be used for molding the part, but sometimes the details are not made available to him.

Concurrent Involvement: The mold designer and the mold maker must get in touch with the material supplier to get information before designing the runners, gates, and vents. For example, it makes a big difference whether a 30% glass filled material or a 30% long fiber glass filled material is used. Apart from this, the mold design must be reviewed between the part designer, the mold designer, the mold maker and the process engineer. When the first shots are produced the mold maker must be present to see how the mold functions. The mold maker must also understand the concepts of Scientific Processing and why a process window is important.

14.3.4 The Material Supplier

In most cases, the material suppliers sell the resin but do not get involved, unless there is a processing issue. Since they are the suppliers of the materials they have all the information such as data sheets, processing information, mold design requirements, part design requirements and are therefore the best source of information.

Conventional Involvement: They provide material specs to the designer, help with material selection, and sell the resin.

Concurrent Involvement: Material suppliers will be able to review gate types, gate sizes, and vent sizes for any given material. For example, sub-gates are popularly and commonly used because they automatically separate the part from the runner; however, they should not be used for highly fiber-filled materials. In case a material is molded for the first time in a particular facility, the molder must invite the material supplier. The material supplier can make sure that the material is processed properly for optimum properties and cycle time. It also helps to involve the material supplier at the mold design stage, again in particular if this is a new material.

14.3.5 The Process Engineer

Unfortunately, often the process engineer is the least involved member of the whole team when in fact he is the most important one. The process engineer is the person who has to deliver the final molded product. Project managers are anxious to see the first molded part in their hand and therefore are usually waiting at the molding machine on the day of the trial. Every other stage of the project, for example mold build, typically has an end date, but the process engineer is under pressure to deliver the parts hours after the mold has arrived into the molding facility and minutes after the mold is mounted in the molding machine. Most likely, the project manager has promised the customer parts by next day delivery or may have also invited the customer to be present for the trial. This puts tremendous pressure on the process engineer who is expected to mold acceptable parts. However, in many cases the process engineer has to take the blame for all the factors that were not taken into consideration and all the mistakes and miscommunications during the entire project if the parts were not acceptable.

Conventional Involvement: In some organizations the process engineer sees the mold or knows about the project on the day it is scheduled and in some cases they are involved at the mold design stage.

Concurrent Requirement: The process engineer must be involved at every stage of the project. Based on his molding experience, he can give a number of suggestions to improve the moldability and the molding process of the part. Process engineers are better judges of features such as vents, gate locations and so on. Mold designers tend to place gates in locations convenient for mold making which are not always the best locations regarding processing. Even non-technical requirements, such as the orientation of the mold, should be reviewed by the process engineer. The choice of the machine must be left to the process engineer. Again, based on his calculations of tonnage, % shot size used, and residence time, he can suggest the best machine for the job at hand.

14.3.6 The Quality Engineer

Typically, the quality engineer is only a 'measurement guy' and often not involved appropriately. The quality engineer past experience with similar parts or plastics can prove valuable. He usually has knowledge about shrinkages and appropriate tolerances during production.

Conventional Involvement: He is one of the invitees to the product release meetings and is usually assigned the task to obtain drawings and work on methods for measuring the final molded parts.

Concurrent Requirement: The drawings must be discussed with the product designer. If an stereolithography model of the part or a prototype model is available, it should be used to detail the inspection plan. Sometimes fixtures are required and these can be planned in advance. Initial Gage Reproducibility & Repeatability (Gage R&R) studies can even be performed on prototype parts. Any impractical tolerances can be mentioned to the designer in advance.

14.3.7 The Sales Team at the Molder

The sales team is usually the one that introduces the product designer to the molder. They are looking to increase the sales of the molder and therefore try to acquire as much business as possible. However, they must understand the technical aspects of the molding in order to get the right set of customers. They must evaluate if an incoming project is suitable to and compatible with the capabilities of the molding operation. If product and molder are not the right match, this could have a negative impact not only on the program but also on the relations between the two parties. All future work between the two parties can get jeopardized.

Conventional Involvement: The sales force brings in a job mainly based on machine tonnage. Then there are other factors that they consider, such as any special requirements. Special requirements include clean rooms, in-mold decoration, or insert molding.

Concurrent involvement: The sales team must understand the capabilities of the molder. This also includes the strength and weaknesses of each department in the organization. If not, he is setting up the company for failure. The rules of machine selection for a particular job must be clearly understood. Tonnage, shot sizes, percentage of shot size used, residence times and other parameters must be known before accepting a job.

14.3.8 Mandatory for All Departments

The final molded part is the result of the efforts of all departments involved. In traditional 'Over the Wall' practices, every department performed their list of things to do and passed the project on to the next department. However, the practice fails to address the issues the receiving departments could have as a result of an action from a department previously involved. Understanding the needs of each department and moreover a 'why these needs' makes the jobs easier, efficient, and ensures on-time delivery of a part that meets the specifications. Every department must understand cross-functional duties and have a basic understanding of each other's job function.

For example, everyone must be trained scientific processing methods and understand why it is important to have a good process window. If the process engineer sends a mold back to the mold maker or the tooling engineer because he does not have a sufficient process window, he has a very valid reason. He can probably mold ten good parts, but that does not mean that he can make half a million good parts, because the process will not be capable. Knowledge of process capability must be commonplace in every department of the molding facility.

14.3.9 Implementing Concurrent Engineering

Concurrent engineering is probably the easiest concept to implement, because all it takes is to get all the involved departments in one room or on a conference call to discuss the project. Table 14.2 provided a simple matrix of typical job functions and activities involved. Set up a meeting between the representatives of each department for the various phases of the project. The following must be taken into account when following the table.

1. The order of certain decisions does not necessarily reflect the order in which certain activities are performed. For example, a machine must be selected at the quoting stage to make sure the molder has the machine for the particular job and not when the job has been accepted and a machine must be selected from the ones that the molder has.
2. The list of activities and job functions is a general list and every company has its own organizational structure. Therefore, every company must generate a customized matrix. A meeting or updates at the end of every stage should be mandatory. Although all job functions may not have a direct role in every stage, their decisions will be based on the information provided to them from the previous stages. Therefore, the status and decisions made regarding the project must be communicated to the whole group.

The molding business is getting extremely competitive in terms of cost and lead times. The time lines from conceptualizing the project to having an actual molded part are shrinking and it is expected that acceptable parts are produced on the first molding attempt. Only companies that can achieve this are likely to survive in the more and more competitive market. Concurrent engineering should be practiced because it provides the extra set of eyes to 'Look out of the Box' and raise potential concerns. These are meetings and times well spent. Regular reviews must be done, especially if there is a change in design, material, time line, etc., and must be communicated to everyone. The final product, good or bad, is a result of the involvement of the whole team.

15 Quality Concepts

This chapter is included to give the reader some basic knowledge and understanding of some of the terminology used in injection molding. At the end of the chapter, a list of some suggested texts is mentioned for readers who would like to further their knowledge.

15.1 Basic Concepts

Population: The population is the complete collection of elements to be studied. All the 10,000 parts molded in a production run is a population.

Sample: A subcollection of elements drawn from a population is a sample: 100 parts drawn from the above 10,000 parts will be a sample.

Statistics: The method of analyzing data form a sample and drawing conclusions about the population based on the data is called statistics. Of the 100 parts molded above, if 10 parts had flash, then the observed defects in the sample will help in estimating the defects in the population.

Variation: Variation is present in all processes. If any process or its output does not show variation, then the measuring method is not sensitive enough to measure the variation. If the weight of each of the 100 parts is 4.15 grams, then a scale with a least count of 0.001 gram must be considered to measure the variation. A stable system only shows effects of natural cause variation. There are no sudden or purposeful changes to the inputs of the process that get reflected in the responses. In special cause variation, there is a sudden or purposeful change in the inputs that is reflected in the responses. For example, power usage during the night before office hours is constant until office personnel begin to arrive. Then there is a sudden increase of power requirement when everyone in the office begins to turn on their computers at 8:00 AM. Figure 15.1 shows natural and special cause variations.

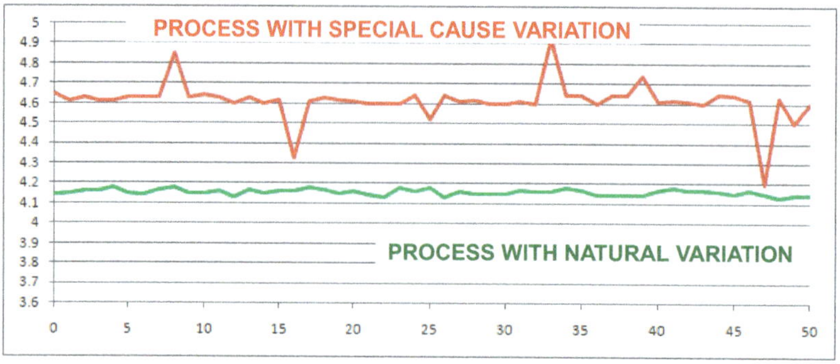

Figure 15.1 Natural and special cause variation

15.2 Histogram

A histogram is a graphical representation of the frequency of each data set. For example, if the weights of 40 parts were recorded and the frequency of each weight was represented graphically, it would be a histogram as shown in Figure 15.2. It shows that there was 1 part that weighed 4.10 grams, 3 parts that weighed 4.11 grams, 2 parts that weighed 4.12 grams, and so on.

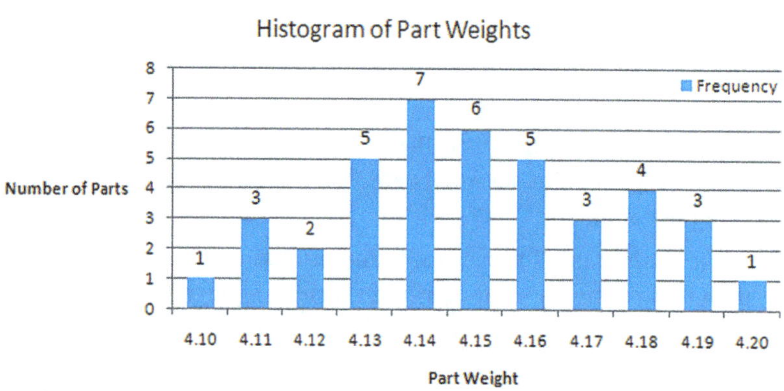

Figure 15.2 Histogram of part weights

15.3 Normal Distribution

In a given population of data, if the data is equally distributed around its mean, the data is said to be normally distributed. If the data for 40 parts were represented by the histogram shown in Figure 15.3, it can be observed that the data are equally distributed on each side of the mean value of 4.15 grams. This is called a normal distribution. The data cluster around the mean value. The curve joining the values is called a bell curve. Figure 15.4 shows examples of normal distributions. The data can also be such that the mean can be skewed to one side as shown in Figure 15.5. Such skewed data is said to have non-normal distribution.

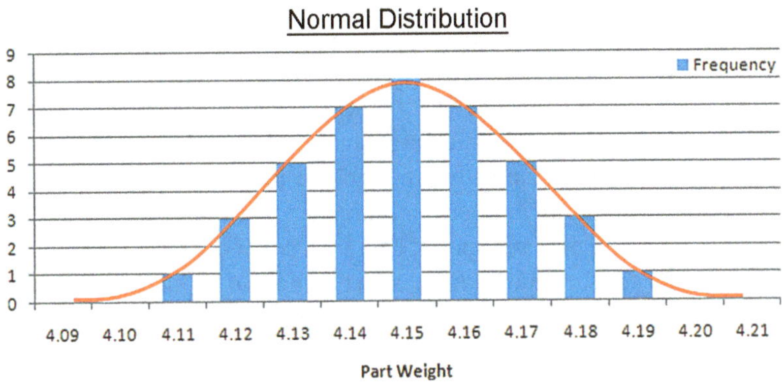

Figure 15.3 Normal distribution

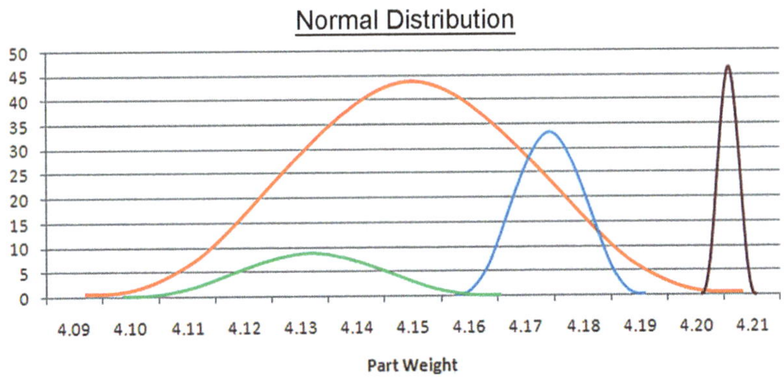

Figure 15.4 Examples of normal distributions

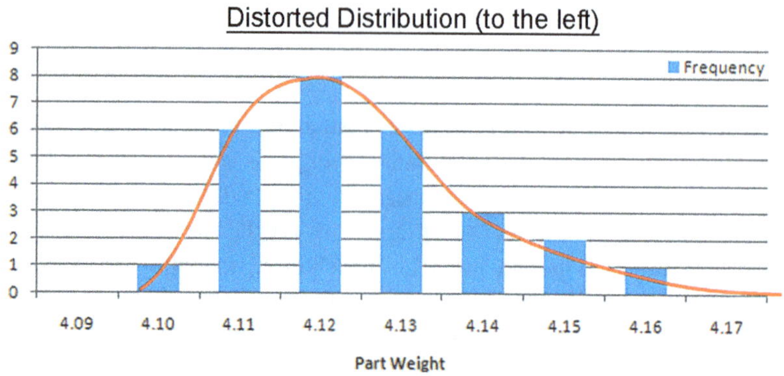

Figure 15.5 Representation of data that is not normally distributed

Significance of normal distribution: Processes that are stable, with only natural cause variations, output data that is normally distributed. The variation is predictable and therefore the output can be predicted. In injection molding, having a stable process with predictable variation translates to quality assurance and the reduction of inspection. Once the production run is completed, the molded parts can be confidently shipped to the customer without any rework or worrying about having returns for rework or scrap. In the presence of any special cause variations, the data can show two distinct sets of data for the same population, each having its own mean and distribution. Such distributions are called bimodal distributions. Clearly, these must be avoided and the source of variation must be investigated.

15.4 Standard Deviation

Refer to Figure 15.6, which shows a normal distribution. Consider the mean to be at a value of 4.15 grams. Because this is a normal distribution, the data are distributed evenly on both sides of the mean. If the area under the curve is divided as shown into 34%, 13.5%, and 2.35% on each side of the mean, it is found that the points intersecting the *x*-axis are all at equal intervals. Each interval is called one standard deviation or one sigma (1 σ). To help understand this concept further, consider that one standard deviation is equal to 0.02 grams (σ = 0.02 grams). The standard deviations on the right side of the mean are positive sigma (+1 σ, +2 σ, +3 σ, and so on) and the standard deviations on the left side of the mean are negative sigma (–1 σ, –2 σ, –3 σ, and so on). Therefore, the area between –1 σ and +1 σ is 68%, the area between –2 σ and +2 σ is 95%, and the area between –3 σ and +3 σ is 99.7%. Therefore, if the data is normally distributed, 99.7% of the data will fall between –3 σ and +3 σ. Standard deviation represents the extent of the variation from the mean. The higher the standard deviation, the higher is the variation.

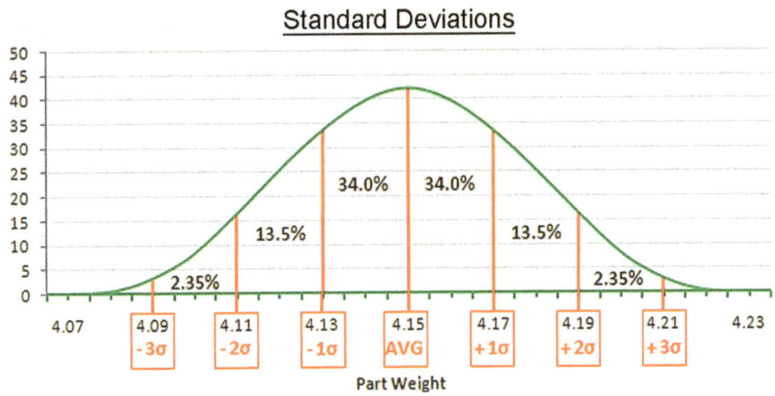

Figure 15.6 Standard deviation and area under the curve

In the example, because 1σ = 0.02 grams, and the average is 4.15 grams, +1 σ = 4.17, +2 σ = 4.19, +3 σ = 4.21, –1 σ = 4.13, –2 σ = 4.11, and –3 σ = 4.09 grams.

15.5 Specification Limits and Standard Deviation

The average part weight in the previous example is 4.15 grams. Assume the nominal value on the part drawing is also 4.15 grams. Refer to Figure 15.7. If the tolerances on the part weight are +/–0.06 grams, the lower specification limit (LSL) is equal to 4.09 grams and the upper specification limit (USL) is equal to 4.21 grams. These tolerances also coincide with –3 σ and +3 σ. Because the data is normally distributed, 99.7% of the parts will be within specifications, and 0.3% of the parts will be out of specifications; 0.15% will be below the LSL, and 0.15% will be above the USL. If the production run was for a 1,000,000 parts, then statistically speaking, 3000 parts will be out of specifications, 1500 parts will be below the LSL, and 1500 parts will be above the USL.

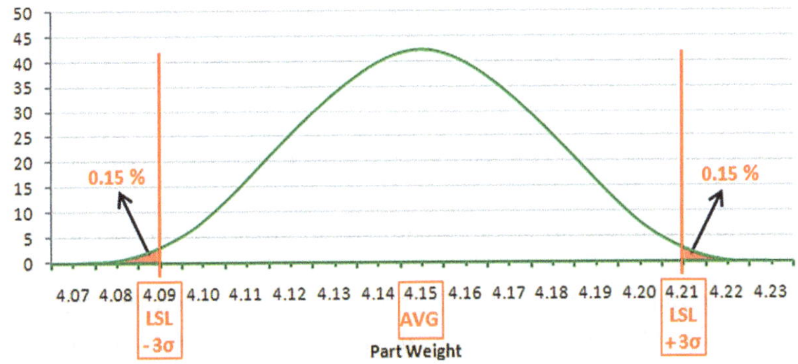

Figure 15.7 Specification limits at –3 σ and + 3 σ

To assure that all the molded parts are within the required tolerances, the specification limits must be well outside –3 σ and +3 σ assuming that the average value of the parts is always at the nominal value on the part print. In day to day production it is rare that the average will always stay at the nominal value because of natural variation. Refer to Figure 15.8. If the specification limits are placed at –6 σ and +6 σ then it is certain that all the parts will fall well within the required specifications. Such a tolerance will easily be able to cushion any natural variations in the process. In recent years this is the reason for the push for the six sigma techniques. It is important that the process is stable and all the variation is natural cause variation. The extent of the variation is debatable. Sometimes, although the data is normally distributed, there can be special cause variation present. For example, ambient temperature during the day and the night may make a significant shift in the distribution. Efforts must be made to minimize this variation by trying to find the source of the variation. It is impossible to eliminate variation.

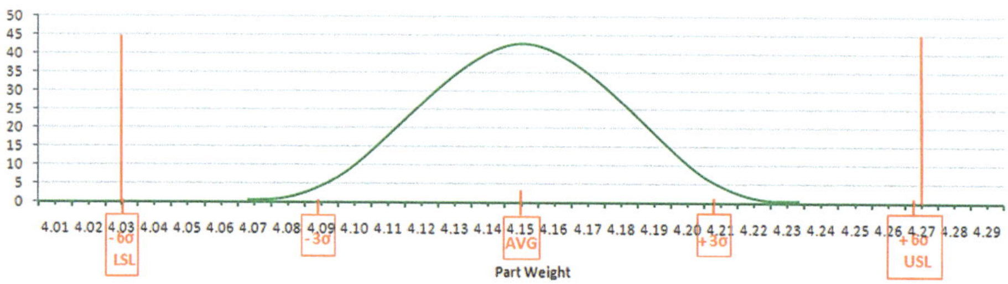

Figure 15.8 Specification limits at –6 σ and +6 σ

15.6 Capability Index

Figure 15.9 shows three processes. All the three have identical distribution and therefore identical standard deviations. Process 2 is the same data that we have been looking at in the earlier sections. The LSL and USL have been placed at –6 σ and +6 σ, with 4.15 grams being the nominal weight. With these limits, all the parts molded with Process 2 will be acceptable. Now consider Process 1. Process 1 has the same distribution and standard deviation as Process 2. However, the average is now shifted to the left to 3.91 grams. All the parts molded with Process 1 will be out of specifications. Process 3 also has the same distribution and standard deviation as Process 2, but the average part weight is 4.29 grams. Some of the parts molded with this process will be acceptable. These are shown as the green shaded area under the curve. All three processes have identical variation of data except their averages are different. But Process 2 is much more capable of producing parts within the required specifications. To make Process 1 and 3 capable, the only change that needs to be made is shifting of their averages towards the nominal.

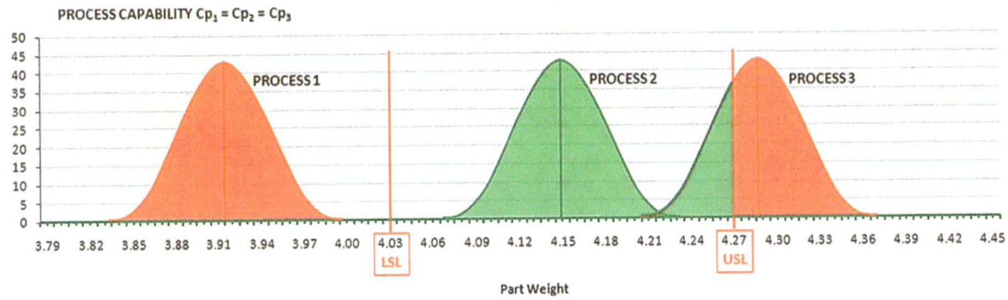

Figure 15.9 Three processes with identical distributions but different averages

Because 99.7% of the data falls between –3 σ and +3 σ, these measures are considered to be the practical limits for the spread of the data. Therefore, the data is spread over 6 σ (3 σ before the nominal and 3 σ after the nominal). If all the parts must be acceptable, then 6 σ must be less than the distance between the LSL and USL, or in other words, the ratio of the difference between the USL and the LSL to 6 σ should be greater than 1. This ratio is called capability index and is denoted as Cp.

$$\text{Capability Index} = \frac{\text{Engineering Requirements}}{\text{Statistical Requirements}} \qquad (15.1)$$

$$Cp = \frac{USL - LSL}{6\sigma}$$

The capability index is not related to the average or nominal value. Having a large Cp value equates to having a low standard deviation or a wide tolerance or both. In Figure 15.9, $Cp_1 = Cp_2 = Cp_3$, they all have identical capability indices.

Case 1: Cp < 1. If Cp < 1, then (USL – LSL) is less than 6 σ. The variation is too large or the specifications limits are too tight. Even if the process is centered, a large number of parts will still be out of specification.

Case 2: Cp = 1. If the Cp = 1 then (USL – LSL) = 6 σ. If centered, the process is capable, but is still molding 0.3% bad parts.

Case 3: Cp > 1. If Cp > 1, then (USL – LSL) is greater than 6 σ. The probability of molding good parts is certainly over 99.7%, depending on the actual value of the Cp. If the Cp value if 1.33, then the probability of molding 100% good parts is very high if the process is centered. If Cp = 2, then the ratio of the (USL – LSL) to the standard deviation is high enough to ensure acceptable parts if the process is centered. The higher the Cp value, the better the chances of molding acceptable parts. Such processes can easily withstand variations and some movements in their averages.

15.7 Process Capability

Process capability is the estimate of how capable the process is in relation to the nearest specification limit. Unlike the capability index, process capability takes into account the average value of the data. Process capability is denoted as Cp_k.

Cp_k is calculated using Equation 15.2. The smaller of the two values is used.

$$Cp_k = \frac{\left(USL - \overline{X}\right)}{3\sigma} \text{ or } Cp_k = \frac{\left(\overline{X} - LSL\right)}{3\sigma} \text{ (whichever is smaller)} \tag{15.2}$$

Refer to Figure 15.10.

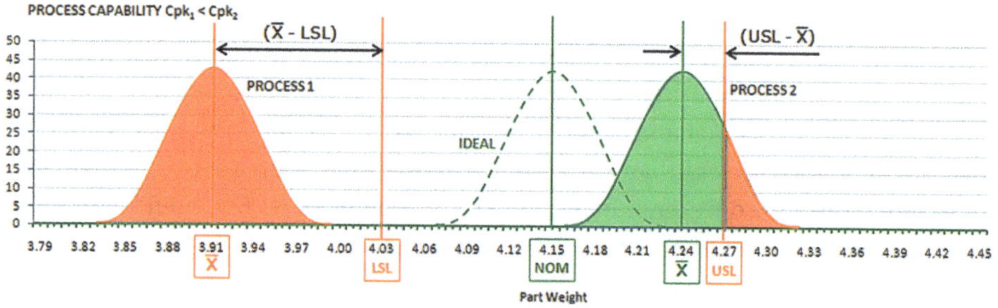

Figure 15.10 Process capability for two processes with identical distributions

The ideal process is shown by the dotted lines. The standard deviation from the previous discussion is 0.02 grams. Consider Process 2. The average of the part weights is 4.27 grams but because of the distribution some of the parts will be out of specifications shown by the red shaded area. If we move the average closer to the nominal, then the chances of molding all the parts acceptably will increase. Cp_k measures how far away the average is from the nominal or in other words, how close is it to one of the specification limits. Applying the formula, we can calculate the Cp_k of Process 2

$$Cp_{k2} = \frac{(4.27 - 4.24)}{3 \times 0.02} = 0.5 \tag{15.3}$$

The numbers for Cp_k follow similar rules as for Cp. A Cp_k value less of than 1.33 suggests that the process is capable of making unacceptable parts at the current location of its average. The confidence of molding acceptable parts increases as the Cp_k increases. A very high Cp_k value suggest that the possibility of molding bad parts is very small.

Consider Process 1. Applying the formula, we can calculate the Cp_k of Process 1.

$$Cp_{k1} = \frac{(3.91 - 4.03)}{3 \times 0.02} = -2 \tag{15.4}$$

Notice that in this case, there is a negative sign associated with the value of the Cp_k. This is in indication that the average is outside the specification limits and, therefore, depending on the distance from the LSL or USL, at least 50% of the parts will be unacceptable.

If the Cp value is high and the Cp_k value is low, this suggests that the process if centered can make acceptable parts. In molding, this is often a steel dimension change in the cavity. If a robust process has been established, the process must not be changed, but the steel must be altered. It is also clear from the previous discussion that if one part is within specifications, then this does not mean that all the parts will be within specifications. There is always a natural variation and therefore a distribution of data. A statistically acceptable sample size of parts must be measured. A minimum acceptable sample size for any statistical sampling is 30.

15.8 Statistical Quality Control (SQC) and Statistical Process Control (SPC)

Application of the above concepts to part quality is SQC. Quality inspectors track the quality of the part to make sure that not only are the parts within specifications, but that the distribution over time is acceptable. The Cp_k values are kept in check for critical dimensions. Control limits are tolerances that are calculated based on the spread of the data. For example, if the USL = 20, LSL= 10, Avg = 15, and the data is spread out over 4 units (2 on each side of 15), then a lower control limit (LCL) of 13 (15 – 2) and an upper control limit (UCL) of 17 (15 + 2) can be applied. If the data starts to drift out of these limits, it can give an early indication of a process shift or an imminent problem. The parts will still be acceptable and a proactive step can be taken to investigate the cause and rectify it. This will ensure continued acceptable production with no stoppage or scrap. In SPC, machine outputs such as cushion values, fill times, and cycle times are tracked following the same methodology. Figure 15.11 displays a run chart showing the different specification limits. The LSL and the USL are usually not shown on control charts, but are included here to help understand the concept.

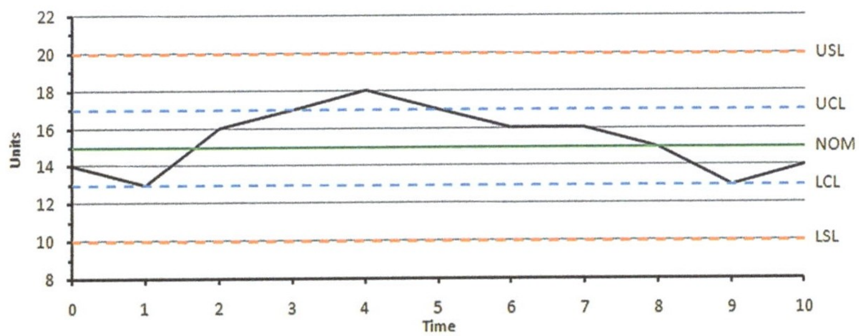

Figure 15.11 Run chart showing the different specification limits

15.9 References

[1] Wheeler, D. J., *Understanding Statistical Process Control* (2010), SPC Press, Knoxville, TN

[2] Triola, M. F., *Elementary Statistics* (2015), Pearson Publishing, NY

[3] Montgomery, D. C., *Introduction to Statistical Quality Control* (2009), John Wiley & Sons Inc., NJ

[4] Juran, J. M., *Juran's Quality Handbook* (2010), McGraw-Hill, NY

[5] Carender, J. W., *Managing Variation for Injection Molding* (2003), Advanced Process Engineering

Appendix A Materials Data Sheet

List of processing related data for commonly injection molded materials – SI system. (Courtesy *www.ides.com*)

Note: These are guidelines only – Individual datasheets must be obtained from the material suppliers or from *www.ides.com*

Long name	Short name	Specific gravity	Mold shrinkage, flow %
Acetal (POM) copolymer	Acetal (POM) copolymer	1.37–1.43	1.6–2.6
Acetal (POM) homopolymer	Acetal (POM) homopolymer	1.42	1.9–2.0
Acrylic, polymethyl methacrylate	Acrylic (PMMA)	1.14–1.20	0.4–0.6
Acrylic, styrene methyl methacrylate copolymer	Acrylic (SMMA)	1.05–1.09	0.4
Acrylonitrile butadiene styrene	ABS	1.04–1.11	0.05–1.4
Acrylonitrile butadiene styrene + PC	ABS+PC	1.10–1.22	1.6
Acrylonitrile styrene acrylate	ASA	1.04–1.12	0.50–0.6
Acrylonitrile styrene scrylate + PC	ASA+PC	1.11–1.24	0.5–0.6
Cellulose acetate	CA	1.23–1.32	0.5
Ethylene tetrafluoroethylene copolymer	ETFE	1.70–1.74	1.8
Perfluoroethylene propylene copolymer	FEP	2.14–2.16	0.7
Perfluoroalkoxy	PFA	2.14–2.15	–
Polyvinylidene fluoride	PVDF	1.75–1.79	2.2–2.5
Methyl methacrylate/ABS	MABS	1.08–1.10	5.5
Polyamide 11	Nylon 11	1.03–1.05	1.2–1.5
Polyamide 12	Nylon 12	0.980–1.23	0.4–1.9
Polyamide 46	Nylon 46	1.17–1.20	0.3–8.3
Polyamide 6	Nylon 6	0.922–1.18	0.01–4.4
Polyamide 610	Nylon 610	1.07–1.17	0.3–2.0
Polyamide 612	Nylon 612	1.06–1.35	0.10–6.0
Polyamide 66	Nylon 66	0.994–1.23	0.02–7.5
Polyamide 66/6 copolymer	Nylon 66/6	1.09–1.15	1.1–1.6
Polyphthalamide	PPA	1.10–1.16	1.1–2.0
Polyarylate	Polyarylate	1.20–1.37	0.81
Polycarbonate	PC	1.17–1.22	0.1–0.16
Polycarbonate + acrylic (PMMA)	PC+Acrylic	1.15–1.29	0.6–0.7
Polycarbonate + PBT	PC+PBT	1.10–32.1	0.5–1.0

Drying temperature (°C)		Drying time (h)	Suggested max moisture (%)	Processing (melt) temp. (°C)		Mold temperature (°C)	
min	max			min	max	min	max
79	121	1.5–5.0	0.15–0.20	189	211	49	96
79	86	2.0–3.5	0.2	199	216	47	91
65	86	3.5–5.0	0.097–0.10	225	271	54	76
65	82	2	–	215	227	40	60
80	91	2.0–3.5	0.010–0.15	223	246	49	76
78	111	3.0–4.0	0.020–0.043	248	288	59	88
80	100	3.0–4.0	0.038–0.10	239	262	55	73
79	110	3.0–4.0	0.020–0.052	250	280	60	80
60	71	2.0–3.0	0.2	220	230	48	60
NA	NA	NA	NA	323	331	135	–
121		3	NA	364		93	–
NA	NA	NA	NA	330	371	180	204
121	150	1.0–2.0	NA	215	249	71	72
70	80	3	NA	220	246	55	65
79	80	4.0–12	0.1	256		52	–
79	100	3.0–10	0.020–0.50	233	275	38	87
89	104	4.0–12	0.050–0.053	310	313	100	114
78	82	2.0–5.5	0.095–0.20	182	314	59	81
79	82	2.0–4.0	0.020–0.20	249	288	52	88
77	82	3.0–4.0	0.020–0.25	243	267	69	80
79	83	3.0–5.5	0.15–0.20	268	296	65	88
80	81	2.5–3.0	0.099–0.20	245	281	67	82
79	135	4.0–7.1	0.045–0.15	311	333	78	151
130		7	NA	NA	NA	100	135
102	128	3.0–4.5	0.019–0.020	282	308	79	102
82	83	4.5	NA	227	253	49	52
94	121	2.0–5.0	0.020–0.022	257	272	62	105

Long name	Short name	Specific gravity	Mold shrink-age, flow %
Polycarbonate + PET	PC+PET	1.20–1.22	0.5–0.9
Polybutylene terephthalate	PBT	1.00–33.6	0.1–0.5
Polyethylene terephthalate	PET	1.32–1.41	0.3–1.8
Polyethylene terephthalate glycol comonomer	PETG	1.25–1.28	0.3–0.5
Polyether imide	PEI	1.26–1.36	0.1–0.2
Polyetheretherketone	PEEK	1.25–1.40	0.1–1.7
Ethylene vinyl acetate copolymer	EVA	0.929–0.962	1.2–1.5
Polyethylene, high density	HDPE	0.932–0.988	0.7–3.0
Polyethylene, low density	LDPE	0.893–0.955	1.3–3.1
Polyethylene, linear low density	LLDPE	0.918–0.948	1.5–2.0
Polyethylene, ultra high molecular weight	UHMWPE	0.920–0.947	0.6–3.0
Polylactic acid	PLA	1.23–1.26	0.3–1.1
Polyphenylene ether	PPE	1.04–1.10	0.6
Polyphenylene sulfide	PPS	1.26–1.75	0.1–0.3
Polypropylene homopolymer	PP Homopolymer	0.901–0.950	1.2–1.8
Polystyrene, general purpose	PS (GPPS)	1.03–1.06	0.4–0.6
Polystyrene, high impact	PS (HIPS)	1.03–1.06	0.4–0.6
Syndiotactic polystyrene	SPS	1.01–1.44	0.3–2.0
Polyether sulfone	PES	1.37–1.38	0.6–1.4
Polysulfone	PSU	1.24–1.25	0.6–1.0
Polyvinyl chloride, chlorinated	CPVC	1.47–1.52	0.6
Polyvinyl chloride, flexible	PVC, Flexible	1.14–1.45	0.9–2.1
Polyvinyl chloride, rigid	PVC, Rigid	0.779–1.47	0.3–0.4
Polyvinyl chloride, semi-rigid	PVC, Semi-Rigid	1.30–1.58	1.1
Styrene acrylonitrile	SAN	1.04–28.1	0.3–0.5
Styrene butadiene block copolymer	SBC	0.850–1.03	0.5–2.4
Styrene butadiene styrene block copolymer	SBS	0.922–1.05	0.4–1.4

Drying temperature (°C)		Drying time (h)	Suggested max moisture (%)	Processing (melt) temp. (°C)		Mold temperature (°C)	
min	max			min	max	min	max
97	118	2.0–8.0	0.019–0.020	267	271	79	81
113	132	3.0–6.0	0.020–0.043	234	265	58	92
120	180	4.0–5.5	0.0030–0.20	256	285	15	130
65	75	3.0–9.0	0.05	218	260	27	40
134	151	5.0–5.5	0.020–0.021	373	374	148	151
135	150	3.0–4.0	0.1	374	384	149	192
60	61	7.0–8.0	NA	98	230	20	180
79	80	1.0	NA	180	251	10	46
79	82	1.0	NA	164	222	15	43
90		1.0	NA	180	240	18	35
NA	NA	1.0	NA	285		45	-
49	51	3.5–4.0	0.010–0.032	199	240	20	105
70	125	3	0.02	260	315	74	90
134	150	4.0–6.0	0.015–0.20	313	323	121	156
75	85	1.0–3.0	0.050–0.20	202	249	2	50
69	82	1.5–2.0	0.02	214	248	30	60
70	78	1.5–2.0	0.1	208	236	29	61
80		3.5	NA	310		70	-
134	177	2.5–6.0	0.020–0.050	355	366	134	160
134	149	3.0–4.0	0.020–0.10	352	366	121	151
NA	NA	NA	NA	203	204	NA	NA
NA	NA	NA	NA	165	200	24	30
66		3	NA	186	206	32	32
NA	NA	NA	NA	188	194	NA	NA
77	80	2.0–4.0	0.020–0.20	204	251	49	65
52	77	0.5–3.0	NA	184	250	39	50
52	52	0.0–3.0	NA	155	232	23	45

Appendix B Conversion Tables for Commonly Used Process Parameters

Speed	1 mm/s = 0.0394 in/s
	1 in/s = 25.4 mm/s
Pressure	1 psi = 0.0069 MPa
	1 psi = 0.0690 bar
	1 bar = 14.50 psi
	1 bar = 0.1 MPa
	1 MPa = 145.04 psi
	1 MPa = 10 bar
Temperature	°Fahrenheit = (1.8 × °Centigrade) + 32
	Example: 50 °C = (1.8 × 50) + 32 = 122 °F
	°Centigrade = (°Fahrenheit – 32)/1.8
	Example: 200 °F = (200 − 32)/1.8 = 93.3 °C
Weight	1 ounce (oz) = 28.35 g
	1 gram (g) = 0.035 oz
Tonnage	1 kN = 0.11 US tons
	1 kN = 0.1 metric ton
	1 US ton = 9.09 kN
	1 US ton = 0.909 metric tons
	1 metric ton = 10 kN
	1 metric ton = 1.1 US tons

Appendix C
Water Flow Tables

Minimum Water Flow (gal/min) Required to Achieve Turbulent Flow

Water temperature (°F)	Pipe diameter in inches				
	¼"	3⁄8"	1⁄5"	¾"	1"
	(0.25 in)	(0.375 in)	(0.5 in)	(0.75 in)	(1 inch)
50	0.41	0.62	0.82	1.23	1.64
60	0.35	0.53	0.71	1.06	1.42
70	0.31	0.46	0.62	0.93	1.23
80	0.27	0.41	0.54	0.81	1.09
90	0.24	0.36	0.48	0.72	0.96
100	0.22	0.32	0.43	0.65	0.86
125	0.17	0.25	0.33	0.50	0.67
150	0.13	0.20	0.27	0.40	0.54
175	0.11	0.17	0.22	0.34	0.45
200	0.09	0.14	0.19	0.28	0.38

Note: This table should be used for water only. The presence of additives, such as rust preventives, will alter the viscosity and therefore the flow rates. It is always better to use more than the recommended flow rates for this reason.

Minimum Water Flow (l/min) Required to Achieve Turbulent Flow

Water temperature (°C)	Pipe diameter in mm				
	8	10	15	20	25
10	1.96	2.45	3.68	4.90	6.13
15	1.71	2.14	3.21	4.29	5.36
20	1.51	1.89	2.83	3.78	4.72
25	1.34	1.68	2.52	3.36	4.20
30	1.20	1.50	2.26	3.01	3.76
35	1.08	1.36	2.03	2.71	3.39
40	0.98	1.23	1.84	2.46	3.07
45	0.90	1.12	1.68	2.24	2.80
55	0.76	0.94	1.42	1.89	2.36
65	0.65	0.81	1.22	1.62	2.03
80	0.53	0.66	0.99	1.32	1.66
95	0.44	0.56	0.83	1.11	1.39

Note: This table should be used for water only. The presence of additives, such as rust preventives, will alter the viscosity and therefore the flow rates. It is always better to use more than the recommended flow rates for this reason.

Appendix D
Part Design Checklist

Following are examples of some of the important questions that must be asked during a part design review:

Part Design Checklist

- Is the gate location acceptable for fill?
- Is the number of gates sufficient for part fill?
- Is the gate location acceptable for cosmetics?
- Is the gate location acceptable for warpage?
- Can the thick areas be cored out?
- Is processing information about the plastic material available?
- Is the shrink factor available?
- Is the engraving in an acceptable position?
- Is the texture acceptable?
- Is the part designed to stay on the B-side of the mold?
- Is sufficient draft provided on the part for ejection?

Appendix E
Mold Design Checklist

Following are examples of some of the important questions that must be asked during a mold design review:

1. **Mold Cooling:**
   a) Is the mold cooling acceptable?
   b) Does the mold require waterlines or oil lines?
   c) What is the size of the waterlines?
   d) Are the waterlines recessed?
2. **Mold Construction:**
   a) Is the 'TOP' of the mold acceptable for part ejection and removal?
   b) Are eyebolt holes provided on the 'TOP' of the mold?
   c) Will the mold fit between the tie bars?
   d) Are the clamp slots/holes acceptable?
   e) Do the waterlines and tie bars interfere with each other?
   f) Are parting line interlocks necessary?
   g) Are pry bar slots necessary?
   h) Is the mold opening sequence acceptable?
   i) Are insulator plates necessary?
   j) Are any spares necessary?
   k) Are the machine tie bars shown on the drawings?
3. **Ejection:**
   a) Is the number of ejector pins sufficient?
   b) Are the sizes of ejector pins sufficient?

c) Does the ejector pattern on the mold match the ejector pattern on the machine?

d) Is the ejection distance sufficient for the part to get ejected out of the mold?

e) Are return springs necessary?

f) Is an early ejector return system necessary?

g) Are support pillars necessary?

h) Is guided ejection necessary?

i) Does the path of any of the ejector pins and the slides interfere with each other?

4. **Venting:**

a) Are vents identified on the mold?

b) Is the last point to fill vented?

c) Are all the corners vented?

d) Are bosses, ribs, and tabs vented?

e) Are deep pockets vented?

f) Are the runners vented?

5. **Gates and Runners:**

a) Is the gate size acceptable?

b) Will the gate leave a gate vestige?

c) Is the runner size acceptable?

d) Is the runner balanced?

e) Will the runners stay on the B-side?

f) In case of a three-plate mold will the runner drop freely?

g) Is there a sprue puller?

h) Do any of the sucker pins interfere with the plastic flow?

Appendix F Mold Qualification Checklist

a) Are the gates acceptable?

b) Are the parts dimensionally acceptable?

c) Has the cooling study been done?

d) Has the gate seal test been done?

e) Has the viscosity test been performed?

f) Is the cavity balance acceptable?

g) Is the cycle time acceptable?

h) Is the ejection acceptable?

i) Is the part fill acceptable?

j) Is the process window acceptable?

k) Is the runner acceptable?

l) Is the venting acceptable?

m) Is the water flow through the mold acceptable?

n) Was the mold safe to be hung in the press?

Appendix G Regrind Tables – Percentage of Regrind in Total Shot

(The runner and part will always have the same percentage of regrind.)

Table G.1 Part Weight = 95%; Runner Weight = 5%

Regrind Generation (g)	Pass (p)				
	1	2	3	4	5
0	100	95.00	95.00	95.00	95.00
1	–	5.00	4.75	4.75	4.75
2	–	–	0.25	0.24	0.24
3	–	–	–	0.01	0.01
4	–	–	–	–	0.00

Table G.2 Part Weight = 90%; Runner Weight = 10%

Regrind Generation (g)	Pass (p)				
	1	2	3	4	5
0	100	90.00	90.00	90.00	90.00
1	–	10.00	9.00	9.00	9.00
2	–	–	1.00	0.90	0.90
3	–	–	–	0.10	0.09
4	–	–	–	–	0.01

Table G.3 Part Weight = 80%; Runner Weight = 20%

Regrind Generation (g)	Pass (p)				
	1	2	3	4	5
0	100	80.00	80.00	80.00	80.00
1	–	20.00	16.00	16.00	16.00
2	–	–	4.00	3.20	3.20
3	–	–	–	0.80	0.64
4	–	–	–	–	0.16

Table G.4 Part Weight = 75%; Runner Weight = 25%

Regrind Generation (g)	Pass (p)				
	1	2	3	4	5
0	100	75.00	75.00	75.00	75.00
1	–	25.00	18.75	18.75	18.75
2	–	–	6.25	4.69	4.69
3	–	–	–	1.56	1.17
4	–	–	–	–	0.39

Table G.5 Part Weight = 50%; Runner Weight = 50%

Regrind Generation (g)	Pass (p)				
	1	2	3	4	5
0	100	50.00	50.00	50.00	50.00
1	–	50.00	25.00	25.00	25.00
2	–	–	25.00	12.50	12.50
3	–	–	–	12.50	6.25
4	–	–	–	–	6.25

Table G.6 Part Weight = 25%; Runner Weight = 75%

Regrind Generation (g)	Pass (p)				
	1	2	3	4	5
0	100	25.00	25.00	25.00	25.00
1	–	75.00	18.75	18.75	18.75
2	–	–	56.25	14.06	14.06
3	–	–	–	42.19	10.55
4	–	–	–	–	31.64

Appendix H Calculation of Melt Density

Following is the formula to calculate the melt density of the plastic in the barrel of the injection molding machine:

Melt density = Shot weight/(0.785 × Screw dia × Screw dia × (Shot size − Cushion value)

Where

Shot weight is in grams and

Screw dia, Shot size and Cushion value are all in centimeters.

Melt density will be in gms/cc

Example:

Shot size = 65.0 mm = 6.50 cm

Cushion value = 5.0 mm = 0.5 cm

Screw Diameter = 35 mm = 3.5 cm

Shot weight including the runners = 40.0 gms

Calculated melt density = 40.0/(0.785 × 35 × 35 × (65.0 − 5.0) = 0.69 gms/cc

Index

A

ABS 106
acetal 110
acrylate 106
acrylics 109
acrylonitrile 106
actual melt temperature 167
actual mold temperature 168
addition polymerization 18
additives 111
aliasing 282, 283
alloys 105
amorphous 24
– polymers 21
anisotropic material 44
anti-aging additives 113
apparent shear rate 61
apparent viscosity 55
array
– orthogonal 278
atactic 23

B

back pressure 161
– optimization 200
balanced array 278
barrel clearance 121
barrel heat profile 29
barrel temperatures 155
– setting 191
Beaumont Technologies 67
blowing agents 114
Box-Behnken design 278
bubbles 80
built-in stress 262
burn marks 81
butadiene 106

C

capability index 380
capillary rheometer 73
cavity balance 222
– procedure 221
– study 219
cavity imbalance
– acceptable level 227
– calculation 226
cavity pressure 219, 220
– curve at the end of fill 357
– curve near the gate 357
– sensing technology 355
– sensor 361
cavity steel adjustment 291
chain scission 113

characterization of polymer viscosity 73
check ring 126
clamp force 121, 132
clamp tonnage 129, 132
classification of polymers 103
coefficient of friction 113
co-injection molding 72
colorants 114
commercial plastics 105
common defects
– cause 326
– prevention 326
concurrent engineering 12, 366
condensation polymerization 18
confounding 282, 283
consistency 143
– cavity to cavity 3
– quality 3
– run-to-run 3
– shot-to-shot 3
constant factors 274
contact type probes 221
contour cooling 336
contour plot 288
control factors 274
control process window (CoPW) 241, 313
control sensor 359
coolant
– type of 337
cooling line 336
cooling phase 149
cooling time 30, 164, 255
– optimization 256
– study 253
copolymers 104
cosmetic process development 312
cosmetic process window (CPW) 239, 312
critical dimension 290
critical factors of molding 9
crystalline 24
– polymers 21
crystallinity 25, 36
crystallite 21
crystallization 113
– temperature 28, 35
cushion value 171, 213
cycle time 173, 256
– breakdown 257
cyclone separator 348

D

decompression 165
decoupled molding℠ 176, 361
degradation 79, 320
degree of crystallinity 22
desiccant dryer 89
design of experiments (DOE) 280
– definition 272
dew point 87, 92
dieseling effect 339
differential scanning calorimeter (DSC) 36
dilatant 57
dimensional process window (DPW) 241, 288, 313
dimensional variation 81
DIOP 112
direct sensor 358
dispersive mixing 124
distributive mixing 123
documentation books 318
drying
– equipment 88
– of plastics 85
– precautions 185
– temperature 85, 86
– time 85, 86, 183, 347

E

effect of regrind 352
elastomers 104
electronic moisture analyzer 91
end-chain degradation 79

end-of-fill cavity pressure integral 363
epoxidized soybean oil (ESO) 112

F

factor 274, 276, 277, 289
factorial experiments 270, 284
family molds 229
feed depth 123
filler 111
– content 262
fill pattern 223
fill progression 213
fill time 170
fines 348
flame retardants 112
flow
– front 58
– imbalance 64
– laminar 336
– promoters 114
– transitional 336
– turbulent 336, 337, 338
fluoropolymers 111
flush mount sensor 358
forced venting 345
fountain flow 58, 69, 216
fountain flow, effect on
– crystallinity 72
– fiber orientation 72
– molecular orientation 72
free radicals 113
free state 45
frozen layer 70
full factorial experiment 284
full part weight 175

G

gage reproducibility & repeatability 371
gassing
– excessive 320
gate freeze study 247
gate seal 247
– cautions and exceptions 249
– procedure 247
– study 246
– time 248
general purpose polystyrene (GPPS) 121
general purpose screw 123
glass transition temperature 33
glassy state 61

H

heater bands 122
heat history 346
heat soaking 317
helix angle 125
high-shear lamina 65
histogram 376
holding phase 149, 150
holding pressure 150, 238, 239
holding time 238
hold pressure 160
hold time 160
homopolymers 104
horizontal clamp molding machine 119
horizontal injection machine 119
hot air dryer 89
hydraulic 'fill only' integral 362
hydraulic pressure curve 356
hydrolytic degradation 79, 182
hydrophilic 77
hydrophobic 77
hygroscopic 77, 83

I

indirect sensor 358
injection
– force 153
– molding, parameter 271
– phase 149, 150
– speed 153
injection molding cycle 149
injection only shot weight 174

injection pressure 159
injection speed 158
injection time 170
injection velocity 158
in-mold rheology study 208
inorganic polymers 104
intensification ratio (IR) 126, 127
– of the screw 122
interaction 280, 281
– column 282
– types of 281
intermolecular forces 19, 34
internal voids 153
isotropic material 44

K

Karl Fischer titration 91
knowledge base 364

L

laminar flow 65
L/D ratio 121, 125
level 275, 276
lot-to-lot variations 209
lower specification limit (LSL) 148, 289
lower tolerance 266
low temperature – high pressure corner 242
low temperature – low pressure corner 241, 242
lubricants 113

M

machine selection 11, 129
machine specifications 120
material purging instructions 317
material selection 9, 10
material supplier
– concurrent involvement 370
– conventional involvement 370
maximum moisture level 77
MeltFlipper® 67
melt flow index (MFI) 74
melt flow rate (MFR) 74
melt homogeneity 29, 121
melting point 35
melt preparation 189
melt processing range 27
melt processing window 35
melt rheology 73
melt rotation technology 67, 69
melt temperature 28, 34, 239
– check 190
melt viscosity reducers 114
moisture determination 90
mold cooling 335
mold design and construction 9, 10
mold designer
– concurrent involvement 369
– conventional involvement 369
molded-in stress 153
mold filling speed 28
mold function qualification
– procedure 265
mold height 131
molding area diagram 239, 243
molding machine 9
– classification 118
molding process 9, 11
– effect on the part properties 347
molding shut down 322
molding startup 321
mold length 130
mold maker
– concurrent involvement 369
– conventional involvement 369
mold open stroke 131
mold qualification
– book 318
– checklist 267, 313, 314
– flowchart 311
mold shut-down procedure 317
mold stack height 130
mold temperature 28, 239
– map 316

mold temperatures 156
mold width 130
molecular weight 19, 25, 347
– distribution (MWD) 19, 20
monitor sensor 359
morphology 21
– role of 26
multishot machine 120

N

natural cause variations 143
naturally balanced runner 66
Newtonian 57
noise factors 274
nominal dimension 266
non-Newtonian 57, 103
non-return valve 126
normal distribution 377
nozzle drool 82
nozzle temperature control 30
nucleating agents 113
number average molecular weight 19
number of experiments 277
nylon overdrying 97

O

operator instructions 317
optical clarity 30
organic polymers 104
orthogonal array 279
Ostwald and de Waele 59
out of specifications 148
oven dryers 88
overdrying 79, 93
– controller 100
– prevention 100

P

packing phase 149
packing pressure 219, 238
packing time 238
pack pressure 159
pack speed 166
pack time 159
parallel cooling 338
Pareto chart 286
part design 9
partial factorial experiment 284
parting line shot 119
parting line vent 345
part quality qualification 313
PBT overdrying 94
peak injection pressure 170
percentage shot size used 137
phthalates 112
piezoelectric sensor 357
Plackett-Burman design 278
plasticating capacity 122
plastic drying 77, 182
plasticizers 112
plastic pressure 122
platen 130
– deflection 130
plugging up of the vents 81
Poisson's effect 45
polar group 83, 85
polyamides 107
polycarbonates 109
polyesters 109
polyethylene 105
polymer 17
– blends 115
– crystallinity 24
– morphology 229
– rheology 55
polyolefins 105
polyoxymethylene 110
polypropylene 105
polystyrenes 108
polyvinyl chloride 109
post-gate cavity pressure integral 362
post-molding shrinkage 153
post-mold shrinkage 261
prediction equation 289

pressure at transfer 171
pressure drop 234
– studies 229
– study procedure 231
pressure graph 359
pressure limited 230, 233, 234
pressure sensor 357
process capability 382
process change log 315
– sheet 324
process consistency 3, 146, 266
process documentation 267, 314
process engineer
– concurrent requirement 371
– conventional involvement 371
processing aids 114
process robustness 145, 266
process selection 291
process sensitivity chart 289
process sheet 314, 315
process window 238, 243, 288
– cautions and exceptions 245
– crystalline materials 242
– study 237
– study procedure 241
product designer
– conventional involvement 368
– required involvement 368
production book 318, 324
progression of the melt 70
purging 320
– compound 322
PVT graph 272

Q

qualification runs 319
qualifications 325
qualitative factor 274
quantitative factor 274
quartz-based sensor 357

R

racetrack effect 66
randomization 302, 303
reducing inspection 293
regrind 346
– batch process 349
– continuous process 349
– generation estimating 350
– generation number 348
– molecular weight 352
– particle size distribution 348
– processing 348
relative humidity 87
residence time 140
resin storage 182
response 275, 276, 289
– qualitative 275
– quantitative 275
Reynolds number (Re) 336, 337
rheological behavior 205
rheological imbalance 66, 223
rheology study 205
– cautions and exceptions 213
– procedure 210
robust process 143, 148, 149, 289
runner
– venting 344
runner weight 175

S

scientific molding 8, 143
scientific processing 143, 145, 179, 205
– definition 6
screening experiment 278
screw 122
– bounce-back 213
– channel depth 123
– compression ratio 124
– compression zone 123
– designs 125
– diameter 121
– feed zone 123

– metering zone 124
– outside diameter 123
– recovery speed 29
– root diameter 123
– slipping 123
– speed 160
– transition zone 123
screw rotation
– procedure 197
screw rotation speed 160
– optimization 195
screw rotation time 173
semicrystalline 22
sensor 356
– locations 363
series cooling 338
setup instructions 317
shear energy 160
shear force 56
shear friction 29
shear heat 122
shear rate 56, 57, 59, 73, 207
shear rate profile 62, 63
shear stress 57
shear thinning 57, 103
shelf life limitations 182
short shot 220, 233, 339
shot control 82
shot size 120, 121, 124, 158
shrinkage 26, 44, 45, 219, 261, 262
– additives 262
– annealing 262
– filler 262
– glass transition temperature 261
– measuring procedure 263
– mold temperature 262
– part thickness 262
– processing conditions 262
– rate of 261
sink 150
solution rheology 73
special cause variations 143
specifications 143
specific volume 34, 219
speed and pressure 153
splay 80
standard deviation 378
startup procedure 321
statistical analysis 286
statistical data 277
statistical process control (SPC) 383
statistical quality control (SQC) 383
stereoregular 23
strain gage sensor 357
stress build-up 67
styrene 106
suck back position 165
surface defects 80
syndiotactic 23

T

Taguchi design 278
target dimension 289
Teflon 111
tensile properties 104
thermal imaging system 221
thermal region 61
thermal transition 32, 36
thermoplastics 104
thermosets 104
toggle machine 130
tolerances
– process change 292
Tomasetti volatile indicator test 90
tonnage 121, 132, 166
– optimization 257
tonnage calculation 137
tooling engineer
– concurrent requirement 369
– conventional involvement 369
tornado chart 286, 287
transfer position 158
troubleshooting 322, 325
– guideline 319, 323
TVI Test 90

U

upper specification limit (USL) 148, 289
upper tolerance 266
UV stabilizers 113

V

vacuum assist 345
vacuum venting 345
valve gate mold 345
velocity 62
– profile 62
vent
– depth 341
– dimensions 340
– locations 344
– reliefs 341
– size 342
venting 339
vertical clamp molding machine 119
vertical injection 119
virgin material 346
viscosity 19, 28, 55, 59, 61, 153, 206, 207
– curve 209
– curve worksheet 211
– effect of temperature on 61
– in polymer melts 58
– linear scale 206
– logarithmic scale 206
– model 73
volumetric shrinkage 150

W

warpage 153, 262
waterline diagram 315, 316
water lines 338
weld line 82
Wheatstone bridge 357

www.ingramcontent.com/pod-product-compliance
Lightning Source LLC
LaVergne TN
LVHW081929200926
848484LV00009B/2396
* 9 7 8 1 5 6 9 9 0 9 0 8 9 *